AF544909

# Springer Tracts in Advanced Robotics

Volume 2

---

Editors: Bruno Siciliano · Oussama Khatib · Frans Groen

# Springer Tracts in Advanced Robotics

## Edited by B. Siciliano, O. Khatib, and F. Groen

**Vol. 21:** Ang Jr., M.H.; Khatib, O. (Eds.)
Experimental Robotics IX – The 9th International Symposium on Experimental Robotics
624 p. 2006 [3-540-28816-3]

**Vol. 20:** Xu, Y.; Ou, Y.
Control of Single Wheel Robots
188 p. 2005 [3-540-28184-3]

**Vol. 19:** Lefebvre, T.; Bruyninckx, H.; De Schutter, J.
Nonlinear Kalman Filtering for Force-Controlled Robot Tasks
280 p. 2005 [3-540-28023-5]

**Vol. 18:** Barbagli, F.; Prattichizzo, D.; Salisbury, K. (Eds.)
Multi-point Interaction with Real and Virtual Objects
281 p. 2005 [3-540-26036-6]

**Vol. 17:** Erdmann, M.; Hsu, D.; Overmars, M.; van der Stappen, F.A (Eds.)
Algorithmic Foundations of Robotics VI
472 p. 2005 [3-540-25728-4]

**Vol. 16:** Cuesta, F.; Ollero, A.
Intelligent Mobile Robot Navigation
224 p. 2005 [3-540-23956-1]

**Vol. 15:** Dario, P.; Chatila R. (Eds.)
Robotics Research – The Eleventh International Symposium
595 p. 2005 [3-540-23214-1]

**Vol. 14:** Prassler, E.; Lawitzky, G.; Stopp, A.; Grunwald, G.; Hägele, M.; Dillmann, R.; Iossifidis. I. (Eds.)
Advances in Human-Robot Interaction
414 p. 2005 [3-540-23211-7]

**Vol. 13:** Chung, W.
Nonholonomic Manipulators
115 p. 2004 [3-540-22108-5]

**Vol. 12:** Iagnemma K.; Dubowsky, S.
Mobile Robots in Rough Terrain – Estimation, Motion Planning, and Control with Application to Planetary Rovers
123 p. 2004 [3-540-21968-4]

**Vol. 11:** Kim, J.-H.; Kim, D.-H.; Kim, Y.-J.; Seow, K.-T.
Soccer Robotics
353 p. 2004 [3-540-21859-9]

**Vol. 10:** Siciliano, B.; De Luca, A.; Melchiorri, C.; Casalino, G. (Eds.)
Advances in Control of Articulated and Mobile Robots
259 p. 2004 [3-540-20783-X]

**Vol. 9:** Yamane, K.
Simulating and Generating Motions of Human Figures
176 p. 2004 [3-540-20317-6]

**Vol. 8:** Baeten, J.; De Schutter, J.
Integrated Visual Servoing and Force Control
198 p. 2004 [3-540-40475-9]

**Vol. 7:** Boissonnat, J.-D.; Burdick, J.; Goldberg, K.; Hutchinson, S. (Eds.)
Algorithmic Foundations of Robotics V
577 p. 2004 [3-540-40476-7]

**Vol. 6:** Jarvis, R.A.; Zelinsky, A. (Eds.)
Robotics Research – The Tenth International Symposium
580 p. 2003 [3-540-00550-1]

**Vol. 5:** Siciliano, B.; Dario, P. (Eds.)
Experimental Robotics VIII
685 p. 2003 [3-540-00305-3]

**Vol. 4:** Bicchi, A.; Christensen, H.I.; Prattichizzo, D. (Eds.)
Control Problems in Robotics
296 p. 2003 [3-540-00251-0]

**Vol. 3:** Natale, C.
Interaction Control of Robot Manipulators – Six-degrees-of-freedom Tasks
120 p. 2003 [3-540-00159-X]

**Vol. 2:** Antonelli, G.
Underwater Robots – Motion and Force Control of Vehicle-Manipulator Systems
209 p. 2003 [3-540-00054-2]

**Vol. 1:** Caccavale, F.; Villani, L. (Eds.)
Fault Diagnosis and Fault Tolerance for Mechatronic Systems – Recent Advances
191 p. 2002 [3-540-44159-X]

Gianluca Antonelli

# Underwater Robots

## Motion and Force Control of Vehicle-Manipulator Systems

## Second Edition

With 133 Figures

Springer

**Professor Bruno Siciliano**, Dipartimento di Informatica e Sistemistica, Università degli Studi di Napoli Federico II, Via Claudio 21, 80125 Napoli, Italy, email: siciliano@unina.it

**Professor Oussama Khatib**, Robotics Laboratory, Department of Computer Science, Stanford University, Stanford, CA 94305-9010, USA, email: khatib@cs.stanford.edu

**Professor Frans Groen**, Department of Computer Science, Universiteit van Amsterdam, Kruislaan 403, 1098 SJ Amsterdam, The Netherlands, email: groen@science.uva.nl

## Author

Dr. Gianluca Antonelli

Università degli Studi di Cassino
Dipartimento di Automazione, Elettromagnetismo,
Ingegneria dell'Informazione e Matematica Industriale
Via di Biasio 43
03043 Cassino
Italy

ISSN print edition: 1610-7438
ISSN electronic edition: 1610-742X

ISBN-10 3-540-31752-X **Springer Berlin Heidelberg New York**
ISBN-13 978-3-540-31752-4 **Springer Berlin Heidelberg New York**

Library of Congress Control Number: 2006920068



**Springer is a part of Springer Science+Business Media**
springer.com



Typesetting: Digital data supplied by author.
Data-conversion and production: PTP-Berlin Protago-TEX-Production GmbH, Germany
Cover-Design: design & production GmbH, Heidelberg
Printed on acid-free paper 89/3141/Yu - 5 4 3 2 1 0

STAR (Springer Tracts in Advanced Robotics) has been promoted under the auspices of EURON (European Robotics Research Network)

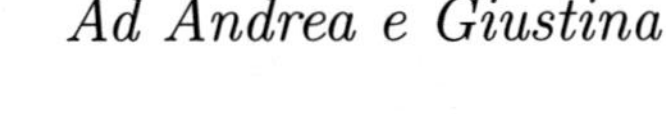

*Ad Andrea e Giustina*

*E la locomotiva sembrava fosse un mostro strano*
*che l'uomo dominava con il pensiero e con la mano...*
*Francesco Guccini, La locomotiva, 1972.*

# Foreword

At the dawn of the new millennium, robotics is undergoing a major transformation in scope and dimension. From a largely dominant industrial focus, robotics is rapidly expanding into the challenges of unstructured environments. Interacting with, assisting, serving, and exploring with humans, the emerging robots will increasingly touch people and their lives.

The goal of the new series of *Springer Tracts in Advanced Robotics (STAR)* is to bring, in a timely fashion, the latest advances and developments in robotics on the basis of their significance and quality. It is our hope that the wider dissemination of research developments will stimulate more exchanges and collaborations among the research community and contribute to further advancement of this rapidly growing field.

The volume by Gianluca Antonelli is the second edition of a successful monograph, which was one of the first volumes to be published in the series. Being focused on an important class of robotic systems, namely underwater vehicle-manipulator systems, this volume improves the previous material while expanding the state-of-the-art in the field. New features deal with fault-tolerant control and coordinated control of autonomous underwater vehicles.

A well-balanced blend of theoretical and experimental results, this volume represents a fine confirmation in our STAR series!

Naples, Italy
October 2005

*Bruno Siciliano,*
STAR Editor

# Foreword

# Acknowledgements

The contributions of a quite large number of people were determinant for the realization of this monograph.

Prof. Stefano Chiaverini was my co-tutor during my PhD at the Università degli Studi di Napoli Federico II, he is now Full Professor at the Università degli Studi di Cassino. Since the beginning of the doctoral experience up to present all the research that I have done is shared with him. As a matter of fact, I consider myself as a co-author of this monograph. Furthermore, Stefano was, somehow, also *responsible* of my decision to join the academic career.

Prof. Lorenzo Sciavicco developed a productive, and friendly, research environment in Napoli that was important for my professional growing. Prof. Bruno Siciliano, my tutor both in the Master and PhD thesis, to whom goes my warmest acknowledgements.

Prof. Fabrizio Caccavale, currently at the Università della Basilicata, Prof. Giuseppe Fusco currently at the Università degli Studi di Cassino, Dr. Tarun Podder, currently at the University of Rochester, Prof. Nilanjan Sarkar, currently at the Vanderbilt University, Prof. Luigi Villani currently at the Università degli Studi di Napoli Federico II, Dr. Michael West, currently at the University of Hawaii; all of them are co-authors of my *wet* papers and deserve a lot of credit for this work.

During the PhD I have been visiting researcher at the Autonomous Systems Laboratory of the University of Hawaii where I carried out some experiments on dynamic control of autonomous underwater vehicles and worked on the interaction control chapter. I would like to acknowledge Prof. Nilanjan Sarkar and Prof. Junku Yuh, my guests during the staying.

For this second edition several colleagues provide me with their illustrative material, I would like to thank Eng. Massimo Caccia, Prof. Giuseppe Casalino, Prof. Tom McLain, Prof. Daniel Stilwell, Eng. Gianmarco Veruggio and Prof. Junku Yuh.

My mother, my father, my brothers Marco and Fabrizio, my wife Giustina and, recently, my son Andrea, they all tolerated, and will have to tolerate for longtime, my *engineeringness*.

# About the Author

**Gianluca Antonelli** was born in Roma, Italy, on December 19, 1970. He received the "Laurea" degree in Electronic Engineering and the "Research Doctorate" degree in Electronic Engineering and Computer Science from the Università degli Studi di Napoli Federico II in 1995 and 2000, respectively. From January 2000 he is with the Università degli Studi di Cassino where he currently is an Associate Professor. He has published more than 60 journals and conference papers; he was awarded with the "EURON Georges Giralt PhD Award", First Edition for the thesis published in the years 1999-2000. From September, 2005 he is an *Associate Editor* of the *IEEE Transactions on Robotics*. His research interests include simulation and control of underwater robotic systems, force/motion control of robot manipulators, path planning and obstacle avoidance for autonomous vehicles, identification, multi-robot systems.

Prof. Gianluca Antonelli
Dipartimento di Automazione, Elettromagnetismo,
Ingegneria dell'Informazione e Matematica Industriale
Università degli Studi di Cassino
via G. Di Biasio 43,
03043, Cassino (FR), Italy
antonelli@unicas.it
http://webuser.unicas.it/antonelli

# Preface to the Second Edition

The purpose of this Second Edition is to add material not covered in the First Edition as well as streamline and improve the previous material.

The organization of the book has been substantially modified, an introductory Chapter containing the state of the art has been considered; the modeling Chapter is substantially unmodified. In Chapter 3 the problem of controlling a 6-Degrees-Of-Freedoms (DOFs) Autonomous Underwater Vehicle (AUV) is investigated. Chapter 4 is a new Chapter devoted at a survey of fault detection/tolerant strategies for ROVs/AUVs, it is mainly based on the Chapter published in [10]. The following Chapter (Chapter 5) reports experimental results obtained with the vehicle ODIN. The following 3 Chapters, from Chapter 6 to Chapter 8 are devoted at presenting kinematic, dynamic and interaction control strategies for Underwater Vehicle Manipulator Systems (UVMSs); new material has been added thanks also to several colleagues who provided me with valuable material, I warmly thank all of them. The content of Chapter 9 is new in this Second Edition and reports preliminary results on the emerging topic of coordinated control of platoon of AUVs. Finally, the bibliography has been updated.

The reader might be interested in knowing what she/he will not find in this book. Since the core of the book is the coordinated control of manipulators mounted on underwater vehicles, control of non-holonomic vehicles is not dealt with; this is an important topic also in view of the large number of existing *torpedo*-like vehicles. Another important aspect concerns the sensorial apparatus, both from the technological point of view and from the algorithmic aspect; most of the AUVs are equipped with redundant sensorial systems required both for localization/navigation purposes and for fault detection/tolerant capabilities. Actuation is mainly obtained by means of thrusters; those are still object of research for the modeling characteristics and might be the object of improvement in terms of dynamic response.

Cassino, Italy *Gianluca Antonelli*
January 2006

# Preface to the First Edition

Underwater Robotics have known in the last years an increasing interest from research and industry. Currently, it is common the use of manned underwater robotics systems to accomplish missions as sea bottom and pipeline survey, cable maintenance, off-shore structures' monitoring and maintenance, collect/release of biological surveys. The strong limit of the use of manned vehicles is the enormous cost and risk in working in such an hostile environment. The aim of the research is to progressively make it possible to perform such missions in a completely autonomous way.

This objective is challenging from the technological as well as from the theoretical aspects since it implies a wide range of technical and research topics. Sending an autonomous vehicle in an unknown and unstructured environment, with limited on-line communication, requires some on board *intelligence* and the ability of the vehicle to react in a reliable way to unexpected situations. Techniques as artificial intelligence, neural network, discrete events, fuzzy logic can be useful in this *high* level mission control. The sensory system of the vehicle must deal with a noisy and unstructured environment; moreover, technologies as GPS are not applicable due to the impossibility to underwater electromagnetic transmission; vision based systems are not fully reliable due to the generally poor visibility. The actuating system is usually composed of thrusters and control surfaces; all of them have a non-linear dynamics and are strongly affected by the hydrodynamic effects.

In this framework the use of a manipulator mounted on a autonomous vehicle plays an important role. From the control point of view, underwater robotics is much more challenging with respect to ground robotics since the former deal with unstructured environments, mobile base, significant external disturbance, low bandwidth of sensory and actuating systems, difficulty in the estimation of the dynamic parameters, highly non-linear dynamics.

Referring to Autonomous Underwater Vehicles (AUVs), i.e., unthetered, unmanned vehicles to be used mainly in survey missions, [294, 321] present the state of the art of several existing AUVs and their control architecture. Currently, there are more than 46 AUV models [321], among others: *ABE* of the Woods Hole Oceanographic Institution (MA, USA), *MARIUS* developed under the Marine Science and Technology Programme of the IV framework of European Commission (Lisbon, Portugal), *ODIN* designed at the Autonomous Systems Laboratory of the University of Hawaii (Honolulu, HI, USA), *OTTER* from the Monterey Bay Acquarium and Stanford University (CA, USA), *Phoenix* and *ARIES* belonging to the Naval Postgraduate School (Monterey, CA, USA), *Twin Burgers* developed at the University of Tokyo (Tokyo, Japan), *Theseus* belonging to ISE Research Ltd (Canada). Reference [92] shows the control architecture of *VORTEX*, a vehicle developed by Inria and Ifremer (France), and *OTTER*. Focusing on the low level motion control of AUVs, most of the proposed control schemes take into account the uncertainty in the model by resorting to an adaptive strategy [83, 91, 126, 130, 138, 314] or a robust approach [90, 93, 145, 201, 259, 310, 311]. In [145] an estimation of the dynamic parameters of the vehicle NPS AUV *Phoenix* is also provided. An overview of control techniques for AUVs is reported in [127].

As a curiosity, in the Figure below there is a draw of one of the first *manned* underwater vehicles. It was found in the *Codice Atlantico* (Codex Atlanticus), written by Leonardo Da Vinci between 1480 and 1518, together with the development of some diver's devices. Legends say that Leonardo worked on the idea of an underwater military machine that he further destroyed by himself the results judged too dangerous. Maybe the first idea of an underwater machine is from Aristotle; following the legend he built a machine: *skaphe andros* (boat-man) that allowed Alexander the Great to stay in deep for at least half a day during the war of Tiro in 325 b. C. This is unrealistic, of course, also considering that the Archimedes's law was still to become a reality (around 250 b. C.).

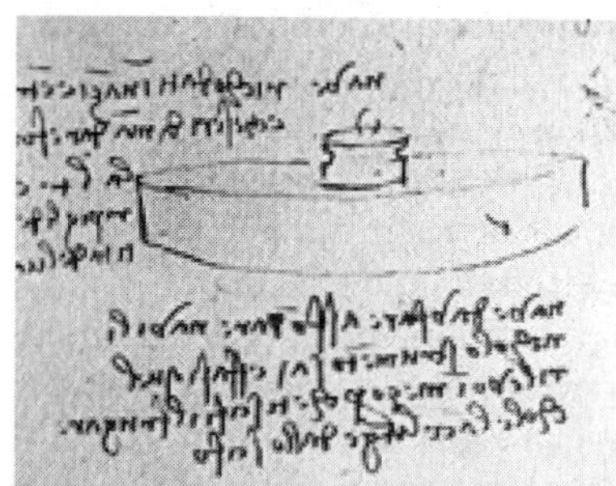

Draw of the manned underwater vehicle developed by Leonardo Da Vinci

The current technology in control of underwater manipulation is limited to the use of a master/slave approach in which a skilled operator has to move a master manipulator that works as *joystick* for the slave manipulator that is performing the task [56, 287]. The limitations of such a technique are evident: the operator must be well trained, underwater communication is hard and a significant delay in the control is experienced. Moreover, if the task has to be performed in deep waters, a manned underwater vehicle close to the unmanned vehicle with the manipulator need to be considered to overcome the communication problems thus leading to enormous cost increasing. Few research centers are equipped with an autonomous Underwater Vehicle-Manipulator System. Among the others:

- *ODIN* and *OTTER* can be provided with a one/two link manipulator to study the interaction of the manipulator and the vehicle in order to execute automatic retrieval tasks [297];
- on *VORTEX* a 7-link manipulator (*PA10*) can be mounted with a large inertia with respect to the vehicle that implies a strong interaction between them;
- *SAUVIM*, a semi-autonomous vehicle with an *Ansaldo* 7-link manipulator is under development at the Autonomous Systems Laboratory of the University of Hawaii; this vehicle, in the final version, will be able to operate at the depth of 4000 m.
- *AMADEUS*, an acronym for Advanced MAnipulation for DEep Underwater Sampling, funded by the European Commission, that involved the Heriot-Watt University (UK), the Università di Genova (Italy), CNR Istituto Automazione Navale, (Italy), the Universitat de Barcelona (Spain), the Institute of Marine Biology of Crete (Greece). The project focused on the co-ordinated control of two tele-operated underwater *Ansaldo* 7-link manipulators and the development of an underwater hand equipped with a slip sensor.

Focusing on the motion control of UVMSs, [56, 159] present a telemanipulated arm; in [192] an *intelligent* underwater manipulator prototype is experimentally validated; [67, 68, 69] present some simulation results on a Composite Dynamics approach for *VORTEX/PA10*; [106] evaluates the dynamic coupling for a specific UVMS; adaptive approaches are presented in [124, 197, 198]. Reference [206] reports some interesting experiments of coordinated control. Very few papers investigated the redundancy resolution of UVMSs by applying inverse kinematics algorithm with different secondary tasks [20, 24, 25, 249, 250].

This book deals with the main control aspects in underwater manipulation tasks and dynamic control of AUVs. First, the mathematical model is discussed; the aspects with significant impact on the control strategy will be remarked. In Chap. 6, kinematic control for underwater manipulation is presented. Kinematic control plays a significant role in unstructured robotics where off-line trajectory planning is not a reliable approach; moreover, the vehicle-manipulator system is often kinematically redundant with respect to the most common tasks and redundancy resolution algorithms can then be applied to exploit such characteristic. Dynamic control is then discussed in Chap. 7; several motion control schemes are analyzed and presented in this book. Some experimental results with the autonomous vehicle *ODIN* (without manipulator) are presented, moreover some theoretical results on adaptive control of AUVs are discussed. In Chap. 8, the interaction with the environment is detailed. Such kind of operation is critical in underwater manipulation for several reasons that do not allow direct implementation of the force control strategies developed for ground robotics. Finally, after having developed some conclusions, a simulation tool for multi-body systems is presented. This software package, developed for testing the control strategies studied along the book, has been designed according to modular requirements that make it possible to generate generic robotic systems in any desired environment.

Napoli, August 2002 *Gianluca Antonelli*

# Notation

In this Chapter, the main acronyms and the notation that will be used in the work are listed.

| | |
|---|---|
| AUV | Autonomous Underwater Vehicle |
| CLIK | Closed Loop Inverse Kinematics |
| DOF | Degree Of Freedom |
| EKF | Extended Kalman Filter |
| FD | Fault Detection |
| FIS | Fuzzy Inference System |
| FTC | Fault Tolerant Controller |
| KF | Kalman Filter |
| ROV | Remotely Operated Vehicle |
| TCM | Thruster Control Matrix |
| UUV | Unmanned Underwater Vehicle |
| UVMS | Underwater Vehicle-Manipulator System |

| | |
|---|---|
| $\Sigma_i, O - \boldsymbol{xyz}$ | inertial frame (see Figure 2.1) |
| $\Sigma_b, O_b - \boldsymbol{x}_b\boldsymbol{y}_b\boldsymbol{z}_b$ | body(vehicle)-fixed frame (see Figure 2.1) |
| $\mathbb{R}$, $\mathbb{N}$ | Real, Natural numbers |
| $\boldsymbol{\eta}_1 = [\,x \quad y \quad z\,]^{\mathrm{T}} \in \mathbb{R}^3$ | body(vehicle) position coordinates in the inertial frame (see Figure 2.1) |
| $\boldsymbol{\eta}_2 = [\,\phi \quad \theta \quad \psi\,]^{\mathrm{T}} \in \mathbb{R}^3$ | body(vehicle) Euler-angle coordinates in the inertial frame (see Figure 2.1) |
| $\mathcal{Q} = \{\boldsymbol{\varepsilon} \in \mathbb{R}^3, \eta \in \mathbb{R}\}$ | quaternion expressing the body(vehicle) orientation with respect to the inertial frame |
| $\boldsymbol{\eta} = [\,\boldsymbol{\eta}_1^{\mathrm{T}} \quad \boldsymbol{\eta}_2^{\mathrm{T}}\,]^{\mathrm{T}} \in \mathbb{R}^6$ | body(vehicle) position/orientation |

| | |
|---|---|
| $\boldsymbol{\eta}_q = [\boldsymbol{\eta}_1^{\mathrm{T}} \quad \boldsymbol{\varepsilon}^{\mathrm{T}} \quad \eta]^{\mathrm{T}} \in \mathbb{R}^7$ | body(vehicle) position/orientation with the orientation expressed by quaternions |
| $\boldsymbol{\nu}_1 = [u \quad v \quad w]^{\mathrm{T}} \in \mathbb{R}^3$ | vector representing the linear velocity of the origin of the body(vehicle)-fixed frame with respect to the origin of the inertial frame expressed in the body(vehicle)-fixed frame (see Figure 2.1) |
| $\boldsymbol{\nu}_2 = [p \quad q \quad r]^{\mathrm{T}} \in \mathbb{R}^3$ | vector representing the angular velocity of the body(vehicle)-fixed frame with respect to the inertial frame expressed in the body(vehicle)-fixed frame (see Figure 2.1) |
| $\boldsymbol{\nu} = [\boldsymbol{\nu}_1^{\mathrm{T}} \quad \boldsymbol{\nu}_2^{\mathrm{T}}]^{\mathrm{T}} \in \mathbb{R}^6$ | vector representing the linear/angular velocity in the body(vehicle)-fixed frame |
| $\boldsymbol{R}_\alpha^\beta \in \mathbb{R}^{3\times 3}$ | rotation matrix expressing the transformation from frame $\alpha$ to frame $\beta$ |
| $\boldsymbol{J}_{k,o}(\boldsymbol{\eta}_2) \in \mathbb{R}^{3\times 3}$ | Jacobian matrix defined in (2.2) |
| $\boldsymbol{J}_{k,oq}(\mathcal{Q}) \in \mathbb{R}^{4\times 3}$ | Jacobian matrix defined in (2.10) |
| $\boldsymbol{J}_e(\boldsymbol{\eta}_2) \in \mathbb{R}^{6\times 6}$ | Jacobian matrix defined in (2.19) |
| $\boldsymbol{J}_{e,q}(\mathcal{Q}) \in \mathbb{R}^{7\times 6}$ | Jacobian matrix defined in (2.23) |
| $\boldsymbol{\tau}_1 = [X \quad Y \quad Z]^{\mathrm{T}} \in \mathbb{R}^3$ | vector representing the resultant forces acting on the rigid body(vehicle) expressed in the body(vehicle)-fixed frame |
| $\boldsymbol{\tau}_2 = [K \quad M \quad N]^{\mathrm{T}} \in \mathbb{R}^3$ | vector representing the resultant moment acting on the rigid body(vehicle) expressed in the body(vehicle)-fixed frame to the pole $O_b$ |
| $\boldsymbol{\tau}_v = [\boldsymbol{\tau}_1^{\mathrm{T}} \quad \boldsymbol{\tau}_2^{\mathrm{T}}]^{\mathrm{T}} \in \mathbb{R}^6$ | generalized forces: forces and moments acting on the vehicle |
| $\boldsymbol{\tau}_v^\star \in \mathbb{R}^6$ | generalized forces in the earth-fixed-frame-based model defined in (2.53) |
| $n$ | degrees of freedom of the manipulator |
| $\boldsymbol{q} \in \mathbb{R}^n$ | joint positions |
| $\boldsymbol{\tau}_q \in \mathbb{R}^n$ | joint torques |
| $\boldsymbol{\tau} = [\boldsymbol{\tau}_v^{\mathrm{T}} \quad \boldsymbol{\tau}_q^{\mathrm{T}}]^{\mathrm{T}} \in \mathbb{R}^{6+n}$ | generalized forces: vehicle forces and moments and joint torques |
| $\boldsymbol{u} \in \mathbb{R}^p$ | control inputs, $\boldsymbol{\tau} = \boldsymbol{B}\boldsymbol{u}$ (see (2.72)) |

| | |
|---|---|
| $\boldsymbol{\zeta} = [\,\boldsymbol{\nu}_1^{\mathrm{T}} \quad \boldsymbol{\nu}_2^{\mathrm{T}} \quad \dot{\boldsymbol{q}}^{\mathrm{T}}\,]^{\mathrm{T}} \in \mathrm{I\!R}^{6+n}$ | *system* velocity |
| $\boldsymbol{\Phi} \in R^{(6+n)\times n_\theta}$ | UVMS regressor defined in (2.73) |
| $\boldsymbol{\theta} \in R^{n_\theta}$ | vector of the dynamic parameters of the UVMS regressor defined in (2.73) |
| $\boldsymbol{\Phi}_v \in R^{6\times n_{\theta,v}}$ | vehicle regressor defined in (2.54) |
| $\boldsymbol{\theta}_v \in R^{n_{\theta,v}}$ | vector of the dynamic parameters of the vehicle regressor defined in (2.54) |
| $\boldsymbol{\eta}_{ee1} = [\,x_E \quad y_E \quad z_E\,]^{\mathrm{T}} \in \mathrm{I\!R}^3$ | position of the end effector in the inertial frame (denoted with $\boldsymbol{x} = [\,x_E \quad y_E \quad z_E\,]^{\mathrm{T}}$ in the interaction control sections) |
| $\boldsymbol{\eta}_{ee2} = [\,\phi_E \quad \theta_E \quad \psi_E\,]^{\mathrm{T}} \in \mathrm{I\!R}^3$ | orientation of the end effector in the inertial frame expressed by Euler angles |
| $\boldsymbol{\nu}_{ee} \in \mathrm{I\!R}^6$ | end-effector linear and angular velocities with respect to the inertial frame expressed in the end-effector frame |
| $\boldsymbol{J}_k(\boldsymbol{R}_B^I) \in \mathrm{I\!R}^{(6+n)\times(6+n)}$ | Jacobian matrix defined in (2.58) |
| $\boldsymbol{J}_w(\boldsymbol{R}_B^I, \boldsymbol{q}) \in \mathrm{I\!R}^{6\times(6+n)}$ | Jacobian matrix defined in (2.67) |
| $\boldsymbol{J}(\boldsymbol{R}_B^I, \boldsymbol{q}) \in \mathrm{I\!R}^{6\times(6+n)}$ | Jacobian matrix used in (2.68) |
| $\boldsymbol{h}_i^i = \left[\,\boldsymbol{f}_i^{i\mathrm{T}} \quad \boldsymbol{\mu}_i^{i\mathrm{T}}\,\right]^{\mathrm{T}} \in \mathrm{I\!R}^6$ | forces and moments exerted by body $i-1$ on body $i$ (see Figure 2.4) |
| $\boldsymbol{h}_e = [\,\boldsymbol{f}_e^{\mathrm{T}} \quad \boldsymbol{\mu}_e^{\mathrm{T}}\,]^{\mathrm{T}} \in \mathrm{I\!R}^6$ | forces and moments at the end effector (see Figure 2.5) |
| $t \in \mathrm{I\!R}$ | time |
| $\lambda_{\min(\max)}(\boldsymbol{X})$ | smallest(largest) eigenvalue of matrix $\boldsymbol{X}$ |
| $\mathrm{diag}\{x_1, \ldots, x_n\}$ | Diagonal matrix filled with $x_i$ in the $i$ row, $i$ column and zero in any other place |
| $\mathrm{blockdiag}\{\boldsymbol{X}_1, \ldots, \boldsymbol{X}_n\}$ | Block diagonal matrix filled with matrices $\boldsymbol{X}_1, \ldots, \boldsymbol{X}_n$ in the main diagonal and zero in any other place |
| $\mathcal{R}(\boldsymbol{X})$ | range of matrix $\boldsymbol{X}$ |
| $\dot{x}$ | time derivative of the variable $x$ |
| $\Vert\boldsymbol{x}\Vert$ | 2-norm of the vector $\boldsymbol{x}$ |
| $\hat{\boldsymbol{x}}\left(\hat{\boldsymbol{X}}\right)$ | estimate of the vector $\boldsymbol{x}$ (matrix $\boldsymbol{X}$) |
| $x_d$ | desired value of the variable $x$ |

| | |
|---|---|
| $\tilde{x}$ | error variable defined as $\tilde{x} = x_d - x$ |
| $\boldsymbol{x}^{\mathrm{T}} \left(\boldsymbol{X}^{\mathrm{T}}\right)$ | transpose of the vector $\boldsymbol{x}$ (matrix $\boldsymbol{X}$) |
| $x_i$ | $i$ th element of the vector $\boldsymbol{x}$ |
| $X_{i,j}$ | element at row $i$, column $j$ of the matrix $\boldsymbol{X}$ |
| $\boldsymbol{X}^{\dagger}$ | Moore-Penrose inversion (pseudoinversion) of matrix $\boldsymbol{X}$<br>If $\boldsymbol{X}$ is low rectangular it is<br>$$\boldsymbol{X}^{\dagger} = \boldsymbol{X}^{\mathrm{T}} \left(\boldsymbol{X}\boldsymbol{X}^{\mathrm{T}}\right)^{-1}$$<br>If $\boldsymbol{X}$ is high rectangular it is<br>$$\boldsymbol{X}^{\dagger} = \left(\boldsymbol{X}^{\mathrm{T}}\boldsymbol{X}\right)^{-1} \boldsymbol{X}^{\mathrm{T}}$$ |
| $\boldsymbol{I}_r$ | $(r \times r)$ identity matrix |
| $\boldsymbol{O}_{r_1 \times r_2}$ | $(r_1 \times r_2)$ null matrix |
| $\boldsymbol{S}(\cdot) \in \mathbb{R}^{3\times 3}$ | matrix performing the cross product between two $(3 \times 1)$ vectors defined in (2.6) |
| $\rho^3$ | water density |
| $\mu$ | fluid dynamic viscosity |
| $R_n$ | Reynolds number |
| $\boldsymbol{g}^I$ | gravity acceleration expressed in the inertial frame |

# Contents

**1. Introduction** ..... 1
1.1 Underwater Vehicles ..... 3
1.2 Sensorial Systems ..... 5
1.3 Actuation ..... 5
1.4 Localization ..... 7
1.5 AUVs' Control ..... 9
1.5.1 Fault Detection/Tolerance for UUVs ..... 11
1.6 UVMS' Coordinated Control ..... 11
1.7 Future Perspectives ..... 11

**2. Modelling of Underwater Robots** ..... 15
2.1 Introduction ..... 15
2.2 Rigid Body's Kinematics ..... 15
2.2.1 Attitude Representation by Euler Angles ..... 16
2.2.2 Attitude Representation by Quaternion ..... 17
2.2.3 Attitude Error Representation ..... 19
2.2.4 6-DOFs Kinematics ..... 21
2.3 Rigid Body's Dynamics ..... 22
2.3.1 Rigid Body's Dynamics in Matrix Form ..... 24
2.4 Hydrodynamic Effects ..... 25
2.4.1 Added Mass and Inertia ..... 26
2.4.2 Damping Effects ..... 28
2.4.3 Current Effects ..... 29
2.5 Gravity and Buoyancy ..... 31
2.6 Thrusters' Dynamics ..... 32
2.7 Underwater Vehicles' Dynamics in Matrix Form ..... 34
2.7.1 Linearity in the Parameters ..... 35
2.8 Kinematics of Manipulators with Mobile Base ..... 36
2.9 Dynamics of Underwater Vehicle-Manipulator Systems ..... 39
2.9.1 Linearity in the Parameters ..... 42
2.10 Contact with the Environment ..... 42
2.11 Identification ..... 43

**3. Dynamic Control of 6-DOF AUVs** .......... 45
3.1 Introduction .......... 45
3.2 Earth-Fixed-Frame-Based, Model-Based Controller .......... 47
3.3 Earth-Fixed-Frame-Based, Non-model-Based Controller .......... 49
3.4 Vehicle-Fixed-Frame-Based, Model-Based Controller .......... 51
3.5 Model-Based Controller Plus Current Compensation .......... 53
3.6 Mixed Earth/Vehicle-Fixed-Frame-Based, Model-Based Controller .......... 55
3.6.1 Stability Analysis .......... 56
3.7 Jacobian-Transpose-Based Controller .......... 57
3.8 Comparison Among Controllers .......... 59
3.8.1 Compensation of the Restoring Generalized Forces .......... 59
3.8.2 Compensation of the Ocean Current .......... 60
3.9 Numerical Comparison Among the Reduced Controllers .......... 60
3.9.1 Results .......... 63
3.9.2 Conclusions and Extension to UVMSs .......... 77

**4. Fault Detection/Tolerance Strategies for AUVs and ROVs** 79
4.1 Introduction .......... 79
4.2 Experienced Failures .......... 80
4.3 Fault Detection Schemes .......... 82
4.4 Fault Tolerant Schemes .......... 86
4.5 Experiments .......... 88
4.6 Conclusions .......... 91

**5. Experiments of Dynamic Control of a 6-DOF AUV** .......... 93
5.1 Introduction .......... 93
5.2 Experimental Set-Up .......... 93
5.3 Experiments of Dynamic Control .......... 94
5.4 Experiments of Fault Tolerance to Thrusters' Fault .......... 101

**6. Kinematic Control of UVMSs** .......... 105
6.1 Introduction .......... 105
6.2 Kinematic Control .......... 106
6.3 The Drag Minimization Algorithm .......... 112
6.4 The Joint Limits Constraints .......... 112
6.5 Singularity-Robust Task Priority .......... 113
6.6 Fuzzy Inverse Kinematics .......... 121
6.7 Conclusions .......... 139

**7. Dynamic Control of UVMSs** .......... 141
7.1 Introduction .......... 141
7.2 Feedforward Decoupling Control .......... 143
7.3 Feedback Linearization .......... 146
7.4 Nonlinear Control for UVMSs with Composite Dynamics .......... 146

7.5 Non-regressor-Based Adaptive Control .................... 149
7.6 Sliding Mode Control .................... 151
7.6.1 Stability Analysis .................... 152
7.6.2 Simulations .................... 154
7.7 Adaptive Control .................... 157
7.7.1 Stability Analysis .................... 158
7.7.2 Simulations .................... 160
7.8 Output Feedback Control .................... 162
7.8.1 Stability Analysis .................... 168
7.8.2 Simulations .................... 172
7.9 Virtual Decomposition Based Control .................... 181
7.9.1 Stability Analysis .................... 186
7.9.2 Simulations .................... 188
7.9.3 Virtual Decomposition with the Proper Adapting Action 194
7.10 Conclusions .................... 198

**8. Interaction Control of UVMSs** .................... 201
8.1 Introduction to Interaction Control of Robots .................... 201
8.2 Dexterous Cooperating Underwater 7-DOF Manipulators .... 203
8.3 Impedance Control .................... 203
8.4 External Force Control .................... 205
8.4.1 Inverse Kinematics .................... 206
8.4.2 Stability Analysis .................... 207
8.4.3 Robustness .................... 208
8.4.4 Loss of Contact .................... 209
8.4.5 Implementation Issues .................... 209
8.4.6 Simulations .................... 210
8.5 Explicit Force Control .................... 213
8.5.1 Robustness .................... 217
8.5.2 Simulations .................... 218
8.6 Conclusions .................... 223

**9. Coordinated Control of Platoons of AUVs** .................... 225
9.1 Introduction .................... 225
9.2 Kinematic Control of AUVs .................... 227
9.2.1 Simulations .................... 232
9.3 Experimental Set-Up at the Virginia Tech .................... 235
9.4 Conclusions .................... 236

**10. Concluding Remarks** .................... 237

A. **Mathematical models** . . . . . 239
A.1 Introduction . . . . . 239
A.2 Phoenix . . . . . 239
A.3 Phoenix+6DOF SMART 3S . . . . . 241
A.4 ODIN . . . . . 242
A.5 9-DOF UVMS . . . . . 243

**References** . . . . . 247

# 1. Introduction

One of the first efforts to design an underwater vehicle is due to Leonardo Da Vinci. It has been found in the *Codice Atlantico* (Codex Atlanticus), written between 1480 and 1518, together with the development of some diver's devices (see Figure 1.1 and 1.2 where the corresponding page of the Codex is reported). Legends say that Leonardo worked on the idea of an underwater military machine and that he further destroyed by himself the results judged too dangerous.

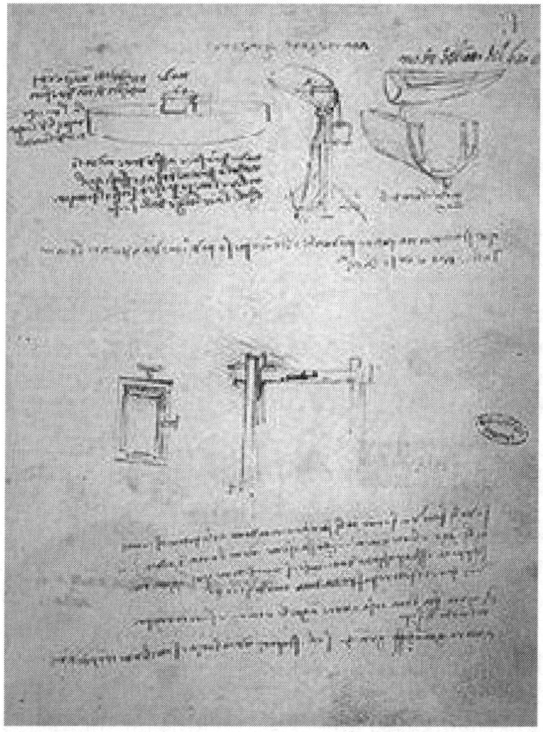

**Fig. 1.1.** Page of the *Codice Atlantico* (around 1500) containing the draw of the manned underwater vehicle developed by Leonardo Da Vinci

Maybe the first idea of an underwater machine is from Aristotle; following the legend he built a machine: *skaphe andros* (boat-man) that allowed Alexander the Great (Alexander III of Macedon, 356 – 323 b. C.) to stay in deep for at least half a day during the war of Tiro in 325 b. C. This is unrealistic, of course, also considering that the Archimedes's law was still to become a reality (around 250 b. C.).

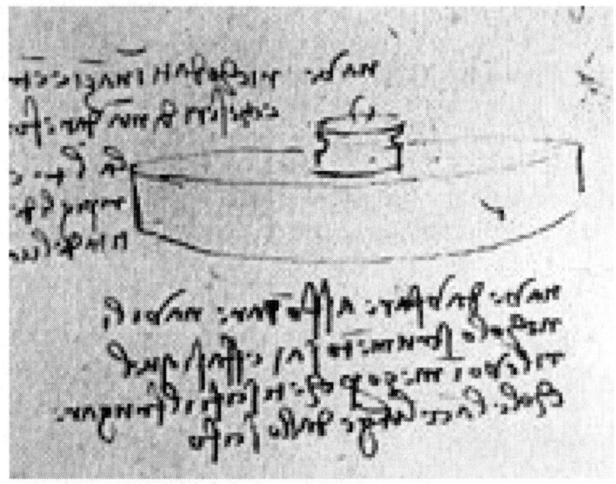

**Fig. 1.2.** Particular of the page of the *Codice Atlantico* containing the draw of the manned underwater vehicle developed by Leonardo Da Vinci

In August, the 4 th, 2005, in the Pacific sea, in front of the Kamchatka, at a depth of 200 meters, a Russian manned submarine, the *AS-28*, got stacked into the cables of a underwater radar; at that moment, seven men were in the vehicle. One day later a British Remotely Operated Vehicle (ROV), *Scorpio*, was there and, after another day of operations, it was possible to cut the cables thus allowing the submarine to surface safely. In addition than exceptional operations like the one mentioned, underwater robots can be used to accomplish missions such as sea bottom and pipeline survey, cable maintenance, off-shore structures' monitoring and maintenance, collect/release of biological surveys. Currently, most of the operations mentioned above are achieved via manned underwater vehicles or remotely operated vehicles; in case of manipulation tasks, moreover, those are performed resorting to remotely operated master-slave systems. The strong limit of the use of manned vehicles is the enormous cost and risk in working in such an hostile environment; the daily operating cost is larger than 8000 € ($\approx$ 10000 $) [323]. The aim of the research is to progressively make it possible to perform such missions in a completely autonomous way.

This objective is challenging from the technological as well as from the theoretical aspects since it implies a wide range of technical and research topics. Sending an autonomous vehicle in an unknown and unstructured environment, with limited on-line communication, requires some on board *intelligence* and the ability of the vehicle to react in a reliable way to unexpected situations. The sensory system of the vehicle must deal with a noisy and unstructured environment; moreover, technologies as GPS (Global Positioning System) are not applicable due to the impossibility to underwater electromagnetic transmission at GPS specific frequencies; vision based systems are not fully reliable due to the generally poor visibility. The actuating system is usually composed of thrusters and control surfaces, both of them have a non-linear dynamics and are strongly affected by the hydrodynamic effects.

The book of T. Fossen [127] is one of the first books dedicated to control problems of marine systems, the case of surface vehicles, in fact, is also taken into account. The same author presents, in [128], an updated and extended version of the topics developed in the first book. Some very interesting talks about state of the art and direction of the underwater robotics were discussed by, e.g., J. Yuh in [315, 317], J. Yuh and M. West in [321], T. Ura in [292]. At the best of our knowledge this is the sole book dedicated to control problems of underwater robotic systems with particular regard with respect to the manipulation [8]; this is an emerging topic in which solid experimental results still need to be achieved.

In this chapter an overview of control problem in underwater robotics is presented; some of these aspects will be further analyzed along this book.

## 1.1 Underwater Vehicles

The therm Remotely Operated Vehicle (ROV) denotes an underwater vehicle physically linked, via the tether, to an operator that can be on a submarine or on a surface ship. The tether is in charge of giving power to the vehicle as well as closing the manned control loop. Autonomous Underwater Vehicles (AUVs), on the other side, are supposed to be completely autonomous, thus relying to onboard power system and *intelligence*. These two types of underwater vehicles share some control problems, in this case one has to refer to them as Unmanned Underwater Vehicles (UUVs). In case of missions that require interaction with the environment, the vehicle can be equipped with one or more manipulators; in this case the system is usually called Underwater Vehicle-Manipulator System (UVMS).

Referring to AUVs, [294, 321] present the state of the art of several existing AUVs and their control architectures. Currently, there are about 100 prototypes in the laboratories all over the world, see e.g., [321]. Among the others: *r2D4* developed at URA laboratory of the University of Tokyo (Tokyo, Japan, `http://underwater.iis.u-tokyo.ac.jp`), *ABE* of the Deep Submerge Laboratory of the Woods Hole Oceanographic Institution

(Massachusetts, USA, `http://www.dsl.whoi.edu`), *Odissey IId* belonging to the AUV Laboratory of the Massachusetts Institute of Technology (Massachusetts, USA, `http://auvlab.mit.edu`), *ODIN III* designed at the Autonomous Systems Laboratory of the University of Hawaii (Hawaii, USA, `http://www.eng.hawaii.edu/~asl`), *Phoenix* and *ARIES*, torpedo-like vehicles developed at the Naval Postgraduate School (California, USA, `http://www.cs.nps.navy.mil/research/auv/`).

Currently, very few companies sell AUVs; among the others: Bluefin Corporations (`http://www.bluefinrobotics.com`) developed, in collaboration with MIT, different AUVs, such as Bluefin 21, for deep operations up to 4500 m; C&C technologies (`www.cctechnol.com`) designed Hugin 3000, able to run autonomously for up to 50 h; the Canadian ISE Research Ltd (`http://www.ise.bc.ca`) developed several AUVs such as, e.g, Explorer or Theseus; Hafmynd, in Iceland, designed a very small AUV named Gavia (`http://www.gavia.is`); the Danish Maridan (`http://www.maridan.dk`) developed the Maridan 600 vehicle.

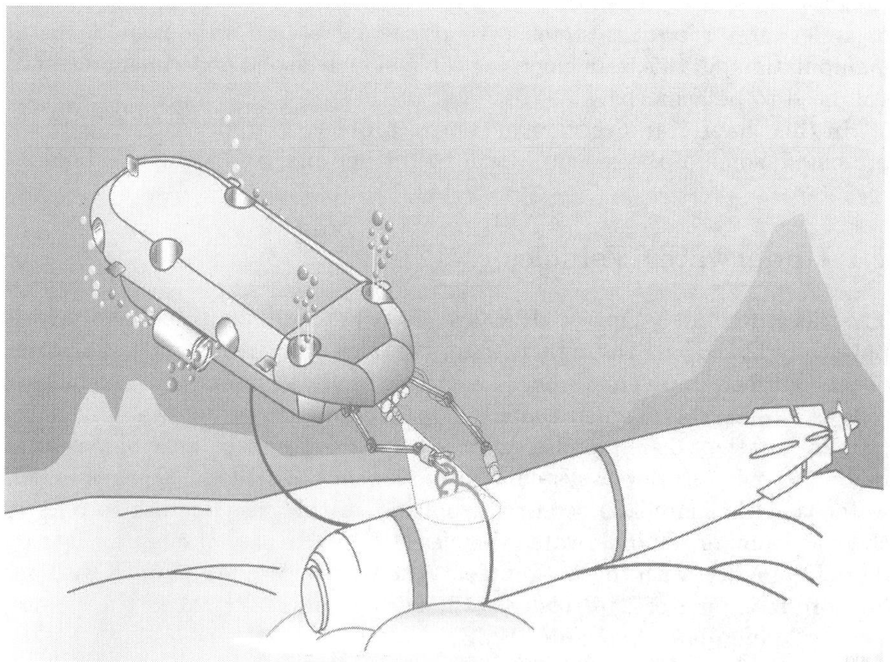

**Fig. 1.3.** Sketch of the underwater vehicle-manipulator system SAUVIM, currently under development at the Autonomous Systems Laboratory of the University of Hawaii (courtesy of J. Yuh)

The UVMSs are still under development; several laboratories built some manipulation devices on underwater structures but very few of them can be considered as capable of autonomous manipulation. *SAUVIM* (see Figure 1.3), a semi-autonomous vehicle with an *Ansaldo* 7-link manipulator is under development at the Autonomous Systems Laboratory of the University of Hawaii; this vehicle, in the final version, will be able to operate at the depth of 4000 ts m; preliminary experiments were performed. *AMADEUS*, an acronym for Advanced MAnipulation for DEep Underwater Sampling, funded by the European Commission, that involved the Heriot-Watt University (UK), the Università di Genova (Italy), the National Research Council-ISSIA (Italy), the Universitat de Barcelona (Spain), the Institute of Marine Biology of Crete (Greece). The project focused on the coordinated control of two teleoperated underwater *Ansaldo* 7-link manipulators and the development of an underwater hand equipped with a slip sensor; Figure 1.4 shows a wet test in a pool. The French company Cybernétix (`http://www.cybernetix.fr`) sells hydraulic manipulators mounted on ROVs that can be remotely operated by means of a joystick or in a master-slave configuration.

## 1.2 Sensorial Systems

The AUVs need to operate in an unstructured hazardous environment; one of the major problems with underwater robotics is in the localization task due to the absence of a single, proprioceptive sensor that measures the vehicle position and the impossibility to use the GPS under the water. The use of redundant multi-sensor system, thus, is common in order to perform sensor fusion tasks and give fault detection and tolerance capability to the vehicle. To give an idea of the sensors used in underwater robotics, Table 1.1 lists the sensors and the corresponding measured variable for Unmanned Underwater Vehicles (UUVs).

As an example, Table 1.2 reports some data of the instrumentations of the ROV developed at the John Hopkins University [271] and Table 1.3 some data of the AUV ODIN III [81, 323].

## 1.3 Actuation

Underwater vehicles are usually controlled by thrusters and/or control surfaces. Control surfaces, such as rudders and sterns, are common in cruise vehicles; those are torpedo-shaped and usually used in cable/pipeline inspection. The main configuration is not changed in the last century, there is a main thruster and at least one rudder and one stern, in Figure 1.5 it is reported the underwater manned vehicle named SLC (*Siluro a Lenta Corsa*, Slow Running Torpedo), or *maiale* (pig), used in the second world war by the *Regia*

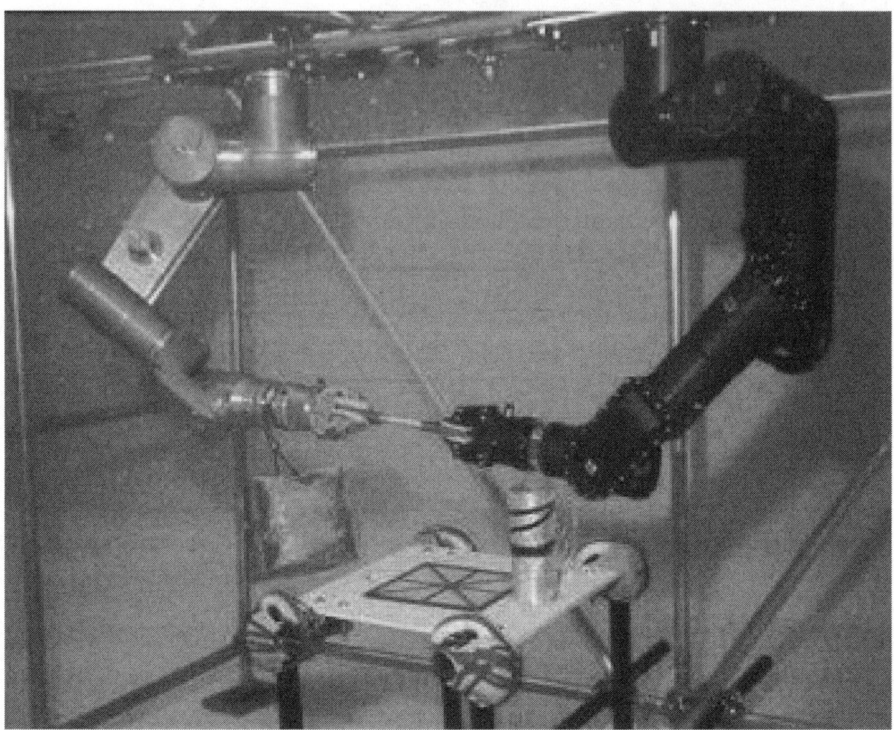

**Fig. 1.4.** Coordinated control of two seven-link Ansaldo manipulators during a wet test in a pool (courtesy of G. Casalino, Genoa Robotics And Automation Laboratory, Università di Genova and G. Veruggio, National Research Council-ISSIA, Italy)

**Table 1.1.** UUV possible instrumentation

| sensor | measured variable |
|---|---|
| Inertial System | linear acceleration and angular velocity |
| Pressure-meter | vehicle depth |
| Frontal sonar | distance from obstacles |
| Vertical sonar | distance from the bottom |
| Ground Speed sonar | relative velocity vehicle/bottom |
| Current-meter | relative velocity vehicle/current |
| Global Positioning System | absolute position at the surface |
| Compass | orientation |
| Acoustic baseline | absolute position in known area |
| Vision systems | relative position/velocity |
| Acoustic Doppler Current Profiler | water current at several positions |

**Table 1.2.** JHUROV instrumentations

| measured variable | sensor | precision | update rate |
|---|---|---|---|
| 3DOF-vehicle position | SHARP acoustic transponder | 0.5 cm | 10 Hz |
| depth | Foxboro/ICT model n. 15 | 2.5 cm | 20 Hz |
| heading | Litton LN200 IMU Gyro | 0.01 deg | 20 Hz |
| roll and pitch | KVH ADGC | 0.1 deg | 10 Hz |
| heading | KVH ADGC | 1 deg | 10 Hz |

**Table 1.3.** ODIN III sensors update

| measured variable | sensor | update rate |
|---|---|---|
| $xy$ vehicle position | 8 sonars | 3 Hz |
| depth | pressure sensor | 30 Hz |
| roll, pitch and yaw | IMU | 30 Hz |

*Marina Italiana* (Royal Italian Navy). Since the force/moment provided by the control surfaces is function of the velocity and it is null in hovering, they are not useful to manipulation missions in which, due to the manipulator interaction, full control of the vehicle is required.

The relationship between the force/moment acting on the vehicle and the control input of the thrusters is highly nonlinear. It is function of some structural variables such as: the density of the water; the tunnel cross-sectional area; the tunnel length; the volumetric flowrate between input-output of the thrusters and the propeller diameter. The state of the dynamic system describing the thrusters is constituted by the propeller revolution, the speed of the fluid going into the propeller and the input torque.

A detailed theoretical and experimental analysis of thrusters' behavior can be found in [40, 147, 176, 289, 300, 309]. In [128] a chapter is dedicated to modelling and control of marine thrusters. Roughly speaking, thrusters are the main cause of limit cycle in vehicle positioning and bandwidth constraint. In [178] the thruster model is explicitly taken into account in the control law. Reference [270] presents experimental results on the performance of model-based control law for AUVs in presence of model mismatching and thrusters' saturation.

## 1.4 Localization

The position and attitude of a free floating vehicle is not measurable by the use of a single, internal sensor. This poses the problem of estimating the AUV's position. As detailed above, several sensors are normally mounted

**Fig. 1.5.** *Pig*, manned vehicle used by the Royal Italian Navy during the second world war in the Mediterranean Sea. The thruster and the group rudder/stern can be observed in the bottom left angle of the photo

on an AUV in order to implement sensor fusion algorithms and obtain an estimation more reliable than by using a single sensor.

Among the possible methods is the use of baseline acoustics, those rely on the use of transmitters/receivers mounted on the vehicle and on known locations needed for triangulations. In same cases one single module can be used and mounted on a surface vehicle the position of which is acquired by means of a GPS. In case of partially *structured* environments, such as harbors, transmitters/receivers at known positions can be easily deployed. In absence of baseline acoustics there is the need to measure a time derivatives of the vehicle position such as the acceleration with the IMU or the velocity with a Ground Speed sonar fused with the vehicle orientation measurements. It is well known in estimation theory, however, that the time integration of a measurement leads to the *Brownian motion*, or *random walk*, i.e., a stochastic model whose variance grows linearly with the elapsed time. After some time, thus, the estimation is useless and a reset of the error is necessary by, e.g., surfacing the vehicle and measuring its real position with a GPS. Finally, when the vehicle uses sonar or video-cameras it can measure several time its relative position with respect to a fixed feature; this information can be used in a, e.g., Extended Kalman Filter (EKF) to improve the estimation of the vehicle's position.

This topic is treated, among the others, in [98, 134, 139, 221, 269, 290, 304].

## 1.5 AUVs' Control

Control of AUVs' is challenging, in fact, even though this problem is kinematically similar to the widely studied one of controlling a free-floating rigid body in a six-dimensional space, the underwater environment makes the dynamics to be faced quite different. An overview of the main control techniques for AUVs can be found, e.g., in [127, 128].

**Fig. 1.6.** Possible scenario of mine countermeasure using an AUV platoon (courtesy of SACLANT Undersea Research Center, North Atlantic Treaty Organization)

A main difference in control of underwater vehicles is related to the type of actuation; cruise vehicles, in fact, are usually actuated by means of one thruster and several control surfaces; they are under-actuated and mainly controlled in the surge, sway and heave directions. On the other hand, if a vehicle is conceived for manipulation tasks it is required that it is actuated in all the DOFs even at very low velocities; 6 or more thrusters are then designed.

An example of cruise vehicle is ARIEL, belonging to the Naval Postgraduate School; a detailed description and its command and control subsystems are provided in [201]. In [145], the control system of the NPS AUV II is given together with experimental results. Control laws for cruise vehicles are usually designed at a nominal velocity since the vehicle is designed for exploration or cable tracking missions, see, e.g., [41] for a pipeline tracking with Twin-Burger 2. The homing operation needs specific algorithms, [116] presents experimental results performed with the vehicle Odyssey IIb, of an homing system based on an electromagnetic guidance rather than an acoustic signal.

Research efforts have been devoted at controlling fully actuated underwater vehicle, in particular at very low velocity or performing a station keeping task. This topic will be discussed in Chapter 3, some experimental results is given in Chapter 5.

Identification of the dynamic parameters of underwater robotic structures is a very challenging task due to the model characteristics, i.e., non-linear and coupled dynamics, difficulty in obtaining effective data; the interested reader can refer to [2, 62, 111, 271].

An emerging topic is also constituted by control of platoon of AUVs, see, e.g., the work of [27, 179, 276]. In Figure 1.6, a possible scenario of mine countermeasure using a platoon of AUVs under study at the SACLANT Undersea Research Center of the North Atlantic Treaty Organization (NATO) is given.

In Figure 1.7, one of the AUVs developed at the Virginia Tech is shown, these vehicles will be very small in size and cheap with most of the components custom-engineered [132].

A brief discussion on control of multi-AUVs is given in Chapter 9.

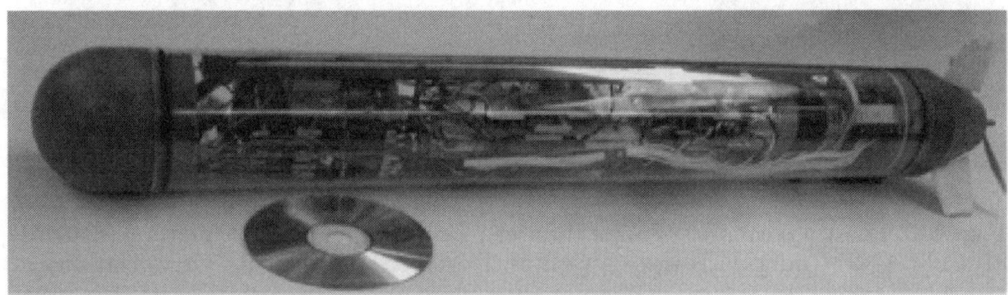

**Fig. 1.7.** One of the vehicles constituting the platoon of AUVs developed at the Virginia Tech (courtesy of D. Stilwell)

### 1.5.1 Fault Detection/Tolerance for UUVs

ROVS and AUVs are complex systems engaged in missions in un-structured, unsafe environments for which the degree of autonomy becomes a crucial issue. In this sense, the capability to detect and tolerate faults is a key to successfully terminate the mission or recuperate the vehicle. An overview of fault detection and fault tolerance algorithms, specifically designed for UUVs is presented in Chapter 4.

In Figure 1.8, the vehicle Romeo operating over thermal vents in the Milos Island, Aegean Sea, Greece, is shown; this vehicle has been built at the RobotLab, National Research Council (CNR-ISSIA), Genova, Italy (`http://www.robotlab.ian.ge.cnr.it/`). This vehicle has been object of several experimental studies on fault detection/tolerance algorithms.

## 1.6 UVMS' Coordinated Control

The use of Autonomous Underwater Vehicles (AUVs) equipped with a manipulator (UVMS) to perform complex underwater tasks give rise to challenging control problems involving nonlinear, coupled, and high-dimensional systems. Currently, the state of the art is represented by tele-operated master/slave architectures; few research centers are equipped with autonomous systems [180, 321].

The *core* of this monograph is dedicated to this topic, in Chapter 6 the kinematic control will be discussed, Chapter 7 presents dynamic control laws for UVMSs and Chapter 8 shows some interaction control schemes.

## 1.7 Future Perspectives

Underwater robotics research is an interesting topic. Current technology allows to safely run long duration missions that involve one single AUV, e.g., as in the case of the Naval Postgraduate School or the Ura laboratory vehicles, or to execute manned-in-the-loop manipulation tasks. There are, however, research topics that need to be further investigated.

The UVMSs need to be studied in the field; from the theoretical aspect, in fact, many of the associated problems have been studied and, possibly, solved: kinematic and dynamic control laws, as well as interaction control laws have been designed and successfully simulated. Few experimental set-up have also been used; these, however, reproduced only oversimplified environments. Interesting results might be achieved by means of a fully actuated autonomous underwater vehicle carrying a 6-DOF manipulator.

The actuating system might be improved in an effort to reduce the limit cycles caused by the thrusters' dynamics at very low velocities; new blade

**Fig. 1.8.** Romeo operating over thermal vents in the Milos Island, Aegean Sea, Greece, during the final demo of the EC-funded project ARAMIS (courtesy of M. Caccia, National Research Council-ISSIA, Italy)

profiles, e.g., might be studied in order to *linearize* the input-output thruster relationships.

The sensory system is also still object of research; recent advances concern the possibility to practically achieve Simultaneously Localization And Mapping (SLAM) with one AUV or perform sensor fusion by the use of platoon of AUVs. In the case of SAUVIM, a passive manipulator is considered with 6 DOFs in charge of measuring the vehicle position/orientation when the UVMS is close to a structure.

In ground and aerial robotics the topic of controlling platoon of vehicles is being studied since longtime; this is an emerging topic also for platoon of AUVs where specific dynamic considerations, communication limitations and control constraints need to be taken into account.

# 2. Modelling of Underwater Robots

*"We have Einstein's space, de Sitter's spaces, expanding universes, contracting universes, vibrating universes, mysterious universes. In fact the pure mathematician may create universes just by writing down an equation, and indeed, if he is an individualist he can have an universe of his own".*
*J.J. Thomson, around 1919.*

## 2.1 Introduction

In this Chapter the mathematical model of UVMSs is derived. Modeling of rigid bodies moving in a fluid or underwater manipulators has been studied in literature by, among others, [137, 156, 157, 174, 182, 189, 203, 242, 255, 256, 285, 286], where a deeper discussion of specific aspects can be found. In [224], the model of two UVMSs holding the same rigid object is derived.

## 2.2 Rigid Body's Kinematics

A rigid body is completely described by its position and orientation with respect to a reference frame $\Sigma_i, O-\boldsymbol{xyz}$ that it is supposed to be earth-fixed and inertial. Let define $\boldsymbol{\eta}_1 \in \mathbb{R}^3$ as

$$\boldsymbol{\eta}_1 = \begin{bmatrix} x \\ y \\ z \end{bmatrix},$$

the vector of the body position coordinates in a earth-fixed reference frame. The vector $\dot{\boldsymbol{\eta}}_1$ is the corresponding time derivative (expressed in the earth-fixed frame). If one defines

$$\boldsymbol{\nu}_1 = \begin{bmatrix} u \\ v \\ w \end{bmatrix}$$

as the linear velocity of the origin of the body-fixed frame $\Sigma_b, O_b - \boldsymbol{x}_b\boldsymbol{y}_b\boldsymbol{z}_b$ with respect to the origin of the earth-fixed frame expressed in the body-fixed frame (from now on: body-fixed linear velocity) the following relation between the defined linear velocities holds:

$$\boldsymbol{\nu}_1 = \boldsymbol{R}_I^B \dot{\boldsymbol{\eta}}_1 \,, \tag{2.1}$$

where $\boldsymbol{R}_I^B$ is the rotation matrix expressing the transformation from the inertial frame to the body-fixed frame.

In the following, two different attitude representations will be introduced: Euler angles and Euler parameters or quaternion. In marine terminology is common the use of Euler angles while several control strategies use the quaternion in order to avoid the representation singularities that might arise by the use of Euler angles.

**Table 2.1.** Common notation for marine vehicle's motion

| | | forces and moments | $\boldsymbol{\nu}_1, \boldsymbol{\nu}_2$ | $\boldsymbol{\eta}_1, \boldsymbol{\eta}_2$ |
|---|---|---|---|---|
| motion in the $x$-direction | surge | $X$ | $u$ | $x$ |
| motion in the $y$-direction | sway | $Y$ | $v$ | $y$ |
| motion in the $z$-direction | heave | $Z$ | $w$ | $z$ |
| rotation about the $x$-axis | roll | $K$ | $p$ | $\phi$ |
| rotation about the $y$-axis | pitch | $M$ | $q$ | $\theta$ |
| rotation about the $z$-axis | yaw | $N$ | $r$ | $\psi$ |

### 2.2.1 Attitude Representation by Euler Angles

Let define $\boldsymbol{\eta}_2 \in \mathrm{IR}^3$ as

$$\boldsymbol{\eta}_2 = \begin{bmatrix} \phi \\ \theta \\ \psi \end{bmatrix}$$

the vector of body Euler-angle coordinates in a earth-fixed reference frame. In the nautical field those are commonly named roll, pitch and yaw angles and corresponds to the elementary rotation around $x$, $y$ and $z$ in fixed frame [254]. The vector $\dot{\boldsymbol{\eta}}_2$ is the corresponding time derivative (expressed in the inertial frame). Let define

$$\boldsymbol{\nu}_2 = \begin{bmatrix} p \\ q \\ r \end{bmatrix}$$

as the angular velocity of the body-fixed frame with respect to the earth-fixed frame expressed in the body-fixed frame (from now on: body-fixed angular velocity). The vector $\dot{\boldsymbol{\eta}}_2$ does not have a physical interpretation and it is related to the body-fixed angular velocity by a proper Jacobian matrix:

$$\boldsymbol{\nu}_2 = \boldsymbol{J}_{k,o}(\boldsymbol{\eta}_2)\dot{\boldsymbol{\eta}}_2 \,. \tag{2.2}$$

The matrix $\boldsymbol{J}_{k,o} \in \mathbb{R}^{3\times3}$ can be expressed in terms of Euler angles as:

$$\boldsymbol{J}_{k,o}(\boldsymbol{\eta}_2) = \begin{bmatrix} 1 & 0 & -s_\theta \\ 0 & c_\phi & c_\theta s_\phi \\ 0 & -s_\phi & c_\theta c_\phi \end{bmatrix}, \tag{2.3}$$

where $c_\alpha$ and $s_\alpha$ are short notations for $\cos(\alpha)$ and $\sin(\alpha)$, respectively. Matrix $\boldsymbol{J}_{k,o}(\boldsymbol{\eta}_2)$ is not invertible for every value of $\boldsymbol{\eta}_2$. In detail, it is

$$\boldsymbol{J}_{k,o}^{-1}(\boldsymbol{\eta}_2) = \frac{1}{c_\theta} \begin{bmatrix} 1 & s_\phi s_\theta & c_\phi s_\theta \\ 0 & c_\phi c_\theta & -c_\theta s_\phi \\ 0 & s_\phi & c_\phi \end{bmatrix}, \tag{2.4}$$

that it is singular for $\theta = (2l+1)\frac{\pi}{2}$ rad, with $l \in \mathbb{N}$, i.e., for a pitch angle of $\pm\frac{\pi}{2}$ rad.

The rotation matrix $\boldsymbol{R}_I^B$, needed in (2.1) to transform the linear velocities, is expressed in terms of Euler angles by the following:

$$\boldsymbol{R}_I^B(\boldsymbol{\eta}_2) = \begin{bmatrix} c_\psi c_\theta & s_\psi c_\theta & -s_\theta \\ -s_\psi c_\phi + c_\psi s_\theta s_\phi & c_\psi c_\phi + s_\psi s_\theta s_\phi & s_\phi c_\theta \\ s_\psi s_\phi + c_\psi s_\theta c_\phi & -c_\psi s_\phi + s_\psi s_\theta c_\phi & c_\phi c_\theta \end{bmatrix}. \tag{2.5}$$

Table 2.1 shows the common notation used for marine vehicles according to the SNAME notation ([272]), Figure 2.1 shows the defined frames and the elementary motions.

### 2.2.2 Attitude Representation by Quaternion

To overcome the possible occurrence of representation singularities it might be convenient to resort to non-minimal attitude representations. One possible choice is given by the quaternion. The term *quaternion* was introduced by Hamilton in 1840, 70 years after the introduction of a four-parameter rigid-body attitude representation by Euler. In the following, a short introduction to quaternion is given.

By defining the mutual orientation between two frames of common origin in terms of the rotation matrix

$$\boldsymbol{R}_{\boldsymbol{k}}(\delta) = \cos\delta \boldsymbol{I}_3 + (1 - \cos\delta)\boldsymbol{k}\boldsymbol{k}^{\mathrm{T}} - \sin\delta \boldsymbol{S}(\boldsymbol{k}),$$

where $\delta$ is the angle and $\boldsymbol{k} \in \mathbb{R}^3$ is the unit vector of the axis expressing the rotation needed to align the two frames, $\boldsymbol{I}_3$ is the $(3\times3)$ identity matrix, $\boldsymbol{S}(\boldsymbol{x})$ is the matrix operator performing the cross product between two $(3\times1)$ vectors

$$\boldsymbol{S}(\boldsymbol{x}) = \begin{bmatrix} 0 & -x_3 & x_2 \\ x_3 & 0 & -x_1 \\ -x_2 & x_1 & 0 \end{bmatrix}, \tag{2.6}$$

the unit quaternion is defined as

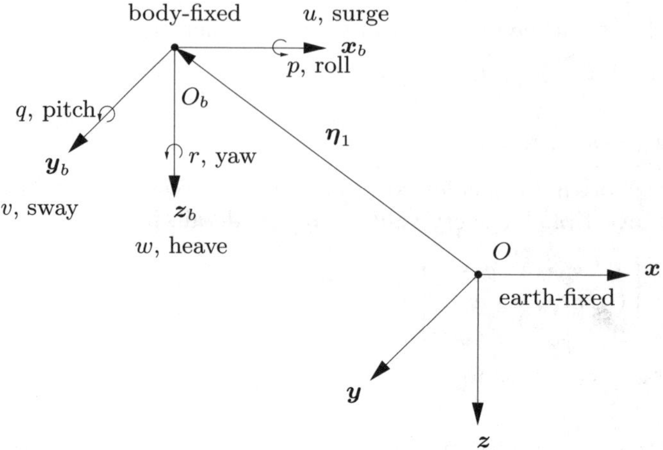


**Fig. 2.1.** Frames and elementary vehicle's motion

$$\mathcal{Q} = \{\boldsymbol{\varepsilon}, \eta\}$$

with

$$\boldsymbol{\varepsilon} = \boldsymbol{k}\sin\frac{\delta}{2},$$

$$\eta = \cos\frac{\delta}{2},$$

where $\eta \geq 0$ for $\delta \in [-\pi, \pi]$ rad. This restriction is necessary for uniqueness of the quaternion associated to a given matrix, in that the two quaternion $\{\boldsymbol{\varepsilon}, \eta\}$ and $\{-\boldsymbol{\varepsilon}, -\eta\}$ represent the same orientation, i.e., the same rotation matrix.

The unit quaternion satisfies the condition

$$\eta^2 + \boldsymbol{\varepsilon}^{\mathrm{T}}\boldsymbol{\varepsilon} = 1. \tag{2.7}$$

The relationship between $\boldsymbol{\nu}_2$ and the time derivative of the quaternion is given by the *quaternion propagation equations*

$$\dot{\boldsymbol{\varepsilon}} = \frac{1}{2}\eta\boldsymbol{\nu}_2 + \frac{1}{2}\boldsymbol{S}(\boldsymbol{\varepsilon})\boldsymbol{\nu}_2, \tag{2.8}$$

$$\dot{\eta} = -\frac{1}{2}\boldsymbol{\varepsilon}^{\mathrm{T}}\boldsymbol{\nu}_2, \tag{2.9}$$

that can be rearranged in the form:

$$\begin{bmatrix} \dot{\boldsymbol{\varepsilon}} \\ \dot{\eta} \end{bmatrix} = \boldsymbol{J}_{k,oq}(\mathcal{Q})\boldsymbol{\nu}_2 = \frac{1}{2}\begin{bmatrix} \eta & -\varepsilon_3 & \varepsilon_2 \\ \varepsilon_3 & \eta & -\varepsilon_1 \\ -\varepsilon_2 & \varepsilon_1 & \eta \\ -\varepsilon_1 & -\varepsilon_2 & -\varepsilon_3 \end{bmatrix}\boldsymbol{\nu}_2. \tag{2.10}$$

The matrix $\boldsymbol{J}_{k,oq}(\mathcal{Q})$ satisfies:

$$\boldsymbol{J}_{k,oq}^{\mathrm{T}} \boldsymbol{J}_{k,oq} = \frac{1}{4} \boldsymbol{I}_3 \,,$$

that allows to invert the mapping (2.10) yielding:

$$\boldsymbol{\nu}_2 = 4 \boldsymbol{J}_{k,oq}^{\mathrm{T}} \begin{bmatrix} \dot{\boldsymbol{\varepsilon}} \\ \dot{\eta} \end{bmatrix} .$$

For completeness the rotation matrix $\boldsymbol{R}_I^B$, needed to compute (2.1), in terms of quaternion is given:

$$\boldsymbol{R}_I^B(\mathcal{Q}) = \begin{bmatrix} 1 - 2(\varepsilon_2^2 + \varepsilon_3^2) & 2(\varepsilon_1\varepsilon_2 + \varepsilon_3\eta) & 2(\varepsilon_1\varepsilon_3 - \varepsilon_2\eta) \\ 2(\varepsilon_1\varepsilon_2 - \varepsilon_3\eta) & 1 - 2(\varepsilon_1^2 + \varepsilon_3^2) & 2(\varepsilon_2\varepsilon_3 + \varepsilon_1\eta) \\ 2(\varepsilon_1\varepsilon_3 + \varepsilon_2\eta) & 2(\varepsilon_2\varepsilon_3 - \varepsilon_1\eta) & 1 - 2(\varepsilon_1^2 + \varepsilon_2^2) \end{bmatrix} . \quad (2.11)$$

### 2.2.3 Attitude Error Representation

Let now define $\boldsymbol{R}_B^I \in \mathbb{R}^{3\times 3}$ as the rotation matrix from the body-fixed frame to the earth-fixed frame, which is also described by the quaternion $\mathcal{Q}$, and $\boldsymbol{R}_d^I \in \mathbb{R}^{3\times 3}$ the rotation matrix from the frame expressing the desired vehicle orientation to the earth-fixed frame, which is also described by the quaternion $\mathcal{Q}_d = \{\boldsymbol{\varepsilon}_d, \eta_d\}$. One possible choice for the rotation matrix necessary to align the two frames is

$$\widetilde{\boldsymbol{R}} = \boldsymbol{R}_B^{I^{\mathrm{T}}} \boldsymbol{R}_d^I = \boldsymbol{R}_I^B \boldsymbol{R}_d^I \,,$$

where $\boldsymbol{R}_I^B = \boldsymbol{R}_B^{I\,\mathrm{T}}$. The quaternion $\widetilde{\mathcal{Q}} = \{\tilde{\boldsymbol{\varepsilon}}, \tilde{\eta}\}$ associated with $\widetilde{\boldsymbol{R}}$ can be obtained directly from $\widetilde{\boldsymbol{R}}$ or computed by composition (quaternion product): $\widetilde{\mathcal{Q}} = \mathcal{Q}^{-1} * \mathcal{Q}_d$, where $\mathcal{Q}^{-1} = \{-\boldsymbol{\varepsilon}, \eta\}$:

$$\tilde{\boldsymbol{\varepsilon}} = \eta \boldsymbol{\varepsilon}_d - \eta_d \boldsymbol{\varepsilon} + \boldsymbol{S}(\boldsymbol{\varepsilon}_d)\boldsymbol{\varepsilon}, \quad (2.12)$$

$$\tilde{\eta} = \eta\eta_d + \boldsymbol{\varepsilon}^{\mathrm{T}} \boldsymbol{\varepsilon}_d \,. \quad (2.13)$$

Since the quaternion associated with $\widetilde{\boldsymbol{R}} = \boldsymbol{I}_3$ (i.e. representing two aligned frames) is $\widetilde{\mathcal{Q}} = \{\boldsymbol{0}, 1\}$, it is sufficient to represent the attitude error as $\tilde{\boldsymbol{\varepsilon}}$.

The quaternion propagation equations can be rewritten also in terms of the error variables:

$$\dot{\tilde{\boldsymbol{\varepsilon}}} = \frac{1}{2} \tilde{\eta} \tilde{\boldsymbol{\nu}}_2 + \frac{1}{2} \boldsymbol{S}(\tilde{\boldsymbol{\varepsilon}}) \tilde{\boldsymbol{\nu}}_2 \,, \quad (2.14)$$

$$\dot{\tilde{\eta}} = -\frac{1}{2} \tilde{\boldsymbol{\varepsilon}}^{\mathrm{T}} \tilde{\boldsymbol{\nu}}_2 \,, \quad (2.15)$$

where $\tilde{\boldsymbol{\nu}}_2 = \boldsymbol{\nu}_{2,d} - \boldsymbol{\nu}_2$ is the angular velocity error expressed in body-fixed frame. Defining

$$\boldsymbol{z} = \begin{bmatrix} \tilde{\boldsymbol{\varepsilon}} \\ \tilde{\eta} \end{bmatrix} ,$$

the relations in (2.14)–(2.15) can be rewritten in the form:

$$\dot{z} = \frac{1}{2}\begin{bmatrix} \tilde{\eta}\boldsymbol{I}_3 + \boldsymbol{S}(\tilde{\varepsilon}) \\ -\tilde{\varepsilon}^{\mathrm{T}} \end{bmatrix}\tilde{\boldsymbol{\nu}}_2 = \boldsymbol{J}_{k,oq}(\boldsymbol{z})\tilde{\boldsymbol{\nu}}_2 \tag{2.16}$$

The equations above are given in terms of the body-fixed angular velocity. In fact, they will be used in the control laws of Chap. 7. The generic expression of the propagation equations is the following:

$$\begin{aligned} \dot{\tilde{\varepsilon}}^a_{ba} &= \frac{1}{2}\boldsymbol{E}(\widetilde{\mathcal{Q}}_{ba})\tilde{\boldsymbol{\omega}}^a_{ba}\,, \\ \dot{\tilde{\eta}}_{ba} &= -\frac{1}{2}\tilde{\varepsilon}^{a\,\mathrm{T}}_{ba}\tilde{\boldsymbol{\omega}}^a_{ba}\,, \end{aligned}$$

with

$$\boldsymbol{E}(\widetilde{\mathcal{Q}}_{ba}) = \tilde{\eta}_{ba}\boldsymbol{I}_3 - \boldsymbol{S}(\tilde{\varepsilon}^a_{ba}).$$

where $\widetilde{\mathcal{Q}}_{ba} = \{\tilde{\varepsilon}^a_{ba}, \tilde{\eta}_{ba}\}$ is the quaternion associated to $\boldsymbol{R}^a_b = \boldsymbol{R}^{\mathrm{T}}_a\boldsymbol{R}_b$ and the angular velocity $\tilde{\boldsymbol{\omega}}^a_{ba} = \boldsymbol{R}^{\mathrm{T}}_a(\boldsymbol{\omega}_b - \boldsymbol{\omega}_a)$ of the frame $\Sigma_b$ relative to the frame $\Sigma_a$, expressed in the frame $\Sigma_a$.

### Quaternion from Rotation Matrix

It can be useful to recall the procedure needed to extract the quaternion from the rotation matrix [127, 261].

Given a generic rotation matrix $\boldsymbol{R}$:

1. compute the trace of $\boldsymbol{R}$ according to:
$$R_{4,4} = \mathrm{tr}(\boldsymbol{R}) = \sum_{j=1}^{3} R_{j,j}$$
2. compute the index $i$ according to:
$$R_{i,i} = \max(R_{1,1}, R_{2,2}, R_{3,3}, R_{4,4})$$
3. define the scalar $c_i$ as:
$$|c_i| = \sqrt{1 + 2R_{i,i} - R_{4,4}}$$
in which the sign can be plus or minus.
4. compute the other three values of $c$ by knowing the following relationships:
$$\begin{aligned} c_4c_1 &= R_{3,2} - R_{2,3} \\ c_4c_2 &= R_{1,3} - R_{3,1} \\ c_4c_3 &= R_{2,1} - R_{1,2} \\ c_2c_3 &= R_{3,2} + R_{2,3} \\ c_3c_1 &= R_{1,3} + R_{3,1} \\ c_1c_2 &= R_{2,1} + R_{1,2} \end{aligned}$$
simply dividing the equations in which $c_i$ is involved by $c_i$ itself.

5. compute the quaternion $\mathcal{Q}$ by the following:

$$[\varepsilon \quad \eta]^{\mathrm{T}} = \frac{1}{2}[c_1 \quad c_2 \quad c_3 \quad c_4]^{\mathrm{T}}.$$

### Quaternion from Euler Angles

The transformation from Euler angles to quaternion is always possible, i.e., it is not affected by the occurrence of representation singularities [127]. This implies that the use of quaternion to control underwater vehicles is compatible with the common use of Euler angles to express the desired trajectory of the vehicle.

The algorithm consists in computing the rotation matrix expressed in Euler angles by (2.5) and using the procedure described in the previous subsection to extract the corresponding quaternion.

## 2.2.4 6-DOFs Kinematics

It is useful to collect the kinematic equations in 6-dimensional matrix forms. Let us define the vector $\boldsymbol{\eta} \in \mathrm{I\!R}^6$ as

$$\boldsymbol{\eta} = \begin{bmatrix} \boldsymbol{\eta}_1 \\ \boldsymbol{\eta}_2 \end{bmatrix} \tag{2.17}$$

and the vector $\boldsymbol{\nu} \in \mathrm{I\!R}^6$ as

$$\boldsymbol{\nu} = \begin{bmatrix} \boldsymbol{\nu}_1 \\ \boldsymbol{\nu}_2 \end{bmatrix}, \tag{2.18}$$

and by defining the matrix $\boldsymbol{J}_e(\boldsymbol{R}_B^I) \in \mathrm{I\!R}^{6\times 6}$

$$\boldsymbol{J}_e(\boldsymbol{R}_B^I) = \begin{bmatrix} \boldsymbol{R}_I^B & \boldsymbol{O}_{3\times 3} \\ \boldsymbol{O}_{3\times 3} & \boldsymbol{J}_{k,o} \end{bmatrix}, \tag{2.19}$$

where the rotation matrix $\boldsymbol{R}_I^B$ given in (2.5) and $\boldsymbol{J}_{k,o}$ is given in (2.3), it is

$$\boldsymbol{\nu} = \boldsymbol{J}_e(\boldsymbol{R}_B^I)\dot{\boldsymbol{\eta}}. \tag{2.20}$$

The inverse mapping, given the block-diagonal structure of $\boldsymbol{J}_e$, is given by:

$$\dot{\boldsymbol{\eta}} = \boldsymbol{J}_e^{-1}(\boldsymbol{R}_B^I)\boldsymbol{\nu} = \begin{bmatrix} \boldsymbol{R}_B^I & \boldsymbol{O}_{3\times 3} \\ \boldsymbol{O}_{3\times 3} & \boldsymbol{J}_{k,o}^{-1} \end{bmatrix}\boldsymbol{\nu}, \tag{2.21}$$

where $\boldsymbol{J}_{k,o}^{-1}$ is given in (2.4).

On the other side, it is possible to represent the orientation by means of quaternions. Let us define the vector $\boldsymbol{\eta}_q \in \mathrm{I\!R}^7$ as

$$\boldsymbol{\eta}_q = \begin{bmatrix} \boldsymbol{\eta}_1 \\ \boldsymbol{\varepsilon} \\ \eta \end{bmatrix} \tag{2.22}$$

and the matrix $\boldsymbol{J}_{e,q}(\boldsymbol{R}_B^I) \in \mathbb{R}^{6\times 7}$

$$\boldsymbol{J}_{e,q}(\boldsymbol{R}_B^I) = \begin{bmatrix} \boldsymbol{R}_I^B & \boldsymbol{O}_{3\times 4} \\ \boldsymbol{O}_{3\times 3} & 4\boldsymbol{J}_{k,oq}^{\mathrm{T}} \end{bmatrix}, \tag{2.23}$$

where $\boldsymbol{J}_{k,oq}$ is given in (2.10); it is

$$\boldsymbol{\nu} = \boldsymbol{J}_{e,q}(\boldsymbol{R}_B^I)\dot{\boldsymbol{\eta}}_e . \tag{2.24}$$

The inverse mapping is given by:

$$\dot{\boldsymbol{\eta}}_e = \begin{bmatrix} \boldsymbol{R}_B^I & \boldsymbol{O}_{3\times 3} \\ \boldsymbol{O}_{4\times 3} & \boldsymbol{J}_{k,oq} \end{bmatrix} \boldsymbol{\nu} . \tag{2.25}$$

## 2.3 Rigid Body's Dynamics

Several approaches can be considered when deriving the equations of motion of a rigid body. In the following, the Newton-Euler formulation will be briefly summarized.

The motion of a generic system of material particles subject to external forces can be described by resorting to the *fundamental principles of dynamics* (Newton's laws of motion). Those relate the resultant force and moment to the time derivative of the linear and angular momentum.

Let $\rho$ be the density of a particle of volume $dV$ of a rigid body $\mathcal{B}$, $\rho dV$ is the corresponding mass denoted by the position vector $\boldsymbol{p}$ in an inertial frame $O - \boldsymbol{xyz}$. Let also $V_{\mathcal{B}}$ be the the body volume and

$$m = \int_{V_{\mathcal{B}}} \rho dV$$

be the total mass. The *center of mass* of $\mathcal{B}$ is defined as

$$\boldsymbol{p}_C = \frac{1}{m} \int_{V_{\mathcal{B}}} \boldsymbol{p} \rho dV.$$

The *linear momentum* of the body $\mathcal{B}$ is defined as the vector

$$\boldsymbol{l} = \int_{V_{\mathcal{B}}} \dot{\boldsymbol{p}} \rho dV = m\dot{\boldsymbol{p}}_C.$$

For a system with constant mass, the Newton's law of motion for the linear part

$$\boldsymbol{f} = \dot{\boldsymbol{l}} = m\frac{d}{dt}\dot{\boldsymbol{p}}_C \tag{2.26}$$

can be rewritten simply by the *Newton's equations of motion*:

$$\boldsymbol{f} = m\ddot{\boldsymbol{p}}_C \tag{2.27}$$

where $\boldsymbol{f}$ is the resultant of the external forces.

Let us define the *Inertia tensor* of the body $\mathcal{B}$ relative to the pole $O$:

$$\boldsymbol{I}_O = \int_{V_\mathcal{B}} \boldsymbol{S}^\mathrm{T}(\boldsymbol{p})\boldsymbol{S}(\boldsymbol{p})\rho dV,$$

where $\boldsymbol{S}$ is the skew-symmetric operator defined in (2.6). The matrix $\boldsymbol{I}_O$ is symmetric and positive definite. The positive diagonal elements $I_{Oxx}$, $I_{Oyy}$, $I_{Ozz}$ are the *inertia moments* with respect to the three coordinate axes of the reference frame. The off diagonal elements are the *products of inertia.*

The relationship between the inertia tensor in two different frames $\boldsymbol{I}_O$ and $\boldsymbol{I}'_O$, related by a rotation matrix $\boldsymbol{R}$, with the same pole $O$, is the following:

$$\boldsymbol{I}_O = \boldsymbol{R}\boldsymbol{I}'_O\boldsymbol{R}^\mathrm{T}.$$

The change of pole is related by the *Steiner's theorem*:

$$\boldsymbol{I}_O = \boldsymbol{I}_C + m\boldsymbol{S}^\mathrm{T}(\boldsymbol{p}_C)\boldsymbol{S}(\boldsymbol{p}_C),$$

where $\boldsymbol{I}_C$ is the inertial tensor relative to the center of mass, when expressed in a frame parallel to the frame in which $\boldsymbol{I}_O$ is defined.

Notice that $O$ can be either a fixed or moving pole. In case of a fixed pole the elements of the inertia tensor are function of time. A suitable choice of the pole might be a point fixed to the rigid body in a way to obtain a constant inertia tensor. Moreover, since the inertia tensor is symmetric positive definite is always possible to find a frame in which the matrix attains a diagonal form, this frame is called *principal frame*, also, if the pole coincides with the center of mass, it is called *central frame*. This is true also if the body does not have a significant geometric symmetry.

Let $\Omega$ be any point in space and $\boldsymbol{p}_\Omega$ the corresponding position vector. $\Omega$ can be either moving or fixed with respect to the reference frame. The *angular momentum* of the body $\mathcal{B}$ relative to the pole $\Omega$ is defined as the vector:

$$\boldsymbol{k}_\Omega = \int_{V_\mathcal{B}} \dot{\boldsymbol{p}} \times (\boldsymbol{p}_\Omega - \boldsymbol{p})\, \rho dV. \tag{2.28}$$

Taking into account the definition of center of mass, (2.28) can be rewritten in the form:

$$\boldsymbol{k}_\Omega = \boldsymbol{I}_C\boldsymbol{\omega} + m\dot{\boldsymbol{p}}_C \times (\boldsymbol{p}_\Omega - \boldsymbol{p}_C)\,, \tag{2.29}$$

where $\boldsymbol{\omega}$ is the angular velocity.

The resultant moment $\boldsymbol{\mu}_\Omega$ with respect to the pole $\Omega$ of a rigid body subject to $n$ external forces $\boldsymbol{f}_1, \ldots, \boldsymbol{f}_n$ is:

$$\boldsymbol{\mu}_\Omega = \sum_{i=1}^{n} \boldsymbol{f}_i \times (\boldsymbol{p}_\Omega - \boldsymbol{p}_i)\,.$$

In case of a system with constant mass and rigid body, the angular part of the Newton's law of motion

$$\boldsymbol{\mu}_\Omega = \dot{\boldsymbol{k}}_\Omega$$

yields the *Euler equations of motion*:

$$\boldsymbol{\mu}_\Omega = \boldsymbol{I}_\Omega \dot{\boldsymbol{\omega}} + \boldsymbol{\omega} \times (\boldsymbol{I}_\Omega \boldsymbol{\omega})\,. \tag{2.30}$$

The right-hand side of the Newton and Euler equations of motion, (2.27) and (2.30), are defined inertial forces and inertial moments, respectively.

### 2.3.1 Rigid Body's Dynamics in Matrix Form

To derive the equations of motion in matrix form it is useful to refer the quantities to a body-fixed frame $O_b - \boldsymbol{x}_b \boldsymbol{y}_b \boldsymbol{z}_b$ using the body-fixed linear and angular velocities that has been introduced in Section 2.2.

The following relationships hold:

$$\boldsymbol{p}_\Omega - \boldsymbol{p}_C = \boldsymbol{R}_B^I \boldsymbol{r}_C^b \tag{2.31}$$

$$\dot{\boldsymbol{R}}_B^I \cdot = \boldsymbol{\omega} \times (\boldsymbol{R}_B^I \cdot) \tag{2.32}$$

$$\boldsymbol{R}_I^B \left( \boldsymbol{\omega} \times \boldsymbol{R}_B^I \cdot \right) = \boldsymbol{\nu}_2 \times \cdot \tag{2.33}$$

$$\boldsymbol{\omega} = \boldsymbol{R}_B^I \boldsymbol{\nu}_2 \tag{2.34}$$

$$\dot{\boldsymbol{\omega}} = \boldsymbol{R}_B^I \dot{\boldsymbol{\nu}}_2 \tag{2.35}$$

$$\dot{\boldsymbol{p}}_C = \boldsymbol{R}_B^I (\boldsymbol{\nu}_1 + \boldsymbol{\nu}_2 \times \boldsymbol{r}_C^b) \tag{2.36}$$

$$\boldsymbol{I}_C = \boldsymbol{R}_B^I \boldsymbol{I}_C^b \boldsymbol{R}_I^B \tag{2.37}$$

where, according to the (2.31), $\boldsymbol{r}_C^b$ is the vector position from the origin of the body-fixed frame to the center of mass expressed in the body-fixed frame ($\dot{\boldsymbol{r}}_C^b = \boldsymbol{0}$ for a rigid body).

Equation (2.26) can be rewritten in terms of the linear body-fixed velocities as

$$\begin{aligned}
\boldsymbol{f} &= m \frac{d}{dt} \left[ \boldsymbol{R}_B^I \left( \boldsymbol{\nu}_1 + \boldsymbol{\nu}_2 \times \boldsymbol{r}_C^b \right) \right] \\
&= m \boldsymbol{R}_B^I \left( \dot{\boldsymbol{\nu}}_1 + \dot{\boldsymbol{\nu}}_2 \times \boldsymbol{r}_C^b + \boldsymbol{\nu}_2 \times \dot{\boldsymbol{r}}_C^b \right) + m \boldsymbol{\omega} \times \boldsymbol{R}_B^I (\boldsymbol{\nu}_1 + \boldsymbol{\nu}_2 \times \boldsymbol{r}_C^b),
\end{aligned}$$

Premultiplying by $\boldsymbol{R}_I^B$ and defining as

$$\boldsymbol{\tau}_1 = \begin{bmatrix} X \\ Y \\ Z \end{bmatrix},$$

the resultant forces acting on the rigid body expressed in a body-fixed frame, and as

$$\boldsymbol{\tau}_2 = \begin{bmatrix} K \\ M \\ N \end{bmatrix},$$

the corresponding resultant moment to the pole $O_b$, one obtains:

$$\boldsymbol{\tau}_1 = m\dot{\boldsymbol{\nu}}_1 + m\dot{\boldsymbol{\nu}}_2 \times \boldsymbol{r}_C^b + m\boldsymbol{\nu}_2 \times \boldsymbol{\nu}_1 + m\boldsymbol{\nu}_2 \times (\boldsymbol{\nu}_2 \times \boldsymbol{r}_C^b).$$

Equation (2.29) is written in an inertial frame. It is possible to rewrite the angular momentum in terms of the body-fixed velocities:

$$\boldsymbol{k}_\Omega = \boldsymbol{R}_B^I \left( \boldsymbol{I}_C^b \boldsymbol{\nu}_2 + m\boldsymbol{\nu}_1 \times \boldsymbol{r}_C^b \right). \tag{2.38}$$

Derivating (2.38) one obtains:

$$\boldsymbol{\tau}_2^I = \boldsymbol{\omega} \times \boldsymbol{R}_B^I \left( \boldsymbol{I}_C^b \boldsymbol{\nu}_2 + m\boldsymbol{\nu}_1 \times \boldsymbol{r}_C^b \right) + \boldsymbol{R}_B^I \left( \boldsymbol{I}_C^b \dot{\boldsymbol{\nu}}_2 + m\dot{\boldsymbol{\nu}}_1 \times \boldsymbol{r}_C^b \right),$$

that, using the relations above, can be written in the form:

$$\boldsymbol{\tau}_2 = \boldsymbol{I}_C^b \dot{\boldsymbol{\nu}}_2 + \boldsymbol{\nu}_2 \times (\boldsymbol{I}_C^b \boldsymbol{\nu}_2) + m\boldsymbol{\nu}_2 \times (\boldsymbol{\nu}_1 \times \boldsymbol{r}_C^b) + m\dot{\boldsymbol{\nu}}_1 \times \boldsymbol{r}_C^b.$$

It is now possible to rewrite the Newton-Euler equations of motion of a rigid body moving in the space. It is:

$$\boldsymbol{M}_{RB}\dot{\boldsymbol{\nu}} + \boldsymbol{C}_{RB}(\boldsymbol{\nu})\boldsymbol{\nu} = \boldsymbol{\tau}_v, \tag{2.39}$$

where

$$\boldsymbol{\tau}_v = \begin{bmatrix} \boldsymbol{\tau}_1 \\ \boldsymbol{\tau}_2 \end{bmatrix}.$$

The matrix $\boldsymbol{M}_{RB}$ is constant, symmetric and positive definite, i.e., $\dot{\boldsymbol{M}}_{RB} = \boldsymbol{O}$, $\boldsymbol{M}_{RB} = \boldsymbol{M}_{RB}^{\mathrm{T}} > \boldsymbol{O}$. Its unique parametrization is in the form:

$$\boldsymbol{M}_{RB} = \begin{bmatrix} m\boldsymbol{I}_3 & -m\boldsymbol{S}(\boldsymbol{r}_C^b) \\ m\boldsymbol{S}(\boldsymbol{r}_C^b) & \boldsymbol{I}_{O_b} \end{bmatrix},$$

where $\boldsymbol{I}_3$ is the $(3 \times 3)$ identity matrix, and $\boldsymbol{I}_{O_b}$ is the inertia tensor expressed in the body-fixed frame.

On the other hand, it does not exist a unique parametrization of the matrix $\boldsymbol{C}_{RB}$, representing the Coriolis and centripetal terms. It can be demonstrated that the matrix $\boldsymbol{C}_{RB}$ can always be parameterized such that it is skew-symmetrical, i.e.,

$$\boldsymbol{C}_{RB}(\boldsymbol{\nu}) = -\boldsymbol{C}_{RB}^{\mathrm{T}}(\boldsymbol{\nu}) \qquad \forall \boldsymbol{\nu} \in \mathbb{R}^6,$$

explicit expressions for $\boldsymbol{C}_{RB}$ can be found, e.g., in [127].

Notice that (2.39) can be greatly simplified if the origin of the body-fixed frame is chosen coincident with the central frame, i.e., $\boldsymbol{r}_C^b = \boldsymbol{0}$ and $\boldsymbol{I}_{O_b}$ is a diagonal matrix.

## 2.4 Hydrodynamic Effects

In this Section the major hydrodynamic effects on a rigid body moving in a fluid will be briefly discussed.

The theory of fluidodynamics is rather complex and it is difficult to develop a reliable model for most of the hydrodynamic effects. A rigorous analysis for incompressible fluids would need to resort to the Navier-Stokes equations (distributed fluid-flow). However, in this book modeling of the hydrodynamic effects in a context of automatic control is considered. In literature, it is well known that kinematic and dynamic coupling between vehicle and manipulator can not be neglected [182, 203, 204, 206], while most of the hydrodynamic effects have no significant influence in the range of the operative velocities.

### 2.4.1 Added Mass and Inertia

When a rigid body is moving in a fluid, the additional inertia of the fluid surrounding the body, that is accelerated by the movement of the body, has to be considered. This effect can be neglected in industrial robotics since the density of the air is much lighter than the density of a moving mechanical system. In underwater applications, however, the density of the water, $\rho \approx 1000\,\text{kg/m}^3$, is comparable with the density of the vehicles. In particular, at 0°, the density of the fresh water is $1002.68\,\text{kg/m}^3$; for sea water with 3.5% of salinity it is $\rho = 1028.48\,\text{kg/m}^3$.

The fluid surrounding the body is accelerated with the body itself, a force is then necessary to achieve this acceleration; the fluid exerts a reaction force which is equal in magnitude and opposite in direction. This reaction force is the added mass contribution. The added mass is not a quantity of fluid to add to the system such that it has an increased mass. Different properties hold with respect to the $(6 \times 6)$ inertia matrix of a rigid body due to the fact that the added mass is function of the body's surface geometry. As an example, the inertia matrix is not necessarily positive definite.

The hydrodynamic force along $\boldsymbol{x}_b$ due to the linear acceleration in the $\boldsymbol{x}_b$-direction is defined as:

$$X_A = -X_{\dot{u}}\dot{u} \qquad \text{where} \qquad X_{\dot{u}} = \frac{\partial X}{\partial \dot{u}},$$

where the symbol $\partial$ denotes the partial derivative. In the same way it is possible to define all the remaining 35 elements that relate the 6 force/moment components $[X \;\; Y \;\; Z \;\; K \;\; M \;\; N]^{\mathrm{T}}$ to the 6 linear/angular acceleration $[\dot{u} \;\; \dot{v} \;\; \dot{w} \;\; \dot{p} \;\; \dot{q} \;\; \dot{r}]^{\mathrm{T}}$. These elements can be grouped in the Added Mass matrix $\boldsymbol{M}_A \in \mathbb{R}^{6\times 6}$. Usually, all the elements of the matrix are different from zero.

There is no specific property of the matrix $\boldsymbol{M}_A$. For certain frequencies and specific bodies, such as catamarans, negative diagonal elements have been documented [127]. However, for completely submerged bodies it can be considered $\boldsymbol{M}_A > \boldsymbol{O}$. Moreover, if the fluid is ideal, the body's velocity is low, there are no currents or waves and frequency independence it holds [214]:

$$\boldsymbol{M}_A = \boldsymbol{M}_A^{\mathrm{T}} > \boldsymbol{O}. \tag{2.40}$$

The added mass has also an *added* Coriolis and centripetal contribution. It can be demonstrated that the matrix expression can always be parameterized such that:

$$\boldsymbol{C}_A(\boldsymbol{\nu}) = -\boldsymbol{C}_A^{\mathrm{T}}(\boldsymbol{\nu}) \qquad \forall \boldsymbol{\nu} \in \mathbb{R}^6.$$

If the body is completely submerged in the water, the velocity is low and it has three planes of symmetry as common for underwater vehicles, the following structure of matrices $\boldsymbol{M}_A$ and $\boldsymbol{C}_A$ can therefore be considered:

$$\boldsymbol{M}_A = -\operatorname{diag}\{X_{\dot{u}}, Y_{\dot{v}}, Z_{\dot{w}}, K_{\dot{p}}, M_{\dot{q}}, N_{\dot{r}}\},$$

$$\boldsymbol{C}_A = \begin{bmatrix} 0 & 0 & 0 & 0 & -Z_{\dot{w}}w & Y_{\dot{v}}v \\ 0 & 0 & 0 & Z_{\dot{w}}w & 0 & -X_{\dot{u}}u \\ 0 & 0 & 0 & -Y_{\dot{v}}v & X_{\dot{u}}u & 0 \\ 0 & -Z_{\dot{w}}w & Y_{\dot{v}}v & 0 & -N_{\dot{r}}r & M_{\dot{q}}q \\ Z_{\dot{w}}w & 0 & -X_{\dot{u}}u & N_{\dot{r}}r & 0 & -K_{\dot{p}}p \\ -Y_{\dot{v}}v & X_{\dot{u}}u & 0 & -M_{\dot{q}}q & K_{\dot{p}}p & 0 \end{bmatrix}.$$

The added mass coefficients can be theoretically derived exploiting the geometry of the rigid body and, eventually, its symmetry [127], by applying the strip theory. For a cylindrical rigid body of mass $\overline{m}$, length $\overline{L}$, with circular section of radius $\overline{r}$, the following added mass coefficients can be derived [127]:

$$\begin{aligned} X_{\dot{u}} &= -0.1\overline{m} \\ Y_{\dot{v}} &= -\pi\rho\overline{r}^2\overline{L} \\ Z_{\dot{w}} &= -\pi\rho\overline{r}^2\overline{L} \\ K_{\dot{p}} &= 0 \\ M_{\dot{q}} &= -\frac{1}{12}\pi\rho\overline{r}^2\overline{L}^3 \\ N_{\dot{r}} &= -\frac{1}{12}\pi\rho\overline{r}^2\overline{L}^3. \end{aligned}$$

Notice that, despite (2.40), in this case it is $\boldsymbol{M}_A \geq \boldsymbol{O}$. This result is due to the geometrical approach to the derivation of $\boldsymbol{M}_A$. As a matter of fact, if a sphere submerged in a fluid is considered, it can be observed that a pure rotational motion of the sphere does not involve any fluid movement, i.e., it is not necessary to add an inertia term due to the fluid. This small discrepancy is just an example of the difficulty in representing with a closed set of equations a distributed phenomenon as fluid movement.

In [220] the added mass coefficients for an ellipsoid are derived.

In [145], and in the Appendix, the coefficients for the experimental vehicle NPS AUV Phoenix are reported. These coefficients have been experimentally derived and, since the vehicle can work at a maximum depth of few meters, i.e., it is not submerged in an unbounded fluid, the structure of $\boldsymbol{M}_A$ is not diagonal. To give an order of magnitude of the added mass terms, the vehicle has a mass of about 5000 kg, the term $X_{\dot{u}} \approx -500$ kg.

A detailed theoretical and experimental discussion on the added mass effect of a cylinder moving in a fluid can be found in [203] where it is shown that the added mass matrix is state-dependent and its coefficients are function of the distance traveled by the cylinder.

### 2.4.2 Damping Effects

The viscosity of the fluid also causes the presence of dissipative drag and lift forces on the body.

A common simplification is to consider only linear and quadratic damping terms and group these terms in a matrix $\boldsymbol{D}_{RB}$ such that:

$$\boldsymbol{D}_{RB}(\boldsymbol{\nu}) > \boldsymbol{O} \qquad \forall \boldsymbol{\nu} \in \mathbb{R}^6.$$

The coefficients of this matrix are also considered to be constant. For a completely submerged body, the following further assumption can be made:

$$\begin{aligned}\boldsymbol{D}_{RB}(\boldsymbol{\nu}) = &-\operatorname{diag}\left\{X_u, Y_v, Z_w, K_p, M_q, N_r\right\} + \\ &-\operatorname{diag}\left\{X_{u|u|}\left|u\right|, Y_{v|v|}\left|v\right|, Z_{w|w|}\left|w\right|, K_{p|p|}\left|p\right|, M_{q|q|}\left|q\right|, N_{r|r|}\left|r\right|\right\}.\end{aligned}$$

Assuming a diagonal structure for the damping matrix implies neglecting the coupling dissipative terms.

The detailed analysis of the dissipative forces is beyond the scope of this work. In the following, only the nature of these forces will be briefly discussed. Introductory analysis of this phenomenon can be found in [127, 157, 208, 220, 255], while in depth discussion in [253, 274].

The viscous effects can be considered as the sum of two forces, the *drag* and the *lift* forces. The former are parallel to the relative velocity between the body and the fluid, while the latter are normal to it. Both drag and lift forces are supposed to act on the center of mass of the body. In order to solve the distributed flow problem, an integral over the entire surface is required to compute the net force/moment acting on the body. Moreover, the model of drag and lift forces is not known and, also for some widely accepted models, the coefficients are not known and variables.

For a sphere moving in a fluid, the drag force can be modeled as [157]:

$$F_{drag} = 0.5\rho U^2 S C_d(R_n),$$

where $\rho$ is the fluid density, $U$ is the velocity of the sphere, $S$ is the frontal area of the sphere, $C_d$ is the adimensional drag coefficients and $R_n$ is the Reynolds number. For a generic body, $S$ is the projection of the frontal area along the flow direction. The drag coefficient is then dependent on the Reynolds number, i.e., on the laminar/turbulent fluid motion:

$$R_n = \frac{\rho\left|U\right|D}{\mu}$$

where $D$ is the characteristic dimension of the body perpendicular to the direction of $U$ and $\mu$ is the dynamic viscosity of the fluid. In Table 2.2 the drag coefficients in function of the Reynolds number for a cylinder are reported [255]. The drag coefficients can be considered as the sum of two physical effects: a frictional contribution of the surface whose normal is perpendicular to the flow velocity, and a pressure contribution of the surface whose normal is parallel to the flow velocity.

**Table 2.2.** Lift and Drag Coefficient for a cylinder

| Reynolds number | regime motion | $C_d$ | $C_l$ |
|---|---|---|---|
| $R_n < 2 \cdot 10^5$ | subcritical flow | 1 | $3 \div 0.6$ |
| $2 \cdot 10^5 < R_n < 5 \cdot 10^5$ | critical flow | $1 \div 0.4$ | 0.6 |
| $5 \cdot 10^5 < R_n < 3 \cdot 10^5$ | transcritical flow | 0.4 | 0.6 |

The lift forces are perpendicular to the flow direction. For an hydrofoil they can be modeled as [157]:

$$F_{lift} = 0.5 \rho U^2 S C_l(R_n, \alpha),$$

where $C_l$ is the adimensional lift coefficient. It can be recognized that it also depends on the angle of attack $\alpha$. In Table 2.2 the lift coefficients in function of the Reynolds number for a cylinder are reported [255].

Vortex induced forces are an oscillatory effect that affects both drag and lift directions. They are caused by the vortex generated by the body that separates the fluid flow. They then cause a periodic *disturbance* that can be the cause of oscillations in cables and some underwater structures. For underwater vehicles it is reasonable to assume that the vortex induced forces are negligible, this, also in view of the adoption of small design surfaces that can reduce this effect. For underwater manipulators with cylindrical links this effects might be experienced.

### 2.4.3 Current Effects

Control of marine vehicles cannot neglect the effects of specific disturbances such as waves, wind and ocean current. In this book wind and waves phenomena will not be discussed since the attention is focused to autonomous vehicles performing a motion or manipulation task in an underwater environment. However, if this task has to be achieved in very shallow waters, those effects can not be neglected.

Ocean currents are mainly caused by: tidal movement; the atmospheric wind system over the sea earth's surface; the heat exchange at the sea surface; the salinity changes and the Coriolis force due to the earth rotation. Currents can be very different due to local climatic and/or geographic characteristics; as an example, in the fjords, the tidal effect can cause currents of up to 3 m/s [127].

The effect of a small current has to be considered also in structured environments such as a pool. In this case, the refresh of the water is strong enough to affect the vehicle dynamics [34].

Let us assume that the ocean current, expressed in the inertial frame, $\boldsymbol{\nu}_c^I$ is constant and irrotational, i.e.,

$$\boldsymbol{\nu}_c^I = \begin{bmatrix} \nu_{c,x} \\ \nu_{c,y} \\ \nu_{c,z} \\ 0 \\ 0 \\ 0 \end{bmatrix}$$

and $\dot{\boldsymbol{\nu}}_c^I = \mathbf{0}$; its effects can be added to the dynamic of a rigid body moving in a fluid simply considering the *relative* velocity in body-fixed frame

$$\boldsymbol{\nu}_r = \boldsymbol{\nu} - \boldsymbol{R}_I^B \boldsymbol{\nu}_c^I \tag{2.41}$$

in the derivation of the Coriolis and centripetal terms and the damping terms.

A simplified modeling of the current effect can be obtained by assuming the current irrotational and constant in the earth-fixed frame, its effect on the vehicle, thus, can be modeled as a constant disturbance in the earth-fixed frame that is further projected onto the vehicle-fixed frame. To this purpose, let define as $\boldsymbol{\theta}_{v,C} \in \mathbb{R}^6$ the vector of constant parameters contributing to the earth-fixed generalized forces due to the current; then, the vehicle-fixed current disturbance can be modelled as

$$\boldsymbol{\tau}_{v,C} = \boldsymbol{\Phi}_{v,C}(\boldsymbol{R}_B^I)\boldsymbol{\theta}_{v,C}, \tag{2.42}$$

where the $(6 \times 6)$ regressor matrix simply expresses the force/moment coordinate transformation between the two frames and it is given by

$$\boldsymbol{\Phi}_{v,C}(\boldsymbol{R}_B^I) = \begin{bmatrix} \boldsymbol{R}_I^B & \boldsymbol{O}_{3\times3} \\ \boldsymbol{O}_{3\times3} & \boldsymbol{R}_I^B \end{bmatrix}. \tag{2.43}$$

Notice that in [14, 35] compensation of the ocean current effects is obtained through a quaternion-based velocity/force mapping instead. Moreover, in some papers [34, 35, 125, 255], the effect of the current is simply modeled as a time-varying, vehicle-fixed, disturbance $\boldsymbol{\tau}_{v,C}$ that would lead to the trivial regressor

$$\boldsymbol{\Phi}'_{v,C} = \boldsymbol{I}_6. \tag{2.44}$$

## 2.5 Gravity and Buoyancy

*"Ses deux mains s'accrochaient à mon cou; elles ne se seraient pas accrochées plus furieusement dans un naufrage. Et je ne comprenais pas si elle voulait que je la sauve, ou bien que je me noie avec elle".*
*Raymond Radiguet, "Le diable au corps" 1923.*

When a rigid body is submerged in a fluid under the effect of the gravity two more forces have to be considered: the gravitational force and the buoyancy. The latter is the only hydrostatic effect, i.e., it is not function of a relative movement between body and fluid.

Let us define as

$$\boldsymbol{g}^I = \begin{bmatrix} 0 \\ 0 \\ 9.81 \end{bmatrix} \text{m/s}^2$$

the acceleration of gravity, $\nabla$ the volume of the body and $m$ its mass. The submerged weight of the body is defined as $W = m \left\| \boldsymbol{g}^I \right\|$ while its buoyancy $B = \rho \nabla \left\| \boldsymbol{g}^I \right\|$.

The gravity force, acting in the center of mass $\boldsymbol{r}_C^B$ is represented in body-fixed frame by:

$$\boldsymbol{f}_G(\boldsymbol{R}_I^B) = \boldsymbol{R}_I^B \begin{bmatrix} 0 \\ 0 \\ W \end{bmatrix},$$

while the buoyancy force, acting in the center of buoyancy $\boldsymbol{r}_B^B$ is represented in body-fixed frame by:

$$\boldsymbol{f}_B(\boldsymbol{R}_I^B) = -\boldsymbol{R}_I^B \begin{bmatrix} 0 \\ 0 \\ B \end{bmatrix}.$$

The $(6 \times 1)$ vector of force/moment due to gravity and buoyancy in body-fixed frame, included in the left hand-side of the equations of motion, is represented by:

$$\boldsymbol{g}_{RB}(\boldsymbol{R}_I^B) = -\begin{bmatrix} \boldsymbol{f}_G(\boldsymbol{R}_I^B) + \boldsymbol{f}_B(\boldsymbol{R}_I^B) \\ \boldsymbol{r}_G^B \times \boldsymbol{f}_G(\boldsymbol{R}_I^B) + \boldsymbol{r}_B^B \times \boldsymbol{f}_G(\boldsymbol{R}_I^B) \end{bmatrix}.$$

In the following, the symbol $\boldsymbol{r}_G^B = [x_G \quad y_G \quad z_G]^{\mathrm{T}}$ (with $\boldsymbol{r}_G^B = \boldsymbol{r}_C^B$) will be used for the center of gravity.

The expression of $\boldsymbol{g}_{RB}$ in terms of Euler angles is represented by:

$$\boldsymbol{g}_{RB}(\boldsymbol{\eta}_2) = \begin{bmatrix} (W-B)s_\theta \\ -(W-B)c_\theta s_\phi \\ -(W-B)c_\theta c_\phi \\ -(y_G W - y_B B)c_\theta c_\phi + (z_G W - z_B B)c_\theta s_\phi \\ (z_G W - z_B B)s_\theta + (x_G W - x_B B)c_\theta c_\phi \\ -(x_G W - x_B B)c_\theta s_\phi - (y_G W - y_B B)s_\theta \end{bmatrix}, \qquad (2.45)$$

while in terms of quaternion is represented by:

$$
\boldsymbol{g}_{RB}(\mathcal{Q})=\begin{bmatrix}
2(\eta\varepsilon_2-\varepsilon_1\varepsilon_3)(W-B)\\
-2(\eta\varepsilon_1+\varepsilon_2\varepsilon_3)(W-B)\\
(-\eta^2+\varepsilon_1^2+\varepsilon_2^2-\varepsilon_3^2)(W-B)\\
(-\eta^2+\varepsilon_1^2+\varepsilon_2^2-\varepsilon_3^2)(y_GW-y_BB)+2(\eta\varepsilon_1+\varepsilon_2\varepsilon_3)(z_GW-z_BB)\\
-(-\eta^2+\varepsilon_1^2+\varepsilon_2^2-\varepsilon_3^2)(x_GW-x_BB)+2(\eta\varepsilon_2-\varepsilon_1\varepsilon_3)(z_GW-z_BB)\\
-2(\eta\varepsilon_1+\varepsilon_2\varepsilon_3)(x_GW-x_BB)-2(\eta\varepsilon_2-\varepsilon_1\varepsilon_3)(y_GW-y_BB)
\end{bmatrix}.
$$

By looking at (2.45), it can be recognized that the difference between gravity and buoyancy ($W-B$) only affects the linear force acting on the vehicle; it is also clear that the restoring linear force is constant in the earth-fixed frame. On the other hand, the two vectors of the first moment of inertia $W\boldsymbol{r}_G^B$ and $B\boldsymbol{r}_B^B$ affect the moment acting on the vehicle and are constant in the vehicle-fixed frame. In summary, the expression of the restoring vector is linear with respect to the vector of four constant parameters

$$
\boldsymbol{\theta}_{v,R}=[\,W-B \quad x_GW-x_BB \quad y_GW-y_BB \quad z_GW-z_BB\,]^{\mathrm{T}} \tag{2.46}
$$

through the $(6\times 4)$ regressor

$$
\boldsymbol{\Phi}_{v,R}(\boldsymbol{R}_B^I)=\begin{bmatrix}\boldsymbol{R}_I^B\boldsymbol{z} & \boldsymbol{O}_{3\times 3}\\ \boldsymbol{0}_{3\times 1} & \boldsymbol{S}\left(\boldsymbol{R}_I^B\boldsymbol{z}\right)\end{bmatrix}, \tag{2.47}
$$

i.e.,

$$
\boldsymbol{g}_{RB}(\boldsymbol{R}_B^I)=\boldsymbol{\Phi}_{v,R}(\boldsymbol{R}_B^I)\boldsymbol{\theta}_{v,R}.
$$

In (2.47) $\boldsymbol{S}(\cdot)$ is the operator performing the cross product. Notice that, alternatively to (2.45), the restoring vector can be written in terms of quaternions; however, this would lead again to the regressor (2.47) and to the vector of dynamic parameters (2.46).

## 2.6 Thrusters' Dynamics

Underwater vehicles are usually controlled by thrusters (Figure 2.2) and/or control surfaces.

Control surfaces, such as rudders and sterns, are common in cruise vehicles; those are torpedo-shaped and usually used in cable/pipeline inspection. Since the force/moment provided by the control surfaces is function of the velocity and it is null in hovering, they are not useful to manipulation missions in which, due to the manipulator interaction, full control of the vehicle is required.

The relationship between the force/moment acting on the vehicle $\boldsymbol{\tau}_v\in\mathbb{R}^6$ and the control input of the thrusters $\boldsymbol{u}_v\in\mathbb{R}^{p_v}$ is highly nonlinear. It is function of some structural variables such as: the density of the water; the tunnel cross-sectional area; the tunnel length; the volumetric flowrate between input-output of the thrusters and the propeller diameter. The state

**Fig. 2.2.** Thruster of SAUVIM (courtesy of J. Yuh, Autonomous Systems Laboratory, University of Hawaii)

of the dynamic system describing the thrusters is constituted by the propeller revolution, the speed of the fluid going into the propeller and the input torque.

A detailed theoretical and experimental analysis of thrusters' behavior can be found in [40, 147, 176, 178, 220, 270, 300, 309]. Roughly speaking, thrusters are the main cause of limit cycle in vehicle positioning and bandwidth constraint.

A common simplification is to consider a linear relationship between $\boldsymbol{\tau}_v$ and $\boldsymbol{u}_v$:

$$\boldsymbol{\tau}_v = \boldsymbol{B}_v \boldsymbol{u}_v, \tag{2.48}$$

where $\boldsymbol{B}_v \in \mathbb{R}^{6 \times p_v}$ is a known constant matrix known as the Thruster Control Matrix (TCM). Along the book, the matrix $\boldsymbol{B}_v$ will be considered square or low rectangular, i.e., $p_v \geq 6$. This means full control of force/moments of the vehicle.

As an example, ODIN has the following TCM:

$$\boldsymbol{B}_v = \begin{bmatrix} * & * & * & * & 0 & 0 & 0 & 0 \\ * & * & * & * & 0 & 0 & 0 & 0 \\ 0 & 0 & 0 & 0 & * & * & * & * \\ 0 & 0 & 0 & 0 & * & * & * & * \\ 0 & 0 & 0 & 0 & * & * & * & * \\ * & * & * & * & 0 & 0 & 0 & 0 \end{bmatrix} \tag{2.49}$$

where $*$ means a non-zero constant factor depending on the thruster allocation. Different TCM can be observed as in, e.g., the vehicle Phantom S3 manufactured by Deep Ocean Engineering that has 4 thrusters:

$$\boldsymbol{B}_v = \begin{bmatrix} * & * & 0 & 0 \\ 0 & 0 & * & * \\ 0 & 0 & * & * \\ 0 & 0 & * & * \\ * & * & 0 & 0 \\ * & * & 0 & 0 \end{bmatrix} \tag{2.50}$$

in which it can be recognized that not all the directions are independently actuated.

On the other hand, if the vehicle is controlled by thrusters, each of which is locally fed back, the effects of the nonlinearities discussed above is very limited and a linear input-output relation between desired force/moment and thruster's torque is experienced. This is the case, e.g., of ODIN [35, 83, 215, 216] where the experimental results show that the linear approximation is reliable.

## 2.7 Underwater Vehicles' Dynamics in Matrix Form

By taking into account the inertial generalized forces, the hydrodynamic effects, the gravity and buoyancy contribution and the thrusters' presence, it is possible to write the equations of motion of an underwater vehicle in matrix form:

$$\boldsymbol{M}_v\dot{\boldsymbol{\nu}} + \boldsymbol{C}_v(\boldsymbol{\nu})\boldsymbol{\nu} + \boldsymbol{D}_{RB}(\boldsymbol{\nu})\boldsymbol{\nu} + \boldsymbol{g}_{RB}(\boldsymbol{R}_B^I) = \boldsymbol{B}_v\boldsymbol{u}_v, \tag{2.51}$$

where $\boldsymbol{M}_v = \boldsymbol{M}_{RB} + \boldsymbol{M}_A$ and $\boldsymbol{C}_v = \boldsymbol{C}_{RB} + \boldsymbol{C}_A$ include also the added mass terms. Taking into account the current, a possible, approximated, model is given by:

$$\boldsymbol{M}_v\dot{\boldsymbol{\nu}} + \boldsymbol{C}_v(\boldsymbol{\nu})\boldsymbol{\nu} + \boldsymbol{D}_{RB}(\boldsymbol{\nu})\boldsymbol{\nu} + \boldsymbol{g}_{RB}(\boldsymbol{R}_B^I) = \boldsymbol{\tau}_v - \boldsymbol{\tau}_{v,C}. \tag{2.52}$$

The following properties hold:

- the inertia matrix is symmetric and positive definite, i.e., $\boldsymbol{M}_v = \boldsymbol{M}_v^{\mathrm{T}} > \boldsymbol{O}$;
- the damping matrix is positive definite, i.e., $\boldsymbol{D}_{RB}(\boldsymbol{\nu}) > \boldsymbol{O}$;

- the matrix $\boldsymbol{C}_v(\boldsymbol{\nu})$ is skew-symmetric, i.e., $\boldsymbol{C}_v(\boldsymbol{\nu}) = -\boldsymbol{C}_v^{\mathrm{T}}(\boldsymbol{\nu}), \forall \boldsymbol{\nu} \in \mathbb{R}^6$.

It is possible to rewrite the dynamic model (2.51) in terms of earth-fixed coordinates; in this case, the state variables are the $(6 \times 1)$ vectors $\boldsymbol{\eta}$, $\dot{\boldsymbol{\eta}}$ and $\ddot{\boldsymbol{\eta}}$. The equations of motion are then obtained, through the kinematic relations (2.1)–(2.2) as

$$\boldsymbol{M}_v^{\star}(\boldsymbol{R}_B^I)\ddot{\boldsymbol{\eta}} + \boldsymbol{C}_v^{\star}(\boldsymbol{R}_B^I, \dot{\boldsymbol{\eta}})\dot{\boldsymbol{\eta}} + \boldsymbol{D}_{RB}^{\star}(\boldsymbol{R}_B^I, \dot{\boldsymbol{\eta}})\dot{\boldsymbol{\eta}} + \boldsymbol{g}_{RB}^{\star}(\boldsymbol{R}_B^I) = \boldsymbol{\tau}_v^{\star}, \quad (2.53)$$

where [127]

$$\begin{aligned}
\boldsymbol{M}_v^{\star} &= \boldsymbol{J}_e^{-\mathrm{T}}(\boldsymbol{R}_B^I)\boldsymbol{M}_v\boldsymbol{J}_e^{-1}(\boldsymbol{R}_B^I)\\
\boldsymbol{C}_v^{\star} &= \boldsymbol{J}_e^{-\mathrm{T}}(\boldsymbol{R}_B^I)\left(\boldsymbol{C}_v(\boldsymbol{\nu}) - \boldsymbol{M}_v\boldsymbol{J}_e^{-1}(\boldsymbol{R}_B^I)\dot{\boldsymbol{J}}(\boldsymbol{R}_B^I)\right)\boldsymbol{J}_e^{-1}(\boldsymbol{R}_B^I)\\
\boldsymbol{D}_{RB}^{\star} &= \boldsymbol{J}_e^{-\mathrm{T}}(\boldsymbol{R}_B^I)\boldsymbol{D}_{RB}(\boldsymbol{\nu})\boldsymbol{J}_e^{-1}(\boldsymbol{R}_B^I)\\
\boldsymbol{g}_{RB}^{\star} &= \boldsymbol{J}_e^{-\mathrm{T}}(\boldsymbol{R}_B^I)\boldsymbol{g}_{RB}(\boldsymbol{R}_B^I)\\
\boldsymbol{\tau}_v^{\star} &= \boldsymbol{J}_e^{-\mathrm{T}}(\boldsymbol{R}_B^I)\boldsymbol{\tau}_v.
\end{aligned}$$

Again, the current can be taken into account by resorting to the relative velocity or, introducing an approximation, considering the following equations of motion:

$$\boldsymbol{M}_v^{\star}(\boldsymbol{R}_B^I)\ddot{\boldsymbol{\eta}} + \boldsymbol{C}_v^{\star}(\boldsymbol{R}_B^I, \dot{\boldsymbol{\eta}})\dot{\boldsymbol{\eta}} + \boldsymbol{D}_{RB}^{\star}(\boldsymbol{R}_B^I, \dot{\boldsymbol{\eta}})\dot{\boldsymbol{\eta}} + \boldsymbol{g}_{RB}^{\star}(\boldsymbol{R}_B^I) = \boldsymbol{\tau}_v^{\star} - \boldsymbol{\tau}_{v,C}^{\star},$$

where $\boldsymbol{\tau}_{v,C}^{\star} \in \mathbb{R}^6$ is the disturbance introduced by the current. It is worth noticing that the earth-fixed and the body-fixed models with the introduction of the current as a simple external disturbance implies different dynamic properties. In particular, this is true if, in case of the design of a control action, the disturbance is considered as constant or slowly varying.

### 2.7.1 Linearity in the Parameters

Relation (2.51) can be written by exploiting the linearity in the parameters property. It must be noted that, while this property is proved for rigid bodies moving in the space [254], for underwater rigid bodies it depends on a suitable representations of the hydrodynamics terms. With a vector of parameters $\boldsymbol{\theta}_v$ of proper dimension it is possible to write the following:

$$\boldsymbol{\Phi}_v(\boldsymbol{R}_B^I, \boldsymbol{\nu}, \dot{\boldsymbol{\nu}})\boldsymbol{\theta}_v = \boldsymbol{\tau}_v. \quad (2.54)$$

The inclusion of the ocean current is straightforward by using the relative velocity as shown in Subsection 2.4.3. However, it might be useful to consider also the regressor form of the two approximations given by considering the current as an external disturbance. In particular, it is of interest to isolate the contribution of the restoring forces and current effects, those are the sole terms giving a non-null contribution to the dynamic with the vehicle still and for this reason will be defined as *persistent dynamic terms*.

Starting from the equation (2.52) let first consider the current as an external disturbance $\boldsymbol{\tau}_{v,C}$ constant in the body-fixed frame, it is possible to write:

$$\boldsymbol{M}_v\dot{\boldsymbol{\nu}} + \boldsymbol{C}_v(\boldsymbol{\nu})\boldsymbol{\nu} + \boldsymbol{D}_{RB}(\boldsymbol{\nu})\boldsymbol{\nu} + \boldsymbol{\Phi}_{v,R}(\boldsymbol{R}_B^I)\boldsymbol{\theta}_{v,R} + \boldsymbol{\Phi}'_{v,C}\boldsymbol{\theta}_{v,C} = \boldsymbol{\tau}_v$$

that can be rewritten as:

$$\boxed{\boldsymbol{M}_v\dot{\boldsymbol{\nu}} + \boldsymbol{C}_v(\boldsymbol{\nu})\boldsymbol{\nu} + \boldsymbol{D}_{RB}(\boldsymbol{\nu})\boldsymbol{\nu} + \boldsymbol{\Phi}_{v,P'}(\boldsymbol{R}_B^I)\boldsymbol{\theta}_{v,P} = \boldsymbol{\tau}_v} \tag{2.55}$$

with the use of the $(6 \times 10)$ regressor:

$$\boldsymbol{\Phi}_{v,P'}(\boldsymbol{R}_B^I) = \begin{bmatrix} \boldsymbol{R}_I^B\boldsymbol{z} & \boldsymbol{O}_{3\times3} & \boldsymbol{I}_3 & \boldsymbol{O}_{3\times3} \\ \boldsymbol{0}_{3\times1} & \boldsymbol{S}\left(\boldsymbol{R}_I^B\boldsymbol{z}\right) & \boldsymbol{O}_{3\times3} & \boldsymbol{I}_3 \end{bmatrix}.$$

On the other side the current can be modeled as constant in the earth-fixed frame and, merged again with the restoring forces contribution, gives the following

$$\boxed{\boldsymbol{M}_v\dot{\boldsymbol{\nu}} + \boldsymbol{C}_v(\boldsymbol{\nu})\boldsymbol{\nu} + \boldsymbol{D}_{RB}(\boldsymbol{\nu})\boldsymbol{\nu} + \boldsymbol{\Phi}_{v,P}(\boldsymbol{R}_B^I)\boldsymbol{\theta}_{v,P} = \boldsymbol{\tau}_v} \tag{2.56}$$

with the use of the $(6 \times 9)$ regressor:

$$\boldsymbol{\Phi}_{v,P}(\boldsymbol{R}_B^I) = \begin{bmatrix} \boldsymbol{O}_{3\times3} & \boldsymbol{R}_I^B & \boldsymbol{O}_{3\times3} \\ \boldsymbol{S}\left(\boldsymbol{R}_I^B\boldsymbol{z}\right) & \boldsymbol{O}_{3\times3} & \boldsymbol{R}_I^B \end{bmatrix}.$$

It is worth noticing that the two regressors have different dimensions. In order to extrapolate the minimum number of independent parameters, i.e., the number of columns of the regressor, it is possible to resort to the numerical method proposed by Gautier [135] based on the Singular Value Decomposition.

Model (2.56) can by rewritten in a sole regressor of proper dimension yielding:

$$\boldsymbol{\Phi}_{v,T}(\boldsymbol{R}_B^I, \boldsymbol{\nu}, \dot{\boldsymbol{\nu}})\boldsymbol{\theta}_{v,T} = \boldsymbol{\tau}_v. \tag{2.57}$$

## 2.8 Kinematics of Manipulators with Mobile Base

In Figure 2.3 a sketch of an Underwater Vehicle-Manipulator System with relevant frames is shown. The frames are assumed to satisfy the Denavit-Hartenberg convention [254]. The position and orientation of the end effector, thus, is easily obtained by the use of homogeneous transformation matrices.

Let $\boldsymbol{q} \in \mathbb{R}^n$ be the vector of joint positions where $n$ is the number of joints. The vector $\dot{\boldsymbol{q}} \in \mathbb{R}^n$ is the corresponding time derivative. Let define $\boldsymbol{\zeta} \in \mathbb{R}^{6+n}$ as

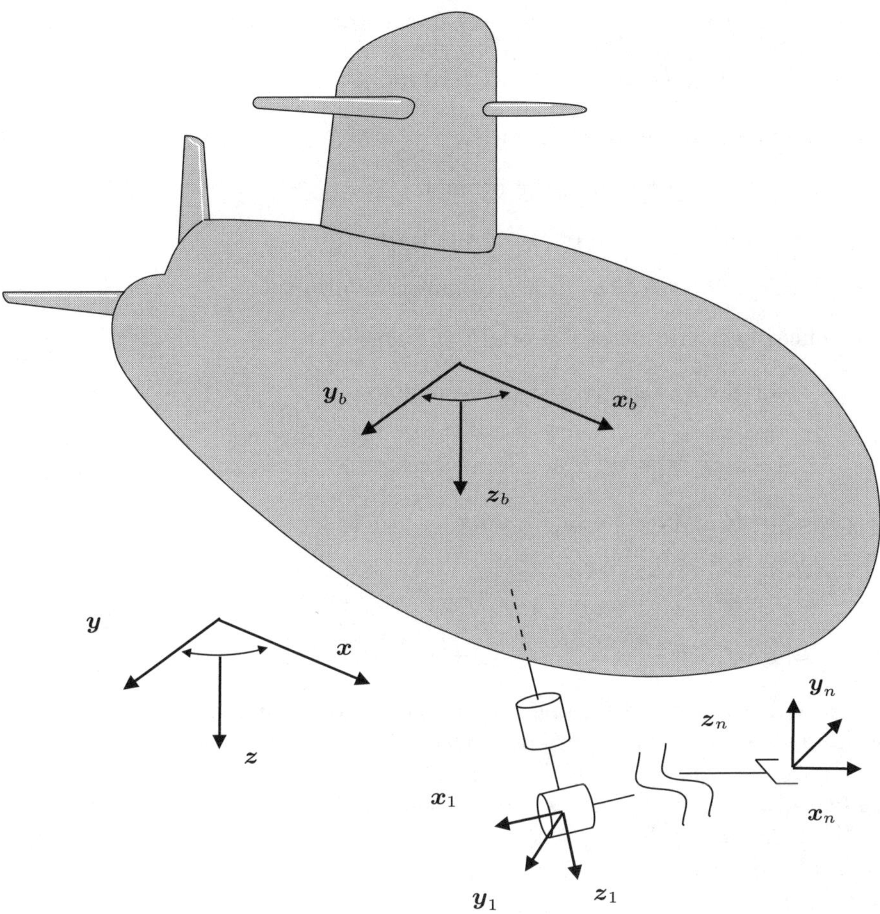


**Fig. 2.3.** Sketch of an Underwater Vehicle-Manipulator System with relevant frames

$$
\boldsymbol{\zeta} = \begin{bmatrix} \boldsymbol{\nu}_1 \\ \boldsymbol{\nu}_2 \\ \dot{\boldsymbol{q}} \end{bmatrix};
$$

It is useful to rewrite the relationship between body-fixed and earth-fixed velocities given in equations (2.1)-(2.2) in a more compact form:

$$
\boldsymbol{\zeta} = \begin{bmatrix} \boldsymbol{\nu}_1 \\ \boldsymbol{\nu}_2 \\ \dot{\boldsymbol{q}} \end{bmatrix} = \begin{bmatrix} \boldsymbol{R}_I^B & \boldsymbol{O}_{3\times3} & \boldsymbol{O}_{3\times n} \\ \boldsymbol{O}_{3\times3} & \boldsymbol{J}_{k,o}(\boldsymbol{R}_I^B) & \boldsymbol{O}_{3\times n} \\ \boldsymbol{O}_{n\times3} & \boldsymbol{O}_{n\times3} & \boldsymbol{I}_n \end{bmatrix} \begin{bmatrix} \dot{\boldsymbol{\eta}}_1 \\ \dot{\boldsymbol{\eta}}_2 \\ \dot{\boldsymbol{q}} \end{bmatrix} = \boldsymbol{J}_k \begin{bmatrix} \dot{\boldsymbol{\eta}}_1 \\ \dot{\boldsymbol{\eta}}_2 \\ \dot{\boldsymbol{q}} \end{bmatrix}, \quad (2.58)
$$

where $\boldsymbol{O}_{n_1\times n_2}$ is the null $(n_1 \times n_2)$ matrix and the matrix $\boldsymbol{J}_{k,o}(\boldsymbol{R}_I^B)$ in terms of Euler angles has been introduced in (2.3).

Knowing $\boldsymbol{\nu}_1$, $\boldsymbol{\nu}_2$, $\dot{\boldsymbol{\nu}}_1$, $\dot{\boldsymbol{\nu}}_2$, (vehicle linear and angular velocities and acceleration in body fixed frame), $\dot{\boldsymbol{q}}$, $\ddot{\boldsymbol{q}}$, (joint velocities and acceleration) it is possible to calculate, for every link, the following variables:

$\boldsymbol{\omega}_i^i$, angular velocity of the frame $i$,

$\dot{\boldsymbol{\omega}}_i^i$, angular acceleration of the frame $i$,

$\boldsymbol{v}_i^i$, linear velocity of the origin of the frame $i$,

$\boldsymbol{v}_{ic}^i$, linear velocity of the center of mass of link $i$,

$\boldsymbol{a}_i^i$, linear acceleration of the origin of frame $i$,

by resorting to the following relationships:

$$\boldsymbol{\omega}_i^i = \boldsymbol{R}_{i-1}^i \left(\boldsymbol{\omega}_{i-1}^{i-1} + \dot{q}_i \boldsymbol{z}_{i-1}\right) \tag{2.59}$$

$$\dot{\boldsymbol{\omega}}_i^i = \boldsymbol{R}_{i-1}^i \left(\dot{\boldsymbol{\omega}}_{i-1}^{i-1} + \boldsymbol{\omega}_{i-1}^{i-1} \times \dot{q}_i \boldsymbol{z}_{i-1} + \ddot{q}_i \boldsymbol{z}_{i-1}\right) \tag{2.60}$$

$$\boldsymbol{v}_i^i = \boldsymbol{R}_{i-1}^i \boldsymbol{v}_{i-1}^{i-1} + \boldsymbol{\omega}_i^i \times \boldsymbol{r}_{i-1,i}^i \tag{2.61}$$

$$\boldsymbol{v}_{ic}^i = \boldsymbol{R}_{i-1}^i \boldsymbol{v}_{i-1}^{i-1} + \boldsymbol{\omega}_i^i \times \boldsymbol{r}_{i-1,c}^i \tag{2.62}$$

$$\boldsymbol{a}_i^i = \boldsymbol{R}_{i-1}^i \boldsymbol{a}_{i-1}^{i-1} + \dot{\boldsymbol{\omega}}_i^i \times \boldsymbol{r}_{i-1,i}^i + \boldsymbol{\omega}_i^i \times (\boldsymbol{\omega}_i^i \times \boldsymbol{r}_{i-1,i}^i) \tag{2.63}$$

where $\boldsymbol{z}_i$ is the versor of frame $i$, $\boldsymbol{r}_{i-1,i}^i$ is the constant vector from the origin of frame $i-1$ toward the origin of frame $i$ expressed in frame $i$.

Since the task of UVMS missions is usually force/position control of the end effector frame, it is necessary to consider the position of the end effector in the inertial frame, $\boldsymbol{\eta}_{ee1} \in \mathbb{R}^3$; this is a function of the system configuration, i.e., $\boldsymbol{\eta}_{ee1}(\boldsymbol{\eta}_1, \boldsymbol{R}_I^B, \boldsymbol{q})$. The vector $\dot{\boldsymbol{\eta}}_{ee1} \in \mathbb{R}^3$ is the corresponding time derivative.

Let us further define $\boldsymbol{\eta}_{ee2} \in \mathbb{R}^3$ as the orientation of the end effector in the inertial frame expressed by Euler angles: also $\boldsymbol{\eta}_{ee2}$ is a function of the system configuration, i.e., $\boldsymbol{\eta}_{ee2}(\boldsymbol{R}_I^B, \boldsymbol{q})$. Again, the vector $\dot{\boldsymbol{\eta}}_{ee2} \in \mathbb{R}^3$ is the corresponding time derivative.

The relation between the end-effector posture $\boldsymbol{\eta}_{ee} = [\boldsymbol{\eta}_{ee1}^{\mathrm{T}} \quad \boldsymbol{\eta}_{ee2}^{\mathrm{T}}]^{\mathrm{T}}$ and the system configuration can be expressed by the following nonlinear equation:

$$\boldsymbol{\eta}_{ee} = \boldsymbol{k}(\boldsymbol{\eta}, \boldsymbol{q}). \tag{2.64}$$

The vectors $\dot{\boldsymbol{\eta}}_{ee1}$ and $\dot{\boldsymbol{\eta}}_{ee2}$ are related to the body-fixed velocities $\boldsymbol{\nu}_{ee}$ via relations analogous to (2.1) and (2.2), i.e.,

$$\boldsymbol{\nu}_{ee1} = \boldsymbol{R}_I^n \dot{\boldsymbol{\eta}}_{ee1} \tag{2.65}$$

$$\boldsymbol{\nu}_{ee2} = \boldsymbol{J}_{k,o}(\boldsymbol{\eta}_{ee2}) \dot{\boldsymbol{\eta}}_{ee2} \tag{2.66}$$

where $\boldsymbol{R}_I^n$ is the rotation matrix from the inertial frame to the end-effector frame (i.e., frame $n$) and $\boldsymbol{J}_{k,o}$ is the matrix defined as in (2.3) with the use of the Euler angles of the end-effector frame. If the end-effector orientation is expressed via quaternion the relation between end-effector angular velocity and time derivative of the quaternion can be easily obtained by the *quaternion propagation equation* (2.10).

The end-effector velocities (expressed in the inertial frame) are related to the body-fixed system velocity by a suitable Jacobian matrix, i.e.,

$$\begin{bmatrix} \dot{\boldsymbol{\eta}}_{ee1} \\ \dot{\boldsymbol{\eta}}_{ee2} \end{bmatrix} = \begin{bmatrix} \boldsymbol{J}_{pos}(\boldsymbol{R}_B^I, \boldsymbol{q}) \\ \boldsymbol{J}_{or}(\boldsymbol{R}_B^I, \boldsymbol{q}) \end{bmatrix} \boldsymbol{\zeta} = \boldsymbol{J}_w(\boldsymbol{R}_B^I, \boldsymbol{q})\boldsymbol{\zeta}. \tag{2.67}$$

In Chapter 6, a different version of the (2.67) will be considered. To have a compact expression to the representation of the attitude error via quaternions, the end-effector velocities (expressed in the earth-fixed frame) are related to the body-fixed system velocity by the following Jacobian matrix:

$$\dot{\boldsymbol{x}}_E = \begin{bmatrix} \dot{\boldsymbol{\eta}}_{ee1} \\ \boldsymbol{R}_n^I \boldsymbol{\nu}_{ee2} \end{bmatrix} = \boldsymbol{J}(\boldsymbol{R}_B^I, \boldsymbol{q})\boldsymbol{\zeta}. \tag{2.68}$$

Notice that the Jacobian has been derived with respect to the angular velocity of the end effector expressed in the earth-fixed frame (the matrix $\boldsymbol{R}_n^I = \boldsymbol{R}_I^{n\,\mathrm{T}}$ is the rotation from the frame $n$ the the earth-fixed frame).

## 2.9 Dynamics of Underwater Vehicle-Manipulator Systems

By knowing the forces acting on a body moving in a fluid it is possible to easily obtain the dynamics of a serial chain of rigid bodies moving in a fluid.

The inertial forces and moments acting on the generic body are represented by:

$$\begin{aligned} \boldsymbol{F}_i^i &= \boldsymbol{M}_i[\boldsymbol{a}_i^i + \dot{\boldsymbol{\omega}}_i^i \times \boldsymbol{r}_{i,c}^i + \boldsymbol{\omega}_i^i \times (\boldsymbol{\omega}_i^i \times \boldsymbol{r}_{i,c}^i)] \\ \boldsymbol{T}_i^i &= \boldsymbol{I}_i^i \dot{\boldsymbol{\omega}}_i^i + \boldsymbol{\omega}_i^i \times (\boldsymbol{I}_i^i \boldsymbol{\omega}_i^i), \end{aligned}$$

where $\boldsymbol{M}_i$ is the $(3 \times 3)$ mass matrix comprehensive of the added mass, $\boldsymbol{I}_i^i$ is the $(3 \times 3)$ inertia matrix plus added inertia with respect to the center of mass, $\boldsymbol{r}_{i,c}^i$ is the vector from the origin of frame $i$ toward the center of mass of link $i$ expressed in frame $i$.

Let us define $\boldsymbol{d}_i^i$ the drag and lift forces acting on the center of mass of link $i$, $\boldsymbol{r}_{i-1,i}^i$ the vector from the origin of frame $i-1$ to the origin of frame $i$ expressed in frame $i$, $\boldsymbol{r}_{i-1,c}^i$ the vector from the origin of frame $i-1$ to the center of mass of link $i$ expressed in frame $i$ and $\boldsymbol{r}_{i-1,b}^i$ the vector from the origin of frame $i-1$ to the center of buoyancy of link $i$ expressed in frame $i$,

$$\boldsymbol{g}^i = \boldsymbol{R}_I^i \boldsymbol{g}^I = \boldsymbol{R}_I^i \begin{bmatrix} 0 \\ 0 \\ 9.81 \end{bmatrix} \quad \mathrm{m/s^2}.$$

The total forces and moments acting on the generic body of the serial chain are given by:

$$\begin{aligned}
\boldsymbol{f}_i^i &= \boldsymbol{R}_{i+1}^i \boldsymbol{f}_{i+1}^{i+1} + \boldsymbol{F}_i^i - m_i \boldsymbol{g}^i + \rho \nabla_i \boldsymbol{g}^i + \boldsymbol{p}_i \\
\boldsymbol{\mu}_i^i &= \boldsymbol{R}_{i+1}^i \boldsymbol{\mu}_{i+1}^{i+1} + \boldsymbol{R}_{i+1}^i \boldsymbol{r}_{i-1,i}^{i+1} \times \boldsymbol{R}_{i+1}^i \boldsymbol{f}_{i+1}^{i+1} + \boldsymbol{r}_{i-1,c}^i \times \boldsymbol{F}_i^i + \boldsymbol{T}_i^i + \\
&\quad + \boldsymbol{r}_{i-1,c}^i \times (-m_i \boldsymbol{g}^i + \boldsymbol{d}_i) + \boldsymbol{r}_{i-1,b}^i \times \rho \nabla_i \boldsymbol{g}^i
\end{aligned}$$

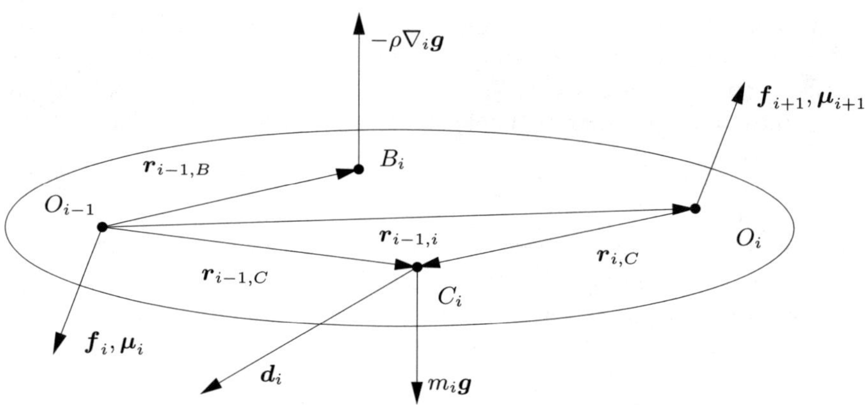


**Fig. 2.4.** Force/moment acting on link $i$

The torque acting on joint $i$ is finally given by:

$$\tau_{q,i} = \boldsymbol{\mu}_i^{i\,\mathrm{T}} \boldsymbol{z}_{i-1}^i + f_{di}\mathrm{sign}(\dot{q}_i) + f_{vi}\dot{q}_i \tag{2.69}$$

with $f_{di}$ and $f_{vi}$ the motor dry and viscous friction coefficients.

Let us define $\boldsymbol{\tau}_q = [\tau_{q,1} \quad \ldots \quad \tau_{q,n}]^{\mathrm{T}} \in \mathbb{R}^n$ the vector of joint torques and $\boldsymbol{\tau} \in \mathbb{R}^{6+n}$

$$\boldsymbol{\tau} = \begin{bmatrix} \boldsymbol{\tau}_v \\ \boldsymbol{\tau}_q \end{bmatrix} \tag{2.70}$$

the vector of force/moment acting on the vehicle as well as joint torques. It is possible to write the equations of motions of an UVMS in a matrix form:

$$\boxed{\boldsymbol{M}(\boldsymbol{q})\dot{\boldsymbol{\zeta}} + \boldsymbol{C}(\boldsymbol{q}, \boldsymbol{\zeta})\boldsymbol{\zeta} + \boldsymbol{D}(\boldsymbol{q}, \boldsymbol{\zeta})\boldsymbol{\zeta} + \boldsymbol{g}(\boldsymbol{q}, \boldsymbol{R}_B^I) = \boldsymbol{\tau}} \tag{2.71}$$

where $\boldsymbol{M} \in \mathbb{R}^{(6+n)\times(6+n)}$ is the inertia matrix including added mass terms, $\boldsymbol{C}(\boldsymbol{q}, \boldsymbol{\zeta})\boldsymbol{\zeta} \in \mathbb{R}^{6+n}$ is the vector of Coriolis and centripetal terms, $\boldsymbol{D}(\boldsymbol{q}, \boldsymbol{\zeta})\boldsymbol{\zeta} \in \mathbb{R}^{6+n}$ is the vector of dissipative effects, $\boldsymbol{g}(\boldsymbol{q}, \boldsymbol{R}_I^B) \in \mathbb{R}^{6+n}$ is the vector of gravity and buoyancy effects. The relationship between the generalized forces $\boldsymbol{\tau}$ and the control input is given by:

$$\boldsymbol{\tau} = \begin{bmatrix} \boldsymbol{\tau}_v \\ \boldsymbol{\tau}_q \end{bmatrix} = \begin{bmatrix} \boldsymbol{B}_v & \boldsymbol{O}_{6\times n} \\ \boldsymbol{O}_{n\times 6} & \boldsymbol{I}_n \end{bmatrix} \boldsymbol{u} = \boldsymbol{B}\boldsymbol{u}, \tag{2.72}$$

where $\boldsymbol{u} \in \mathbb{R}^{p_v+n}$ is the vector of the control input. Notice that, while for the vehicle a generic number $p_v \geq 6$ of control inputs is assumed, for the manipulator it is supposed that $n$ joint motors are available.

It can be proven that:

- The inertia matrix $\boldsymbol{M}$ of the system is symmetric and positive definite:
$$\boldsymbol{M} = \boldsymbol{M}^{\mathrm{T}} > \boldsymbol{O}$$
moreover, it satisfies the inequality
$$\lambda_{\min}(\boldsymbol{M}) \leq \|\boldsymbol{M}\| \leq \lambda_{\max}(\boldsymbol{M}),$$
where $\lambda_{\min}(\boldsymbol{M})$ ($\lambda_{\max}(\boldsymbol{M})$) is the minimum (maximum) eigenvalue of $\boldsymbol{M}$.
- For a suitable choice of the parametrization of $\boldsymbol{C}$ and if all the single bodies of the system are symmetric, $\dot{\boldsymbol{M}} - 2\boldsymbol{C}$ is skew-symmetric [67]
$$\boldsymbol{\zeta}^{\mathrm{T}} \left( \dot{\boldsymbol{M}} - 2\boldsymbol{C} \right) \boldsymbol{\zeta} = 0$$
which implies
$$\dot{\boldsymbol{M}} = \boldsymbol{C} + \boldsymbol{C}^{\mathrm{T}}$$
moreover, the inequality
$$\|\boldsymbol{C}(\boldsymbol{a}, \boldsymbol{b})\boldsymbol{c}\| \leq C_M \|\boldsymbol{b}\| \, \|\boldsymbol{c}\|$$
and the equality
$$\boldsymbol{C}(\boldsymbol{a}, \alpha_1 \boldsymbol{b} + \alpha_2 \boldsymbol{c}) = \alpha_1 \boldsymbol{C}(\boldsymbol{a}, \boldsymbol{b}) + \alpha_2 \boldsymbol{C}(\boldsymbol{a}, \boldsymbol{c})$$
hold.
- The matrix $\boldsymbol{D}$ is positive definite
$$\boldsymbol{D} > \boldsymbol{O}$$
and satisfies
$$\|\boldsymbol{D}(\boldsymbol{q}, \boldsymbol{a}) - \boldsymbol{D}(\boldsymbol{q}, \boldsymbol{b})\| \leq D_M \|\boldsymbol{a} - \boldsymbol{b}\| \,.$$

In [255], it can be found the mathematical model written with respect to the earth-fixed-frame-based vehicle position and the manipulator end-effector. However, it must be noted that, in that case, a 6-dimensional manipulator is considered in order to have square Jacobian to work with; moreover, kinematic singularities need to be avoided.

Reference [174] reports some interesting dynamic considerations about the interaction between the vehicle and the manipulator. The analysis performed allows to divide the dynamics in separate meaningful terms.

### 2.9.1 Linearity in the Parameters

UVMS have a property that is common to most mechanical systems, e.g., serial chain manipulators: linearity in the dynamic parameters. Using a suitable mathematical model for the hydrodynamic forces, (2.71) can be rewritten in a matrix form that exploits this property:

$$\boldsymbol{\Phi}(\boldsymbol{q}, \boldsymbol{R}_B^I, \boldsymbol{\zeta}, \dot{\boldsymbol{\zeta}})\boldsymbol{\theta} = \boldsymbol{\tau} \tag{2.73}$$

with $\boldsymbol{\Phi} \in R^{(6+n)\times n_\theta}$, being $n_\theta$ the total number of parameters. Notice that $n_\theta$ depends on the model used for the hydrodynamic generalized forces and joint friction terms. For a single rigid body the number of dynamic parameter $n_{\theta,v}$ is a number greater than 100 [127]. For an UVMS it is $n_\theta = (n+1) \cdot n_{\theta,v}$, that gives an idea of the complexity of such systems.

Differently from ground fixed manipulators, in this case the number of parameters can not be reduced because, due to the 6 degrees of freedom (DOFs) of the sole vehicle, all the dynamic parameters provide an individual contribution to the motion.

## 2.10 Contact with the Environment

If the end effector of a robotic system is in contact with the environment, the force/moment at the tip of the manipulator acts on the whole system according to the equation ([254])

$$\boxed{\boldsymbol{M}(\boldsymbol{q})\dot{\boldsymbol{\zeta}} + \boldsymbol{C}(\boldsymbol{q}, \boldsymbol{\zeta})\boldsymbol{\zeta} + \boldsymbol{D}(\boldsymbol{q}, \boldsymbol{\zeta})\boldsymbol{\zeta} + \boldsymbol{g}(\boldsymbol{q}, \boldsymbol{R}_B^I) = \boldsymbol{\tau} + \boldsymbol{J}_w^{\mathrm{T}}(\boldsymbol{q}, \boldsymbol{R}_B^I)\boldsymbol{h}_e,} \tag{2.74}$$

where $\boldsymbol{J}_w$ is the Jacobian matrix defined in (2.67) and the vector $\boldsymbol{h}_e \in \mathbb{R}^6$ is defined as

$$\boldsymbol{h}_e = \begin{bmatrix} \boldsymbol{f}_e \\ \boldsymbol{\mu}_e \end{bmatrix}$$

i.e., the vector of force/moments at the end effector expressed in the inertial frame. If it is assumed that only linear forces act on the end effector equation (2.74) becomes

$$\boxed{\boldsymbol{M}(\boldsymbol{q})\dot{\boldsymbol{\zeta}} + \boldsymbol{C}(\boldsymbol{q}, \boldsymbol{\zeta})\boldsymbol{\zeta} + \boldsymbol{D}(\boldsymbol{q}, \boldsymbol{\zeta})\boldsymbol{\zeta} + \boldsymbol{g}(\boldsymbol{q}, \boldsymbol{R}_B^I) = \boldsymbol{\tau} + \boldsymbol{J}_{pos}^{\mathrm{T}}(\boldsymbol{q}, \boldsymbol{R}_B^I)\boldsymbol{f}_e} \tag{2.75}$$

Contact between the manipulator and the environment is usually difficult to model. In the following the simple model constituted by a frictionless and elastically compliant plane will be considered. The force at the end effector is then related to the deformation of the environment by the following simplified model [79] (see Figure 2.5)

$$\boldsymbol{f}_e = \boldsymbol{K}(\boldsymbol{x} - \boldsymbol{x}_e), \tag{2.76}$$

where $\boldsymbol{x}$ is the position of the end effector expressed in the inertial frame, $\boldsymbol{x}_e$ characterizes the constant position of the unperturbed environment expressed in the inertial frame and

$$\boldsymbol{K} = k\boldsymbol{n}\boldsymbol{n}^{\mathrm{T}}, \tag{2.77}$$

with $k > 0$, is the stiffness matrix being $\boldsymbol{n}$ the vector normal to the plane [194].

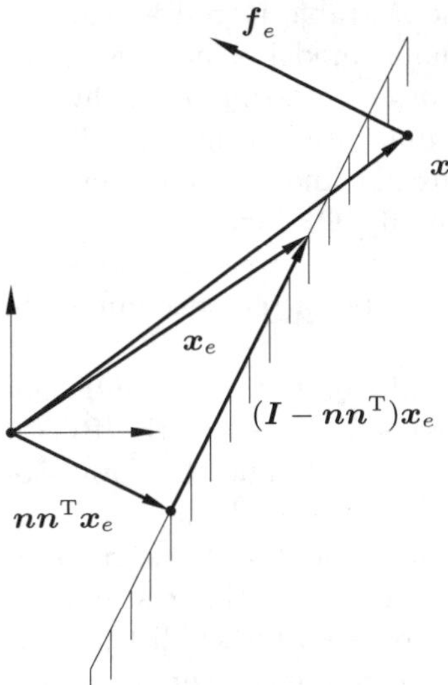


**Fig. 2.5.** Planar view of the chosen model for the contact force

In our case it is $\boldsymbol{x} = \boldsymbol{\eta}_{ee1}$; however, in the force control chapter, the notation $\boldsymbol{x}$ will be maintained.

## 2.11 Identification

Identification of the dynamic parameters of underwater robotic structures is a very challenging task. The mathematical model shares its main characteristics with the model of a ground-fixed industrial manipulator, e.g., it is non-linear and coupled. In case of underwater structure, however, the hydrodynamic terms are approximation of the physical effects. The actuation system of the vehicle is achieved mainly by the thrusters the models of which are still object of research. Finally, accurate measurement of the whole configuration is not easy. For these reasons, while from the mathematical aspect the problem is not new, from the practical point of view it is very difficult to set-up a systematic and reliable identification procedure for UVMSs. At the best of our knowledge, there is no significant results in the identification of full

UVMSs model. Few experimental results, moreover, concern the sole vehicle; often driven in few DOFs.

Since most of the fault detection algorithms rely on the accuracy of the mathematical model, this is also the reason why in this domain too, there are few experimental results (see Chapter 4).

In [2] the hydrodynamic damping terms of the vehicle Roby 2 developed at the Naval Automation Institute, National Research Council, Italy, (now CNR-ISSIA) have been experimentally estimated and further used to develop fault detection/tolerance strategies. The vehicle is stable in roll and pitch, hence, considering a constant depth, the sole planar model is identified.

In [227] some sea trials have been set-up in order to estimate the hydrodynamic derivatives of an 1/3-scale PAP-104 mine countermeasures ROV. The paper assumes that the added mass is already known, moreover, the identification concerns the planar motion for the 6-DOFs model. The position of the vehicle is measured by means of a redundant acoustic system; all the measurements are fused in an EKF in order to obtain the optimum state estimation. Experimental results are given.

The work [271] reports some experimental results on the single DOF models for the ROV developed at the John Hopkins University (JHUROV). First, the mathematical model is written so as to underline the Input-to-State-Stability; then a stable, on-line, adaptive identification technique is derived. The latter method is compared with a classic, off-line, Least-Squares approach. The approximation required by the proposed technique is that the equations of motion are decoupled, diagonal, there is no tether disturbance and the added mass is constant. Interesting experimental results are reported.

The interested reader can refer also to [62, 111].

# 3. Dynamic Control of 6-DOF AUVs

## 3.1 Introduction

In this Chapter, the problem of controlling an Autonomous Underwater Vehicle in 6-DOFs is approached.

To effectively compensate the hydrodynamic effects several adaptive (integral) control laws have been proposed in the literature (see, e.g., [91, 93, 97, 152, 319]). In [126], a number of adaptive control actions are proposed, where the presence of an external disturbance is taken into account and its counteraction is obtained by means of a switching term; simulation on the simplified three-DOF horizontal model of NEROV are given. In [121], a body-fixed-frame based adaptive control law is developed. In [314], an adaptive control law based on Euler angle representation of the orientation has been proposed for the control of an AUV; planar simulations are provided to show the effectiveness of the proposed approach. Reference [184] proposes a self-adaptive neuro-fuzzy inference system that makes use of a 5-layer-structured neural network to improve the function approximation. In [175] a fuzzy membership function based-neural network is proposed; the control's membership functions derivation is achieved by a back propagation network.

Generally speaking, the performance achievable from the application of adaptive control laws has been validated only by means of reduced-order simulations; on the other hand, six-DOF experimental results are seldom presented in the literature [299]. References [34, 35, 122, 126] describe six-DOF control laws in which the orientation is described by the use of quaternions. The papers [34, 35, 84, 215, 216, 323] report six-DOF experimental results on the underwater vehicle ODIN (Omni-Directional Intelligent Navigator).

An experimental work is given in [269, 270] by the use of the Johns Hopkins University ROV on a single DOF. Different simple control laws are tested on the vehicle in presence of model mismatching and thruster saturation and their performance is evaluated.

Among the other hydrodynamic effects acting on a rigid body moving in a fluid, the restoring generalized forces (gravity plus buoyancy) and the ocean current are of major concern in designing a motion control law for underwater vehicles, since they are responsible of steady-state position and orientation errors. However, while the restoring generalized forces are usually dealt with in the framework of adaptive dynamic compensations, only few papers take into account the effect of the ocean current. The works [34, 35, 125] consider a six-DOF control problem in which the ocean current is compensated in vehicle-fixed coordinates; since the current effects are modeled as an external disturbance acting on the vehicle, there is no need for additional sensors. In [145, 243, 244] a different approach is proposed for the three-DOF surge control of the vehicle Phoenix: the current, or more generally, the sea wave, is modeled by an Auto-Regressive dynamic model and an extended Kalman filter is designed to estimate the relative velocity between vehicle and water; the estimated relative velocity is then used by a sliding-mode controller to drive the vehicle. In this case, additional sensors besides those typically available on-board are required. The controller developed in [145] has been also used for the NPS ARIES AUV [201]. Reference [133] reports an algorithm for the underwater navigation of a torpedo-like vehicle in presence of unknown current where the current itself is estimated by resorting to a range measurement from a single location.

A common feature of all the adaptive control laws proposed in the literature is that they are designed starting from dynamic models written either in the earth-fixed frame or in the vehicle-fixed frame. Nevertheless, some hydrodynamic effects are seen as constant in the earth-fixed frame (e.g., the restoring linear force) while some others are constant in the vehicle-fixed frame (e.g., the restoring moment).

In this Chapter a comparison among the controllers developed in [13, 121, 122, 125, 280, 320] is shown. The controllers have been designed for 6-DOFs control of AUVs and they do not need the measurement of the ocean current (when it is taken into account). The analysis will mainly concern the controllers capacity to compensate for the persistent dynamic effects, e.g., the restoring forces and the ocean current. For each controller a reduced version is derived and eventually modified so as to achieve null steady state error under modeling uncertainty and presence of ocean current. It is worth noticing that the reduced controller is not given by the Authors of the corresponding paper; this has to be taken into account while observing the simulation results. The reduced controller will be developed in order to achieve a PD action plus the adaptive/integral compensation of the persistent effects, i.e., it can be considered as the equivalent of an adaptive PD+gravity compensation for industrial manipulator. In other words, the velocity and acceleration based dynamic terms of the model will not be compensated for. Numerical simulations using the model of ODIN (Omni-Directional Intelligent Navigator) [320], have been run to verify the theoretical results.

For easy of readings, Table 3.1 reports the label associated with each controller.

**Table 3.1.** Labels of the discussed controllers

| label | Authors | Sect. | frame |
|---|---|---|---|
| **A** | Fjellstad & Fossen | 3.2 | earth |
| **B** | Yuh *et al.* | 3.3 | earth |
| **C** | Fjellstad & Fossen | 3.4 | vehicle |
| **D** | Fossen & Balchen | 3.5 | earth/vehicle |
| **E** | Antonelli *et al.* | 3.6 | earth/vehicle |
| **F** | Sun & Cheah | 3.7 | earth/vehicle |

## 3.2 Earth-Fixed-Frame-Based, Model-Based Controller

In 1994, O. Fjellstad and T. Fossen [122] propose an earth-fixed-frame-based, model-based controller that makes use of the 4-parameter unit quaternion (Euler parameter) to reach a singularity-free representation of the attitude. The controller is obtained by extending the results obtained in [267] for robot manipulators.

By defining

$$\tilde{\boldsymbol{p}} = \boldsymbol{p}_d - \boldsymbol{p} \tag{3.1}$$

where $\boldsymbol{p} = [\,\boldsymbol{\eta}_1^{\mathrm{T}} \quad \mathcal{Q}^{\mathrm{T}}\,]^{\mathrm{T}} \in \mathrm{I\!R}^7$ is the quaternion-based position/attitude vector of the vehicle and $\boldsymbol{p}_d = [\,\boldsymbol{\eta}_{1,d}^{\mathrm{T}} \quad \mathcal{Q}_d^{\mathrm{T}}\,]^{\mathrm{T}} \in \mathrm{I\!R}^7$ is its desired value. The following $(7 \times 1)$ vector can be further defined:

$$\boldsymbol{s} = \boldsymbol{K}_D\dot{\tilde{\boldsymbol{p}}} + \boldsymbol{K}_P\tilde{\boldsymbol{p}} + \boldsymbol{K}_I\int_0^t \tilde{\boldsymbol{p}}(\tau)d\tau = \dot{\boldsymbol{p}}_r - \boldsymbol{K}_D\dot{\boldsymbol{p}} \tag{3.2}$$

that implies the vector $\dot{\boldsymbol{p}}_r \in \mathrm{I\!R}^7$ defined as

$$\dot{\boldsymbol{p}}_r = \boldsymbol{K}_D\dot{\boldsymbol{p}}_d + \boldsymbol{K}_P\tilde{\boldsymbol{p}} + \boldsymbol{K}_I\int_0^t \tilde{\boldsymbol{p}}(\tau)d\tau \tag{3.3}$$

where $\boldsymbol{K}_D$, $\boldsymbol{K}_P$ and $\boldsymbol{K}_I$ are $(7 \times 7)$ positive definite matrices of gains.

The following control law is proposed

$$\boxed{\boldsymbol{\tau}_v^{\star} = \boldsymbol{M}_v^{\star}\ddot{\boldsymbol{p}}_r + \boldsymbol{C}_v^{\star}\dot{\boldsymbol{p}}_r + \boldsymbol{D}_{RB}^{\star}\dot{\boldsymbol{p}}_r + \boldsymbol{g}_{RB}^{\star} + \boldsymbol{\Lambda}\boldsymbol{s}\,,} \tag{3.4}$$

where $\boldsymbol{\Lambda}$ is a $(7 \times 7)$ positive definite matrix of gains. Notice that the above control law refers to a quaternion-based dynamic model in earth-fixed coordinates which can be obtained from (2.53) by using the matrix $\boldsymbol{J}_{k,oq} \in \mathrm{I\!R}^{4\times 3}$ instead of the matrix $\boldsymbol{J}_{k,o} \in \mathrm{I\!R}^{3\times 3}$ in the construction of the Jacobian $\boldsymbol{J}_e(\boldsymbol{R}_B^I)$. Also notice that, with respect to the model detailed in Section 2.7 the dimension of $\boldsymbol{J}_e(\boldsymbol{R}_B^I)$ are different.

Let now consider the positive semi-definite function:

$$V = \frac{1}{2}\boldsymbol{s}^{\mathrm{T}}\boldsymbol{M}_v^{\star}\boldsymbol{s} > 0, \qquad \forall \boldsymbol{s} \neq \alpha \begin{bmatrix} \mathbf{0} \\ \mathcal{Q} \end{bmatrix}, \qquad \alpha \in \mathrm{I\!R} \tag{3.5}$$

after straightforward calculation, its time derivative is given by:

$$\dot{V} = -\boldsymbol{s}^{\mathrm{T}}\left[\boldsymbol{\Lambda} + \boldsymbol{D}_{RB}^{\star}\right]\boldsymbol{s} < 0, \qquad \forall \boldsymbol{s} \neq \mathbf{0} \tag{3.6}$$

Since $V$ is only positive semi-definite and the system is non-autonomous the stability can not be derived by applying the Lyapunov's theorem. By further assuming that $\dot{\boldsymbol{p}}_r$ is twice differentiable, then $\ddot{V}$ is bounded and $\dot{V}$ is uniformly continuous. Hence, application of the Barbălat's Lemma allows to prove global convergence of $\boldsymbol{s} \to \mathbf{0}$ as $t \to \infty$. Due to the definition of the vector $\boldsymbol{s}$, its convergence to zero also implies convergence of $\tilde{\boldsymbol{p}}$ to the null value.

In case of perfect knowledge of the dynamic model, moreover, the convergence of the error to zero can be demonstrated even for $\boldsymbol{K}_I = \boldsymbol{O}$, i.e., without integral action.

**Compensation of the persistent effects.** If a reduced version of the controller is implemented, e.g., by neglecting the model-based terms in (3.4), the restoring moment is not compensated efficiently. In fact, let consider a vehicle in the two static postures shown in Figure 3.1 and let suppose that the vehicle, starting from the left configuration, is driven to the right configuration and, after a while, back to the left configuration. In the left configuration the integral action in (3.3) does not give any contribution to the control moment as expected, because the vectors of gravity and buoyancy are aligned. Furthermore, in the right configuration the integral action will compensate exactly at the steady state for the moment generated by the misalignment between gravity and buoyancy. When the vehicle is driven back to the left configuration, a null steady-state compensation error is possible after the integral action is discharged; this poses a severe limitation to the control bandwidth that can be achieved. A similar argument holds in the typical practical situation in which the compensation implemented through the vector $\boldsymbol{g}_{RB}^{\star}$ is not exact.

On the other hand, since the error variables are defined in the earth-fixed frame, the controller is appropriate to counteract the current effect. This point will be clarified in next Subsections, when discussing the drawbacks of the controllers **C** and **D** with respect to the current compensation.

Finally, an adaptive version of this controller is not straightforward. In fact, since the dynamic model (2.53) does not depend on the absolute vehicle position, a steady null linear velocity of the vehicle with a non-null position error would *not excite* a corrective adaptive control action. As a result, null position error at rest cannot be guaranteed in presence of ocean current. From the theoretical point of view, this drawback can be avoided by defining the velocity error using the current measurement. However, from the practical point of view, this approach cannot achieve fine positioning of the vehicle since local vortices can make current measurement too noisy.

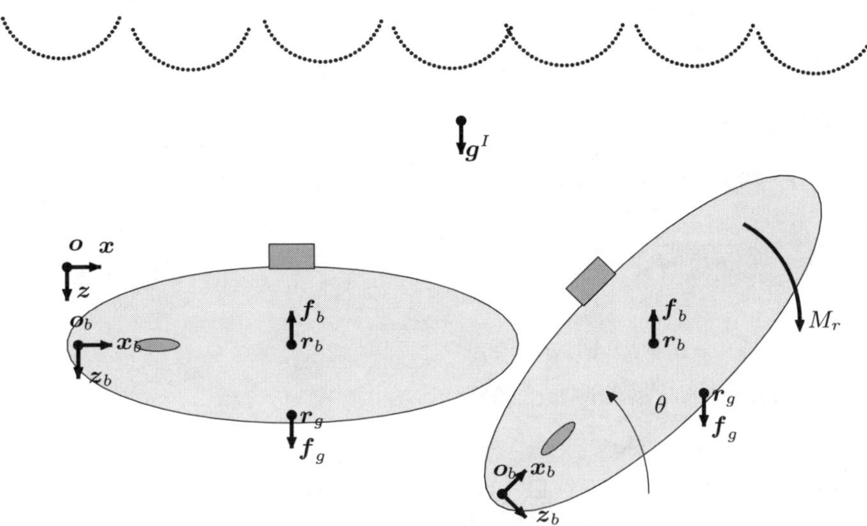


**Fig. 3.1.** Planar view parallel to the $xz$ earth-fixed frame of two different configurations and corresponding restoring forces and moments: $\boldsymbol{r}_G$ ($\boldsymbol{r}_B$) is the center of gravity (buoyancy), $\boldsymbol{f}_G$ ($\boldsymbol{f}_B$) is the gravity (buoyancy) force, and $M_R$ is the $y$ component of the restoring moment

**Reduced Controller.** According to the motivation given in the introduction of this Section, the following reduced version of the control law is considered

$$\boldsymbol{\tau}_v^\star = \hat{\boldsymbol{g}}_{RB}^\star(\boldsymbol{R}_B^I) + \dot{\boldsymbol{p}}_r \,, \tag{3.7}$$

where $\hat{\boldsymbol{g}}_{RB}^\star(\boldsymbol{R}_B^I)$ is a model-based estimate of the restoring generalized force acting on the vehicle. It might be useful to substituting the definition of $\dot{\boldsymbol{p}}_r$ yielding

$$\boldsymbol{\tau}_v^\star = \hat{\boldsymbol{g}}_{RB}^\star(\boldsymbol{R}_B^I) + \boldsymbol{K}_D\dot{\tilde{\boldsymbol{p}}}_d + \boldsymbol{K}_P\tilde{\boldsymbol{p}} + \boldsymbol{K}_I\int_0^t \tilde{\boldsymbol{p}}(\tau)d\tau \tag{3.8}$$

that has a clear interpretation.

## 3.3 Earth-Fixed-Frame-Based, Non-model-Based Controller

In 1999, J. Yuh proposes an earth-fixed-frame-based, non-model-based controller [320]. The control input is given by:

$$\boldsymbol{\tau}_v^\star = \boldsymbol{K}_1\ddot{\boldsymbol{\eta}}_d + \boldsymbol{K}_2\dot{\boldsymbol{\eta}} + \boldsymbol{K}_3 + \boldsymbol{K}_4\dot{\tilde{\boldsymbol{\eta}}} + \boldsymbol{K}_5\tilde{\boldsymbol{\eta}} = \sum_{i=1}^{5} \boldsymbol{K}_i\boldsymbol{\phi}_i\,, \tag{3.9}$$

where $\tilde{\boldsymbol{\eta}} = \boldsymbol{\eta}_d - \boldsymbol{\eta}$, the gains $\boldsymbol{K}_i \in \mathbb{R}^{6\times6}$ are computed as

$$\boldsymbol{K}_i = \frac{\hat{\gamma}_i \boldsymbol{s}_e \boldsymbol{\phi}_i^{\mathrm{T}}}{\|\boldsymbol{s}_e\|\,\|\boldsymbol{\phi}_i\|} \qquad i = 1, \ldots, 5\,,$$

where

$$\boldsymbol{s}_e = \dot{\tilde{\boldsymbol{\eta}}} + \sigma\tilde{\boldsymbol{\eta}} \qquad \text{with } \sigma > 0\,,$$

and the factors $\hat{\gamma}_i$'s are updated by

$$\dot{\hat{\gamma}}_i = f_i\,\|\boldsymbol{s}_e\|\,\|\boldsymbol{\phi}_i\| \qquad \text{with } f_i > 0 \qquad i = 1, \ldots, 5\,.$$

Please notice that $\boldsymbol{\phi}_3 = \boldsymbol{k} \in \mathbb{R}^6$, i.e., a positive constant vector.

Experimental results in 6-DOFs on the use of (3.9) are reported in [215, 216, 323]; these have proven the effectiveness of this controller starting from the surface (with null initial gains) where a smooth version of the controller has been implemented.

The stability analysis can be performed using Lyapunov-like arguments starting from the function

$$V = \frac{1}{2}\tilde{\boldsymbol{\eta}}^{\mathrm{T}}\tilde{\boldsymbol{\eta}} + \frac{1}{2}\sum_{i=1}^{5}\frac{1}{f_i}\left(\gamma_i - \hat{\gamma}_i\right)^2$$

that, differentiated with respect to the time yields a negative semi-definite scalar function; details can be found in [323].

Recently, in [323], Zhao and Yuh propose a model-based version of this control law. The eventual knowledge of the vehicle dynamics is exploited by implementing a *disturbance observer* in charge of partially compensate for the system dynamics. Interesting experimental results with the vehicle ODIN are reported where the tuning of the gains has been achieved on the single DOFs independently. This has been made possible in view of the specific shape of the vehicle, close to a sphere, that reduces the coupling effects and the difference among the directions. It is worth noticing that the considerations below are valid also for this version of the controller.

**Compensation of the persistent effects.** Since the tracking errors are defined in the earth-fixed frame, similar considerations to those developed for the law **A** can be done with respect to compensation of both restoring generalized forces and ocean current effects.

**Reduced Controller.** Being the sole non-model-based law, the controller presented in [320] is characterized by an error behavior much different from that obtained by the other considered controllers and a fair comparison is made difficult. For this reason, a reduced controller has not been retrieved.

## 3.4 Vehicle-Fixed-Frame-Based, Model-Based Controller

Several controllers have been proposed in literature which are based on vehicle-fixed-frame error variables. Among them, the controller proposed in 1994 by O. Fjellstad and T. Fossen [121]; full DOFs experimental results have been reported by G. Antonelli *et al.* in [34, 35].

Let consider the vehicle-fixed variables:

$$\tilde{\boldsymbol{y}} = \begin{bmatrix} \boldsymbol{R}_I^B \tilde{\boldsymbol{\eta}}_1 \\ \tilde{\boldsymbol{\varepsilon}} \end{bmatrix}$$

$$\tilde{\boldsymbol{\nu}} = \boldsymbol{\nu}_d - \boldsymbol{\nu} \,,$$

where $\tilde{\boldsymbol{\eta}}_1 = \boldsymbol{\eta}_{1,d} - \boldsymbol{\eta}_1$, being $\boldsymbol{\eta}_{1,d}$ the desired position, and $\tilde{\boldsymbol{\varepsilon}}$ is the quaternion based attitude error,

$$\boldsymbol{s}_v = \tilde{\boldsymbol{\nu}} + \boldsymbol{\Lambda} \tilde{\boldsymbol{y}} \,, \tag{3.10}$$

with $\boldsymbol{\Lambda} = \text{blockdiag}\{\lambda_p \boldsymbol{I}_3, \lambda_o \boldsymbol{I}_3\}$, $\boldsymbol{\Lambda} > \boldsymbol{O}$.

$$\boldsymbol{\nu}_a = \boldsymbol{\nu}_d + \boldsymbol{\Lambda} \tilde{\boldsymbol{y}} \,. \tag{3.11}$$

Reminding the vehicle regressor $\boldsymbol{\Phi}_v \in R^{6 \times n_{\theta,v}}$ defined in (2.54) and the corresponding vector of dynamic parameters $\boldsymbol{\theta}_v \in R^{n_{\theta,v}}$, the control law is given by:

$$\boxed{\boldsymbol{\tau}_v = \boldsymbol{\Phi}_v(\boldsymbol{R}_B^I, \boldsymbol{\nu}, \boldsymbol{\nu}_a, \dot{\boldsymbol{\nu}}_a) \hat{\boldsymbol{\theta}}_v + \boldsymbol{K}_D \boldsymbol{s}_v \,,} \tag{3.12}$$

where $\boldsymbol{K}_D$ is a $(6 \times 6)$ positive definite matrix. The parameter estimate $\hat{\boldsymbol{\theta}}_v$ is updated by

$$\boxed{\dot{\hat{\boldsymbol{\theta}}}_v = \boldsymbol{K}_\theta^{-1} \boldsymbol{\Phi}_v^{\mathrm{T}}(\boldsymbol{R}_B^I, \boldsymbol{\nu}, \boldsymbol{\nu}_a, \dot{\boldsymbol{\nu}}_a) \boldsymbol{s}_v \,,} \tag{3.13}$$

where $\boldsymbol{K}_\theta$ is a suitable positive definite matrix of appropriate dimension.

The stability analysis is achieved by considering the following Lyapunov candidate function:

$$V = \frac{1}{2} \boldsymbol{s}_v^{\mathrm{T}} \boldsymbol{M}_v \boldsymbol{s}_v + \frac{1}{2} \tilde{\boldsymbol{\theta}}_v^{\mathrm{T}} \boldsymbol{K}_\theta \tilde{\boldsymbol{\theta}}_v > 0, \qquad \forall \boldsymbol{s} \neq \boldsymbol{0},\ \tilde{\boldsymbol{\theta}}_v \neq \boldsymbol{0} \tag{3.14}$$

the time derivative of which, by applying the proposed control law, is given by

$$\dot{V} = -\boldsymbol{s}_v^{\mathrm{T}} \left[ \boldsymbol{K}_D + \boldsymbol{D}_{RB} \right] \boldsymbol{s}_v \tag{3.15}$$

It is now possible to prove the system stability in a Lyapunov-Like sense using the Barbălat's Lemma. Since

- $V$ is lower bounded

- $\dot{V}(\boldsymbol{s}_v, \tilde{\boldsymbol{\theta}}_v) \leq 0$
- $\dot{V}(\boldsymbol{s}_v, \tilde{\boldsymbol{\theta}}_v)$ is uniformly continuous

then

- $\dot{V}(\boldsymbol{s}_v, \tilde{\boldsymbol{\theta}}_v) \rightarrow 0$ as $t \rightarrow \infty$.

Thus $\boldsymbol{s}_v \rightarrow \mathbf{0}$ as $t \rightarrow \infty$. In view of the definition of $\boldsymbol{s}_v$, this implies that $\tilde{\boldsymbol{\nu}} \rightarrow \mathbf{0}$ as $t \rightarrow \infty$; in addition, due to the properties of the quaternion, it results that $\tilde{\eta} \rightarrow 1$ as $t \rightarrow \infty$. However, as usual in adaptive control schemes, it is not possible to prove asymptotic stability of the whole state since $\tilde{\boldsymbol{\theta}}_v$ is only guaranteed to be bounded.

Notice that, in [121], further discussion is developed by considering different choices for the matrix $\boldsymbol{\Lambda}$.

In 1997 [89] G. Conte and A. Serrani develop a Lyapunov-based control for AUVs. The designed controller, by introducing a representation of the model uncertainties, is made robust using Lyapunov techniques. It is worth noticing that this approach does not take into account explicitly for an adaptive/integral action to compensate for the current effect. The position error is represented within a vehicle-fixed representation with a feedback term similar to the vector $\boldsymbol{s}_v$ used in this Section.

**Compensation of the persistent effects.** Following the same reasoning as previously done for the law **A**, it can be deducted that also in this case the adaptive compensation cannot guarantee null position error at rest in the presence of ocean current. In fact, the dynamic model (2.51) shows that a steady null linear velocity of the vehicle with a non-null position error does not excite a corrective adaptive control action. Again, ocean current measurement cannot overcome this problem in practice.

To achieve a null position error in presence of ocean current, an integral action on body-fixed-frame error variables was considered in [34, 35]. However, this integral action has the following drawback: let suppose that the vehicle is at rest in the left configuration of Figure 3.2 in presence of a water current aligned to the earth-fixed $\boldsymbol{y}$ axis; the control action builds the current compensation term which, in this particular configuration, turns out to be parallel to the $\boldsymbol{y}_b$ axis. If the vehicle is now *quickly* rotated to right configuration, the built compensation term rotates together with the vehicle keeping its alignment to the $\boldsymbol{y}_b$ axis; however, this vehicle-fixed axis has now become parallel to the $\boldsymbol{x}$ axis of the earth-fixed frame. Therefore, the built compensation term acts as a disturbance until the integral action has re-built proper current compensation for the right configuration. It is clear at this point that this drawback does not arise for the controller **E** and for the controllers **A** and **B**, since they build ocean current compensation in an earth-fixed frame.

**Reduced Controller.** Similarly to the other cases, a reduced form of the controller has been derived. An integral action on body-fixed-frame error

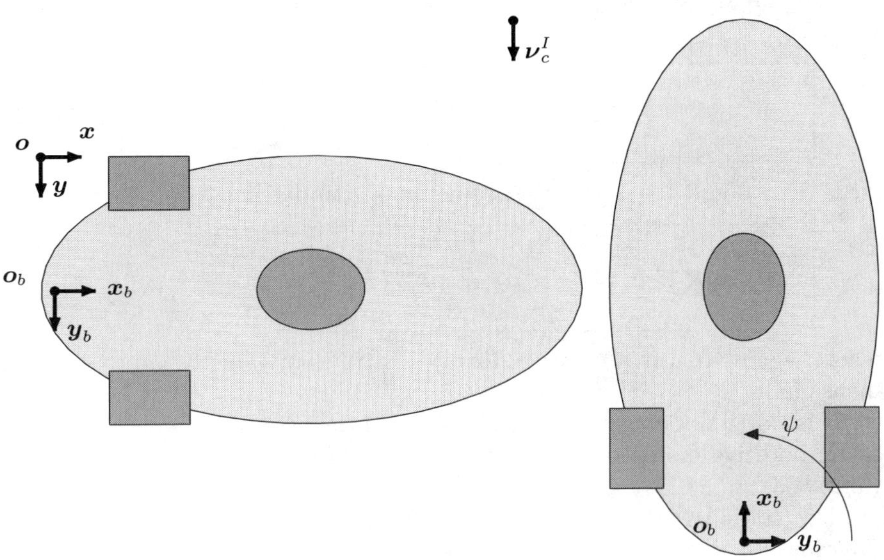


**Fig. 3.2.** Planar view parallel to the $xy$ earth-fixed frame of two different configurations under the constant current $\boldsymbol{\nu}_c^I$

variables has also been considered as in [35] to counteract the current effects, i.e.,

$$\boldsymbol{\tau}_v = \boldsymbol{\Phi}_{v,P'}(\boldsymbol{R}_B^I)\hat{\boldsymbol{\theta}}_v + \boldsymbol{K}_D \boldsymbol{s}_v \tag{3.16}$$

$$\dot{\hat{\boldsymbol{\theta}}}_v = \boldsymbol{K}_\theta^{-1}\boldsymbol{\Phi}_{v,P'}^{\mathrm{T}} \boldsymbol{s}_v \,. \tag{3.17}$$

## 3.5 Model-Based Controller Plus Current Compensation

T. Fossen and J. Balchen, in 1991 [125], propose a control law that explicitly takes into account the ocean current without need for the current measurement.

Being:

$$\boldsymbol{s}_e = \dot{\tilde{\boldsymbol{\eta}}} + \sigma\tilde{\boldsymbol{\eta}} \qquad \text{with } \sigma > 0\,, \tag{3.18}$$

the controller is given by:

$$\boxed{\boldsymbol{\tau}_v = \boldsymbol{\Phi}_v(\boldsymbol{R}_B^I, \boldsymbol{\nu}, \boldsymbol{\nu}_a, \dot{\boldsymbol{\nu}}_a)\hat{\boldsymbol{\theta}}_v + \hat{\boldsymbol{v}} + \boldsymbol{J}_e^{\mathrm{T}}(\boldsymbol{R}_B^I)\boldsymbol{K}_D \boldsymbol{s}_e} \tag{3.19}$$

where $\boldsymbol{K}_D \in \mathbb{R}^{6\times 6}$, is a positive definite matrix of gains. Notice that the PD action is similar to the transpose of the Jacobian approach developed for industrial manipulators [254]. The current compensation $\hat{\boldsymbol{v}}$ is updated by the following

$$\dot{\hat{\boldsymbol{v}}} = \boldsymbol{W}^{-1}\boldsymbol{J}_e^{-1}(\boldsymbol{R}_B^I)\boldsymbol{s}_e \tag{3.20}$$

with $\boldsymbol{W} > 0$ and the dynamic parameters are updated by

$$\dot{\hat{\boldsymbol{\theta}}}_v = \boldsymbol{K}_\theta^{-1}\boldsymbol{\Phi}_v^{\mathrm{T}}(\boldsymbol{R}_B^I, \boldsymbol{\nu}, \boldsymbol{\nu}_a, \dot{\boldsymbol{\nu}}_a)\boldsymbol{J}_e^{-1}(\boldsymbol{R}_B^I)\boldsymbol{s}_e\,, \tag{3.21}$$

where, again, $\boldsymbol{K}_\theta$, is a positive definite matrices of gains of appropriate dimensions.

It is worth noticing that with this control the model considers the current as an additive disturbance, constant in the body-fixed frame (see Subsection 2.4.3).

The stability analysis is developed by defining as Lyapunov candidate function

$$V = \boldsymbol{s}_e^{\mathrm{T}}\boldsymbol{M}_v^\star\boldsymbol{s}_e + \frac{1}{2}\,\tilde{\boldsymbol{\theta}}_v^{\mathrm{T}}\boldsymbol{K}_\theta\tilde{\boldsymbol{\theta}}_v + \frac{1}{2}\,\tilde{\boldsymbol{v}}^{\mathrm{T}}\boldsymbol{W}\tilde{\boldsymbol{v}} \tag{3.22}$$

that is positive definite $\forall \boldsymbol{s}_e \neq \boldsymbol{0},\ \tilde{\boldsymbol{\theta}}_v \neq \boldsymbol{0},\ \tilde{\boldsymbol{v}} \neq \boldsymbol{0}$.

By applying the proposed control law, and following the guidelines in [125], it is possible to demonstrate that

$$\dot{V} = -\boldsymbol{s}_e^{\mathrm{T}}\left[\boldsymbol{K}_D + \boldsymbol{D}_{RB}^\star\right]\boldsymbol{s}_e \leq 0 \tag{3.23}$$

It is now possible to prove again the system stability in a Lyapunov-Like sense using the Barbălat's Lemma. Since

- $V$ is lower bounded
- $\dot{V}(\boldsymbol{s}_e, \tilde{\boldsymbol{\theta}}_v, \tilde{\boldsymbol{v}}) \leq 0$
- $\dot{V}(\boldsymbol{s}_e, \tilde{\boldsymbol{\theta}}_v, \tilde{\boldsymbol{v}})$ is uniformly continuous

then

- $\dot{V}(\boldsymbol{s}_e, \tilde{\boldsymbol{\theta}}_v, \tilde{\boldsymbol{v}}) \to 0$ as $t \to \infty$.

Thus $\boldsymbol{s}_e \to \boldsymbol{0}$ as $t \to \infty$. In view of the definition of $\boldsymbol{s}_e$, this implies that $\tilde{\boldsymbol{\eta}} \to \boldsymbol{0}$ as $t \to \infty$. However, the vectors $\tilde{\boldsymbol{\theta}}_v$ and $\tilde{\boldsymbol{v}}$ are only guaranteed to be bounded.

**Compensation of the persistent effects.** Since the adaptive compensation term $\boldsymbol{\Phi}_v(\boldsymbol{R}_B^I, \boldsymbol{\nu}, \boldsymbol{\nu}_a, \dot{\boldsymbol{\nu}}_a)\hat{\boldsymbol{\theta}}_v$ operates in vehicle-fixed coordinates, the control law is suited for effective compensation of the restoring moment.

On the other hand, despite the error vector $\boldsymbol{s}_e$ is based on earth-fixed quantities, the current-compensation term $\hat{\boldsymbol{v}}$ is built through the integral action (3.20) which works in vehicle-fixed coordinates; this implies that, as for counteraction of the ocean current, this control law suffers from the same drawback as that discussed for the law **C**.

**Reduced Controller.** The model-based compensation has been reduced to the restoring generalized force alone:

$$\boldsymbol{\tau}_v = \boldsymbol{\Phi}_{v,R}(\boldsymbol{R}_B^I)\hat{\boldsymbol{\theta}}_{v,R} + \hat{\boldsymbol{v}} + \boldsymbol{J}_e^{\mathrm{T}}(\boldsymbol{R}_B^I)\boldsymbol{K}_D\boldsymbol{s}_e \tag{3.24}$$

$$\dot{\hat{\boldsymbol{v}}} = \boldsymbol{W}^{-1}\boldsymbol{J}_e^{-1}(\boldsymbol{R}_B^I)\boldsymbol{s}_e \tag{3.25}$$

$$\dot{\hat{\boldsymbol{\theta}}}_{v,R} = \boldsymbol{K}_\theta^{-1}\boldsymbol{\Phi}_{v,R}^{\mathrm{T}}(\boldsymbol{R}_B^I)\boldsymbol{J}_e^{-1}(\boldsymbol{R}_B^I)\boldsymbol{s}_e\,, \tag{3.26}$$

## 3.6 Mixed Earth/Vehicle-Fixed-Frame-Based, Model-Based Controller

In 2001, G. Antonelli, F. Caccavale, S. Chiaverini and G. Fusco [13, 14] propose and adaptive tracking control law that takes into account the different nature of the hydrodynamic effects acting on the AUV (ROV); this is achieved by suitably building each dynamic compensation action in a proper (either inertial or vehicle-fixed) reference frame. In fact, since adaptive or integral control laws asymptotically achieve compensation of the *constant* disturbance terms, it is convenient to build the compensation action in a reference frame with respect to which the disturbance term itself is seen as much as possible as constant. The analysis has been extended in [9, 16].

Let consider the vehicle-fixed variables:

$$\tilde{\boldsymbol{y}} = \begin{bmatrix} \boldsymbol{R}_I^B\tilde{\boldsymbol{\eta}}_1 \\ \tilde{\boldsymbol{\varepsilon}} \end{bmatrix}$$

$$\tilde{\boldsymbol{\nu}} = \boldsymbol{\nu}_d - \boldsymbol{\nu}\,,$$

where $\tilde{\boldsymbol{\eta}}_1 = \boldsymbol{\eta}_{1,d} - \boldsymbol{\eta}_1$, being $\boldsymbol{\eta}_{1,d}$ the desired position, and $\tilde{\boldsymbol{\varepsilon}}$ is the quaternion based attitude error. Let define as $\hat{\boldsymbol{\theta}}_v$ the vector of parameters to be adapted,

$$\boldsymbol{s}_v = \tilde{\boldsymbol{\nu}} + \boldsymbol{\Lambda}\tilde{\boldsymbol{y}}\,, \tag{3.27}$$

with $\boldsymbol{\Lambda} = \mathrm{blockdiag}\{\lambda_p\boldsymbol{I}_3, \lambda_o\boldsymbol{I}_3\}$, $\boldsymbol{\Lambda} > \boldsymbol{O}$. The control law is given by:

$$\boxed{\boldsymbol{\tau}_v = \boldsymbol{K}_D\boldsymbol{s}_v + \boldsymbol{K}\tilde{\boldsymbol{y}} + \boldsymbol{\Phi}_{v,T}\hat{\boldsymbol{\theta}}_v} \tag{3.28}$$

where $\boldsymbol{K}_D \in \mathbb{R}^{6\times 6}$ and $\boldsymbol{K} = \mathrm{blockdiag}\{k_p\boldsymbol{I}_3, k_o\boldsymbol{I}_3\}$ are positive definite matrices of the gains to be designed, and

$$\boxed{\dot{\hat{\boldsymbol{\theta}}}_v = \boldsymbol{K}_\theta^{-1}\boldsymbol{\Phi}_{v,T}^{\mathrm{T}}\boldsymbol{s}_v\,,} \tag{3.29}$$

where $\boldsymbol{K}_\theta$ is also a positive definite matrix of proper dimensions.

### 3.6.1 Stability Analysis

Let us consider the following scalar function

$$V = \frac{1}{2}\boldsymbol{s}_v^{\mathrm{T}}\boldsymbol{M}_v\boldsymbol{s}_v + \frac{1}{2}\tilde{\boldsymbol{\theta}}_v^{\mathrm{T}}\boldsymbol{K}_\theta\tilde{\boldsymbol{\theta}}_v + \frac{1}{2}k_p\tilde{\boldsymbol{\eta}}_1^{\mathrm{T}}\tilde{\boldsymbol{\eta}}_1 + k_o\tilde{\boldsymbol{z}}^{\mathrm{T}}\tilde{\boldsymbol{z}}\,, \tag{3.30}$$

where $\tilde{\boldsymbol{z}} = [1 \quad \boldsymbol{0}^{\mathrm{T}}]^{\mathrm{T}} - \boldsymbol{z} = [1 - \tilde{\eta} \quad -\tilde{\boldsymbol{\varepsilon}}^{\mathrm{T}}]^{\mathrm{T}}$. Due to the positive definiteness of $\boldsymbol{M}_v$, $\boldsymbol{K}_\theta$, and $\boldsymbol{K}$, the scalar function $V(\tilde{\boldsymbol{\eta}}_1, \tilde{\boldsymbol{z}}, \boldsymbol{s}_v, \tilde{\boldsymbol{\theta}}_v)$ is positive definite.

Let us define the following partition for the variable $\boldsymbol{s}_v$ that will be useful later:

$$\boldsymbol{s}_v = \begin{bmatrix} \boldsymbol{s}_p \\ \boldsymbol{s}_o \end{bmatrix}, \tag{3.31}$$

with $\boldsymbol{s}_p \in \mathrm{I\!R}^3$ and $\boldsymbol{s}_o \in \mathrm{I\!R}^3$. In view of (3.31) and the definition of $\boldsymbol{s}_v$, it is

$$\tilde{\boldsymbol{\nu}}_1 = \boldsymbol{s}_p - \lambda_p\boldsymbol{R}_I^B\tilde{\boldsymbol{\eta}}_1 \tag{3.32}$$

$$\tilde{\boldsymbol{\nu}}_2 = \boldsymbol{s}_o - \lambda_o\tilde{\boldsymbol{\varepsilon}}\,. \tag{3.33}$$

Differentiating $V$ with respect to time yields:

$$\begin{aligned}
\dot{V} &= \boldsymbol{s}_v^{\mathrm{T}}\boldsymbol{M}_v\dot{\boldsymbol{s}}_v + \tilde{\boldsymbol{\theta}}_v^{\mathrm{T}}\boldsymbol{K}_\theta\dot{\tilde{\boldsymbol{\theta}}}_v + k_p\tilde{\boldsymbol{\eta}}_1^T\boldsymbol{R}_B^I\tilde{\boldsymbol{\nu}}_1 - 2k_o\tilde{\boldsymbol{z}}^T\boldsymbol{J}_{k,oq}\tilde{\boldsymbol{\nu}}_2 \\
&= \boldsymbol{s}_v^{\mathrm{T}}\boldsymbol{M}_v\left(\dot{\boldsymbol{\nu}}_d - \dot{\boldsymbol{\nu}} + \boldsymbol{\Lambda}\dot{\tilde{\boldsymbol{y}}}\right) + \tilde{\boldsymbol{\theta}}_v^{\mathrm{T}}\boldsymbol{K}_\theta\dot{\tilde{\boldsymbol{\theta}}}_v + \\
&\quad + k_p\tilde{\boldsymbol{\eta}}_1^T\boldsymbol{R}_B^I\left(\boldsymbol{s}_p - \lambda_p\boldsymbol{R}_I^B\tilde{\boldsymbol{\eta}}_1\right) + \\
&\quad - k_o\left[1 - \tilde{\eta} \quad -\tilde{\boldsymbol{\varepsilon}}^{\mathrm{T}}\right]\begin{bmatrix} -\tilde{\boldsymbol{\varepsilon}}^{\mathrm{T}} \\ \tilde{\eta}\boldsymbol{I}_3 + \boldsymbol{S}(\tilde{\boldsymbol{\varepsilon}}) \end{bmatrix}(\boldsymbol{s}_o - \lambda_o\tilde{\boldsymbol{\varepsilon}})\,.
\end{aligned}$$

Then, defining

$$\boldsymbol{\nu}_a = \boldsymbol{\nu}_d + \boldsymbol{\Lambda}\tilde{\boldsymbol{y}} \tag{3.34}$$

yields (dependencies are dropped out to increase readability):

$$\begin{aligned}
\dot{V} &= \boldsymbol{s}_v^{\mathrm{T}}\left[\boldsymbol{M}_v\dot{\boldsymbol{\nu}}_a - \boldsymbol{\tau}_v + \boldsymbol{C}_v\boldsymbol{\nu} + \boldsymbol{D}_{RB}\boldsymbol{\nu} + \boldsymbol{g}_{RB} + \boldsymbol{\tau}_{v,C}\right] + \\
&\quad - \tilde{\boldsymbol{\theta}}_v^{\mathrm{T}}\boldsymbol{K}_\theta\dot{\hat{\boldsymbol{\theta}}}_v + k_p\tilde{\boldsymbol{\eta}}_1^{\mathrm{T}}\boldsymbol{R}_B^I\boldsymbol{s}_p - k_p\lambda_p\tilde{\boldsymbol{\eta}}_1^{\mathrm{T}}\tilde{\boldsymbol{\eta}}_1 + \\
&\quad + k_o\tilde{\boldsymbol{\varepsilon}}^{\mathrm{T}}\boldsymbol{s}_o - \lambda_o k_o\tilde{\boldsymbol{\varepsilon}}^{\mathrm{T}}\tilde{\boldsymbol{\varepsilon}},
\end{aligned} \tag{3.35}$$

that can be rewritten as:

$$\begin{aligned}
\dot{V} &= \boldsymbol{s}_v^{\mathrm{T}}\left[\boldsymbol{\Phi}_{v,T}\boldsymbol{\theta}_v - \boldsymbol{\tau}_v\right] - \boldsymbol{s}_v^{\mathrm{T}}\boldsymbol{D}_{RB}\boldsymbol{s}_v - \tilde{\boldsymbol{\theta}}_v^{\mathrm{T}}\boldsymbol{K}_\theta\dot{\hat{\boldsymbol{\theta}}}_v + \\
&\quad + \boldsymbol{s}_v^{\mathrm{T}}\boldsymbol{K}\tilde{\boldsymbol{y}} - k_p\lambda_p\tilde{\boldsymbol{\eta}}_1^{\mathrm{T}}\tilde{\boldsymbol{\eta}}_1 - k_o\lambda_o\tilde{\boldsymbol{\varepsilon}}^{\mathrm{T}}\tilde{\boldsymbol{\varepsilon}}\,.
\end{aligned}$$

By considering the control law (3.28) and the parameters update (3.29), it is:

$$\dot{V} = -\boldsymbol{s}_v^{\mathrm{T}}(\boldsymbol{K}_D + \boldsymbol{D}_{RB})\boldsymbol{s}_v - k_p\lambda_p\tilde{\boldsymbol{\eta}}_1^{\mathrm{T}}\tilde{\boldsymbol{\eta}}_1 - k_o\lambda_o\tilde{\boldsymbol{\varepsilon}}^{\mathrm{T}}\tilde{\boldsymbol{\varepsilon}}$$

that is negative semi-definite over the state space $\{\tilde{\boldsymbol{\eta}}_1, \tilde{\boldsymbol{z}}, \boldsymbol{s}_v, \tilde{\boldsymbol{\theta}}_v\}$.

It is now possible to prove the system stability in a Lyapunov-Like sense using the Barbălat's Lemma. Since

- $V$ is lower bounded
- $\dot{V}(\tilde{\boldsymbol{\eta}}_1, \tilde{\boldsymbol{z}}, \boldsymbol{s}_v, \tilde{\boldsymbol{\theta}}_v) \le 0$
- $\dot{V}(\tilde{\boldsymbol{\eta}}_1, \tilde{\boldsymbol{z}}, \boldsymbol{s}_v, \tilde{\boldsymbol{\theta}}_v)$ is uniformly continuous

then

- $\dot{V}(\tilde{\boldsymbol{\eta}}_1, \tilde{\boldsymbol{z}}, \boldsymbol{s}_v, \tilde{\boldsymbol{\theta}}_v) \to 0$ as $t \to \infty$.

Thus $\tilde{\boldsymbol{\eta}}_1, \tilde{\boldsymbol{\varepsilon}}, \boldsymbol{s}_v \to \mathbf{0}$ as $t \to \infty$. In view of the definition of $\boldsymbol{s}_v$, this implies that $\tilde{\boldsymbol{\nu}} \to \mathbf{0}$ as $t \to \infty$; in addition, due to the properties of the quaternion, it results that $\tilde{\eta} \to 1$ as $t \to \infty$. However, as usual in adaptive control schemes, it is not possible to prove asymptotic stability of the whole state since $\tilde{\boldsymbol{\theta}}_v$ is only guaranteed to be bounded.

**Compensation of the persistent effects.** The control law has been designed explicitly to take into account the persistent dynamic effects in their proper frame. For this reason it does not suffer from the drawbacks of compensating the restoring forces in the earth-fixed frame, or, dually, the current parameters in the body-fixed frame.

**Reduced Controller.** According to the aim to design a PD-like action plus an adaptive compensation action, the reduced version of the control law is then given by:

$$\boldsymbol{\tau}_v = \boldsymbol{K}_D \boldsymbol{s}_v + \boldsymbol{\Phi}_{v,P} \hat{\boldsymbol{\theta}}_{v,P} \tag{3.36}$$

$$\dot{\hat{\boldsymbol{\theta}}}_{v,P} = \boldsymbol{K}_\theta^{-1} \boldsymbol{\Phi}_{v,P}^{\mathrm{T}} \boldsymbol{s}_v \,, \tag{3.37}$$

where $\boldsymbol{K}_D \in \mathbb{R}^{6\times 6}$ and $\boldsymbol{K}_\theta \in \mathbb{R}^{9\times 9}$ are positive definite matrices of gains to be designed.

## 3.7 Jacobian-Transpose-Based Controller

Sun and Cheah, in 2003 [280], present an adaptive set-point control inspired by the manipulator control approach proposed in [38].

The vector $\tilde{\boldsymbol{\eta}} \in \mathbb{R}^6$ represents, as usual, the error in the earth-fixed frame. Let further define the function $\boldsymbol{F}$ defined on the single components such as:

- It holds: $F_i(x_i) > 0 \quad \forall x_i \neq 0, \qquad F_i(0) = 0$;
- The function $F_i(x_i)$ is twice continuously differentiable;
- The partial derivative $f_i = \partial F_i / \partial x_i$ is strictly increasing in $x_i$ for $|x_i| < \gamma_i$ for a given $\gamma_i$ and saturated for $|x_i| \ge \gamma_i$;
- There exists a positive constant $c_i$ such that $F_i(x_i) > c_i^2 f_i$

The function $\boldsymbol{f} \in \mathbb{R}^6$ collecting the elements $f_i$ is basically a saturated function. Guidelines in its selection are given in [38].

The proposed controller is given by:

$$\boldsymbol{\tau}_v = \boldsymbol{K}_P \boldsymbol{J}_e^{\mathrm{T}}(\boldsymbol{R}_B^I)\boldsymbol{f}(\tilde{\boldsymbol{\eta}}) - \boldsymbol{K}_D\boldsymbol{\nu} + \boldsymbol{\Phi}_{v,R}(\boldsymbol{R}_B^I)\hat{\boldsymbol{\theta}}_{v,R} \tag{3.38}$$

where $\boldsymbol{K}_P$ and $\boldsymbol{K}_D$ are positive definite matrices, selected as scalar in [280] and

$$\dot{\hat{\boldsymbol{\theta}}}_{v,R} = \boldsymbol{K}_\theta^{-1}\boldsymbol{\Phi}_{v,R}^{\mathrm{T}}(\boldsymbol{R}_B^I)\left(\boldsymbol{\nu} + \alpha\boldsymbol{J}_e^{\mathrm{T}}(\boldsymbol{R}_B^I)\boldsymbol{f}(\tilde{\boldsymbol{\eta}})\right) \tag{3.39}$$

where $\alpha > 0$ is a scalar gain.

It can be noticed that the proportional action is similar to the action proposed by Fossen and Balchen in Section 3.5. The derivative action is, on the contrary, based on the vehicle-fixed variable $\boldsymbol{\nu}$ that is different from the derivative action developed by Fossen and Balchen based on the term $\boldsymbol{J}_e^{\mathrm{T}}\dot{\tilde{\boldsymbol{\eta}}}$. Finally they both have an adaptive action that is already reduced to the sole restoring terms in [280].

It is worth noticing that the Authors propose scalar gains $\boldsymbol{K}_P$, $\boldsymbol{K}_D$ and $\alpha$ for the controller. There is a main problem related with the fact that the position and orientation variables have different unit measures; using the same gains might force the designer to tune the performance to the lower bandwidth.

**Compensation of the persistent effects.** The controller proposed by the Authors compensates for the gravity in the vehicle-fixed frame. However, the update law is based on the transpose of the Jacobian, and not on its inverse; the mapping, thus, is not exact and a coupling among the error directions is experienced. This has as a consequence that this control law is not suitable for the restoring force compensation. It is worth noticing, moreover, that, differently from the industrial manipulator case, inversion of the Jacobian is not computational demanding since, being $\boldsymbol{J}_e$ a simple $(6 \times 6)$ matrix (see eq. (2.19)), its inverse $\boldsymbol{J}_e^{-1}$ can be symbolically computed; from a computational aspect, thus, there is no difference in using $\boldsymbol{J}_e^{\mathrm{T}}$ or $\boldsymbol{J}_e^{-1}$. By visual inspection of $\boldsymbol{J}_e^{\mathrm{T}}$ and $\boldsymbol{J}_e^{-1}$ it can be observed that the difference is in the rotational part $\boldsymbol{J}_{k,o}$, being the transpose of a rotation matrix equal to its inverse. In particular, using the common definition that roll pitch an yaw means the use of elementary rotation around $x$, $y$ and $z$ in fixed frame [254], the corresponding matrix $\boldsymbol{J}_{k,o}$ is not function of the yaw angle, in case of null pitch angle it is $\boldsymbol{J}_{k,o}^{\mathrm{T}} = \boldsymbol{J}_{k,o}^{-1}$, and it is singular for a pitch angle of $\pm\pi/2$; close to that singularity the numerical difference between $\boldsymbol{J}_{k,o}^{\mathrm{T}}$ and $\boldsymbol{J}_{k,o}^{-1}$ increases.

The ocean current is not taken into account. Using the controller (3.38)–(3.39) under the effect of the current would lead to an error different from zero at steady state.

**Reduced Controller.** The aim of the Authors is to propose already a controller that matches the definition of *reduced* controller that has been given here. The absence of compensation for the current, however, makes the controller not appealing for practical implementation. The proposed reduced controller, modified to take into account a tracking problem, is given by:

$$\boldsymbol{\tau}_v = \boldsymbol{K}_P \boldsymbol{J}_e^{\mathrm{T}}(\boldsymbol{R}_B^I)\tilde{\boldsymbol{\eta}} + \boldsymbol{K}_D \tilde{\boldsymbol{\nu}} + \boldsymbol{\Phi}_{v,P} \hat{\boldsymbol{\theta}}_{v,P} \tag{3.40}$$

$$\dot{\hat{\boldsymbol{\theta}}}_{v,P} = \boldsymbol{K}_\theta^{-1} \boldsymbol{\Phi}_{v,P}^{\mathrm{T}} \left( \tilde{\boldsymbol{\nu}} + \boldsymbol{\Lambda} \boldsymbol{J}_e^{\mathrm{T}}(\boldsymbol{R}_B^I)\tilde{\boldsymbol{\eta}} \right) \tag{3.41}$$

where $\boldsymbol{\Lambda}$, $\boldsymbol{K}_P$ and $\boldsymbol{K}_D$ are positive definite matrices, selected at least as block-diagonal matrices to keep different dynamics for the position and the orientation.

It is worth noticing that the reduced controller derived is different from the original controller, that would not reach a null steady state error under the effect of the current. Also, the regressor $\boldsymbol{\Phi}_{v,P}$ embeds the gravity regressor as proposed by Sun and Cheah but the drawback in the restoring compensation still exist due to the not proper update law of the parameters.

## 3.8 Comparison Among Controllers

For easy of readings, Table 3.1 reports the label associated with each controller. The controllers developed by the researchers are quite different one each other, in this Section a qualitative comparison with respect to the compensation performed by the controllers with respect to the persistent dynamic effects is provided.

### 3.8.1 Compensation of the Restoring Generalized Forces

The controller **A** is totally conceived in the earth-fixed frame and it is model-based. In case of perfect knowledge of the restoring-related dynamic parameters, thus, there is not need to compensate for the restoring generalized forces. This is, however, unpractical in a real situation; in case of partial compensation of the vector $\boldsymbol{g}_{RB}^{\star}$, in fact, this controller experiences the drawback discussed in Section 3.2. A similar drawback is shared from the controller **B** that, again, is totally based on earth-fixed variables, the integral actions, thus, does not compensate optimally for the restoring action. The controller **F** compensates the restoring force in the vehicle-fixed frame; the update law, however, is not based on the exact mapping between orientation and moments resulting in a not clean adaptive action. The effect, however, is as significant as the vehicle works with a large pitch angle. The remaining 3 controllers properly adapt the restoring-related parameters in the vehicle-fixed frame.

### 3.8.2 Compensation of the Ocean Current

The presence of an ocean current is seldom taken into account in the literature. The effect of the current, however, can be really significant ([127, 323]). In this discussion a main assumption is made: roughly speaking the current is considered as constant in the earth-fixed frame and its effect on the vehicle is modeled as a force, proportional to the current magnitude, that *pushes* the vehicle. The considerations below, thus, are based on this assumption. It is worth noticing that in the simulation the dynamic effect of the current is correctly taken into account by computing the relative velocity in the vehicle model. The accuracy of this assumption, thus, seems to be verified.

Controller **C** and **F** does not consider at all the current, as a result the developed controller does not reach a null steady state error. Controller **D** compensates the current with a term designed on the purpose; this term adapts on vehicle-fixed parameters, as discussed in Section 3.5 this action experiences a drawback. The remaining controllers correctly compensate for the current adapting/integrating on a set of earth-fixed based parameters.

Table 3.2 summarizes the discussed properties of the 6 controllers with respect to the persistent dynamic terms. It is worth noticing that only the controller **E** correctly compensates for both actions.

**Table 3.2.** Summary of the behavior of the controllers with respect to compensation of the hydrodynamic persistent terms

| control law | current effect | restoring forces | null error with current |
|---|---|---|---|
| **A** | **fit** | unfit | **yes** |
| **B** | **fit** | unfit | **yes** |
| **C** | unfit | **fit** | no |
| **D** | unfit | **fit** | **yes** |
| **E** | **fit** | **fit** | **yes** |
| **F** | unfit | unfit | no |

## 3.9 Numerical Comparison Among the Reduced Controllers

A performance comparison among the controllers discussed has been developed by numerical simulations. The simulations have been run on the 6-DOF mathematical model of ODIN, an AUV built at the Autonomous Systems Laboratory of the University of Hawaii (see the Appendix for a description and the model used in the simulation).

An aspect to be considered for proper reading of this study is that the control problem at hand is highly nonlinear, coupled and the controller to be compared are intrinsically different one from the other. Any effort to chose the gains so as to ensure *similar* performance to the controllers has been made; nevertheless, this is impossible in a strict sense. For this reason, the presented results have to be interpreted mainly looking at the error behavior rather then focusing on direct numeric comparison. On the other hand, it was chosen to not emphasize vehicle-related effects that would not be present in general applications. In particular, ODIN has a small metacentric height, yielding low restoring moments; simulations of the controllers to vehicles of larger metacentric height would make even more evident the drawbacks in the compensation of the restoring forces.

**Test trajectory** In order to demonstrate the effects discussed in this paper, the simplest task to be considered is one involving successive changes of the vehicle orientation in presence of ocean current. The simulation length can be divided in different period of 60 s duration. The vehicle is firstly put in the water at the position

$$\boldsymbol{\eta}_d(t=0) = \begin{bmatrix} 0 & 0 & 1 & 0 & 0 & 0 \end{bmatrix}^{\mathrm{T}} \quad [\mathrm{m/deg}]$$

without knowledge of the current but with an estimation of the restoring parameters. The first 60 s are used to adapt the effect of the current. In the successive period the vehicle is required to move in roll and pitch from 0 deg to 10 deg and −15 deg, respectively and come back to the original configuration. In the successive two periods the vehicle is required to move of 90 deg in yaw and come back to the initial position. Finally, 60 s of steady state are given. Figure 3.3 plots the desired trajectory.

**Initial conditions** Table 3.3 reports the initial condition for the adaptive/integral parameters of the controllers, it can be observed that the same values have been used in the estimation of the restoring forces. This is also true for the controller **A**, where the gravity estimation $\hat{\boldsymbol{g}}^{\star}_{RB}$, even if not adaptive, is obtained resorting to

$$\hat{\boldsymbol{\theta}}_{v,R} = \begin{bmatrix} -9 & 0 & 0 & 50 \end{bmatrix}^{\mathrm{T}} .$$

From the model in the Appendix it can be noticed that the true value of the restoring parameters is

$$\boldsymbol{\theta}_{v,R} = \begin{bmatrix} -8.0438 & 0 & 0 & 61.3125 \end{bmatrix}^{\mathrm{T}} .$$

The vehicle initial position is also

$$\boldsymbol{\eta} = \begin{bmatrix} 0 & 0 & 1 & 0 & 0 & 0 \end{bmatrix}^{\mathrm{T}} \quad [\mathrm{m,deg}] ,$$

meaning that the vehicle is supposed to start its motion under the water, at 1 m depth, moreover, there is no initial estimation of the ocean current.

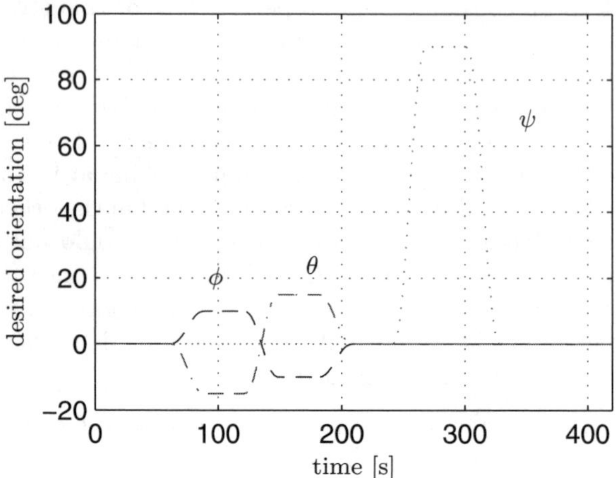


**Fig. 3.3.** Desired trajectory; the desired position is constant

**Table 3.3.** Initial conditions for all the controllers in their reduced version and number of parameter to adapt/integrate

| law | number of par. | $\hat{\boldsymbol{\theta}}(t=0)$ |
|---|---|---|
| **A** | 7 | $[0\ \ 0\ \ 0\ \ 0\ \ 0\ \ 0\ \ 1]^{\mathrm{T}}$ (initial value of the integral) |
| **B** | – | not simulated |
| **C** | 10 | $[-9\ \ 0\ \ 50\ \ 0\ \ 0\ \ 0\ \ 0\ \ 0\ \ 0\ \ 0]^{\mathrm{T}}$ |
| **D** | 10 | $[-9\ \ 0\ \ 50\ \ 0\ \ 0\ \ 0\ \ 0\ \ 0\ \ 0\ \ 0]^{\mathrm{T}}$ |
| **E** | 9 | $[0\ \ 50\ \ 0\ \ 0\ \ 0\ \ 9\ \ 0\ \ 0\ \ 0]^{\mathrm{T}}$ |
| **F** | 9 | $[0\ \ 50\ \ 0\ \ 0\ \ 0\ \ 9\ \ 0\ \ 0\ \ 0]^{\mathrm{T}}$ |

**Ocean current** The simulations have been run considering the following constant current

$$\boldsymbol{\nu}_c^I = [0\ \ 0.3\ \ 0\ \ 0\ \ 0\ \ 0]^{\mathrm{T}} \quad \mathrm{m/s}\,,$$

**Gains selection** All the simulated controllers are non-linear. Moreover, significant differences arise among them, this is, in fact, the object of this discussion. For these reasons it is very difficult to perform a fair numerical comparison and the selection of the gains is very delicate. The gains have been tuned so as to give to the controller similar control effort in terms of force/moment magnitude and eventual presence of chattering. In few cases, since the controller feeds back the same variable, it has been possible to chose the same control gain, it is the case, e.g., of the controllers **C** and **E**. In other cases, such as, e.g., the controller **F**, the use of the transpose of the Jacobian implicitly change the effective gain acting on the error, the equivalence of the gains, thus, is only apparent.

A separate discussion needs to be done for the controller **B**. Its specific structure makes it really hard to tune the parameters following these simple considerations. As recognized in [323], moreover, the parameter tuning of their controller has not been simple and handy for being used in the experiments with the vehicle ODIN III. For this reason, while the controller has been object of the theoretical discussion, the simulation will not be reported.

The following gains have been used for the controller **A**:

$$\begin{aligned} \boldsymbol{K}_P &= \text{blockdiag}\{8.8\boldsymbol{I}_3, 20\boldsymbol{I}_4\}\,, \\ \boldsymbol{K}_D &= \text{blockdiag}\{110\boldsymbol{I}_3, 40\boldsymbol{I}_4\}\,, \\ \boldsymbol{K}_I &= \text{blockdiag}\{0.2\boldsymbol{I}_3, 1\boldsymbol{I}_4\}\,. \end{aligned}$$

The following gains have been used for the controller **C**:

$$\begin{aligned} \boldsymbol{K}_D &= \text{blockdiag}\{110\boldsymbol{I}_3, 40\boldsymbol{I}_3\}\,, \\ \boldsymbol{\Lambda} &= \text{blockdiag}\{0.08\boldsymbol{I}_3, 0.9\boldsymbol{I}_3\}\,, \\ \boldsymbol{K}_\theta^{-1} &= 2\boldsymbol{I}_9\,. \end{aligned}$$

The following gains have been used for the controller **D**:

$$\begin{aligned} \boldsymbol{K}_D &= \text{blockdiag}\{110\boldsymbol{I}_3, 40\boldsymbol{I}_3\}\,, \\ \boldsymbol{\Lambda} &= \text{blockdiag}\{0.08\boldsymbol{I}_3, 0.9\boldsymbol{I}_3\}\,, \\ \boldsymbol{K}_\theta^{-1} &= 2\boldsymbol{I}_4\,, \\ \boldsymbol{W}^{-1} &= 2\boldsymbol{I}_6\,. \end{aligned}$$

The following gains have been used for the controller **E**:

$$\begin{aligned} \boldsymbol{K}_D &= \text{blockdiag}\{110\boldsymbol{I}_3, 40\boldsymbol{I}_3\}\,, \\ \boldsymbol{\Lambda} &= \text{blockdiag}\{0.08\boldsymbol{I}_3, 0.9\boldsymbol{I}_3\}\,, \\ \boldsymbol{K}_{\theta_P}^{-1} &= 2\boldsymbol{I}_9\,. \end{aligned}$$

The following gains have been used for the controller **F**:

$$\begin{aligned} \boldsymbol{K}_D &= \text{blockdiag}\{8.8\boldsymbol{I}_3, 36\boldsymbol{I}_3\}\,, \\ \boldsymbol{K}_V &= \text{blockdiag}\{110\boldsymbol{I}_3, 40\boldsymbol{I}_3\}\,, \\ \boldsymbol{\Lambda} &= \text{blockdiag}\{0.08\boldsymbol{I}_3, 0.9\boldsymbol{I}_3\}\,, \\ \boldsymbol{K}_{\theta_P}^{-1} &= 2\boldsymbol{I}_9\,. \end{aligned}$$

### 3.9.1 Results

The code to reproduce the simulation and to plot or compare all the outputs is given in `http://webuser.unicas.it/antonelli/auv.zip` together with a `readme.txt` file of explanation.

In Figure 3.4 the position and orientation for controller **A** are given; it can be noticed that in the first 60 seconds the control needs to adapt with respect to the current and a movement along $y$ is observed, also, it can be observed that, during the horizontal rotation ($t \in [240, 360]$ s), the controller correctly compensates for the current and a very small coupling is experienced. The drawback in the restoring compensation of this controller is not significant in this numerical case study. Figures 3.5 reports the required control force and moment.

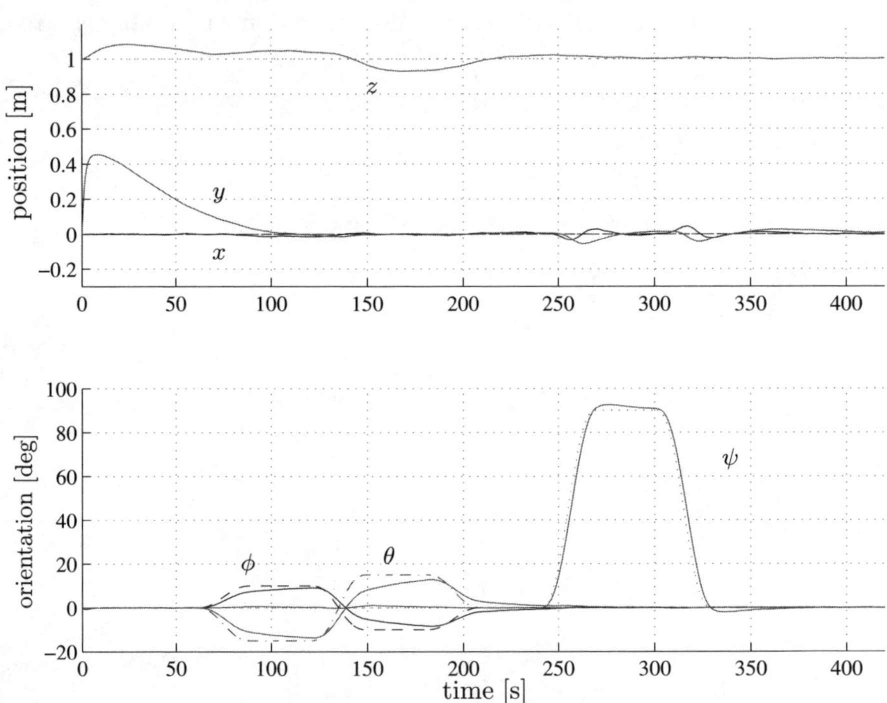


**Fig. 3.4.** Simulated position and orientation for the reduced controller **A**

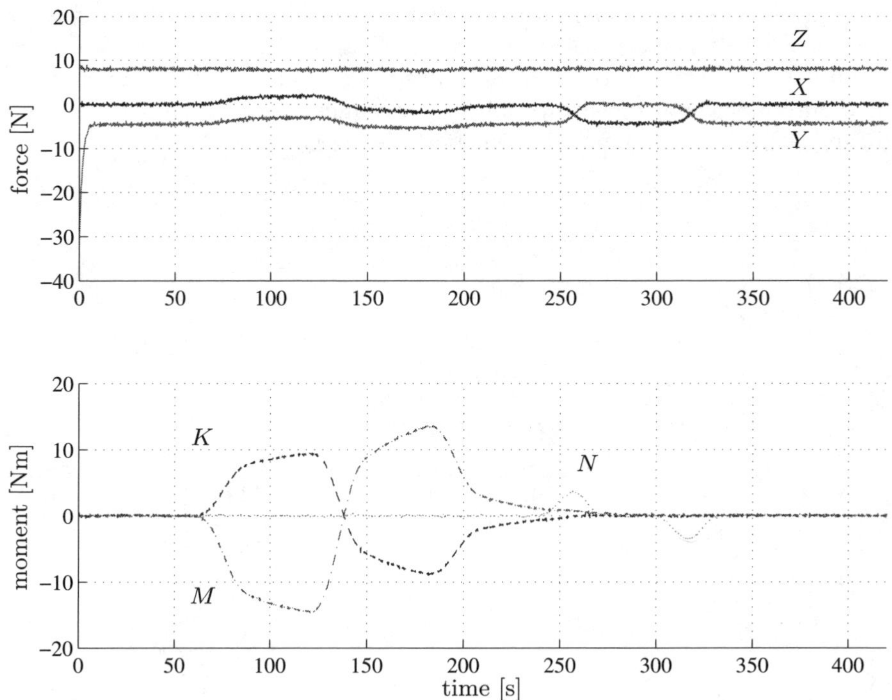


**Fig. 3.5.** Simulated force and moment for the reduced controller **A**

Figure 3.6 and 3.7 report the position, orientation and control effort for controller **C**. It can be observed a strong coupling during the commanded yaw rotation caused by the controller itself. This can be further appreciated in Figure 3.8 where the projection of the vehicle position on the $xy$ plane during the rotation compared with controller **A** is given.

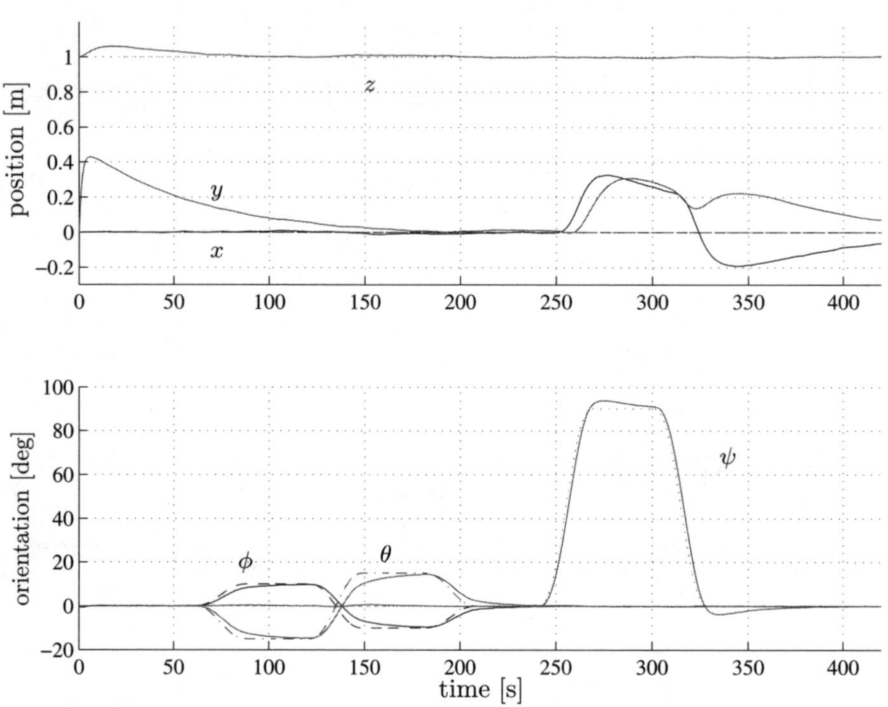


**Fig. 3.6.** Simulated position and orientation for the reduced controller **C**

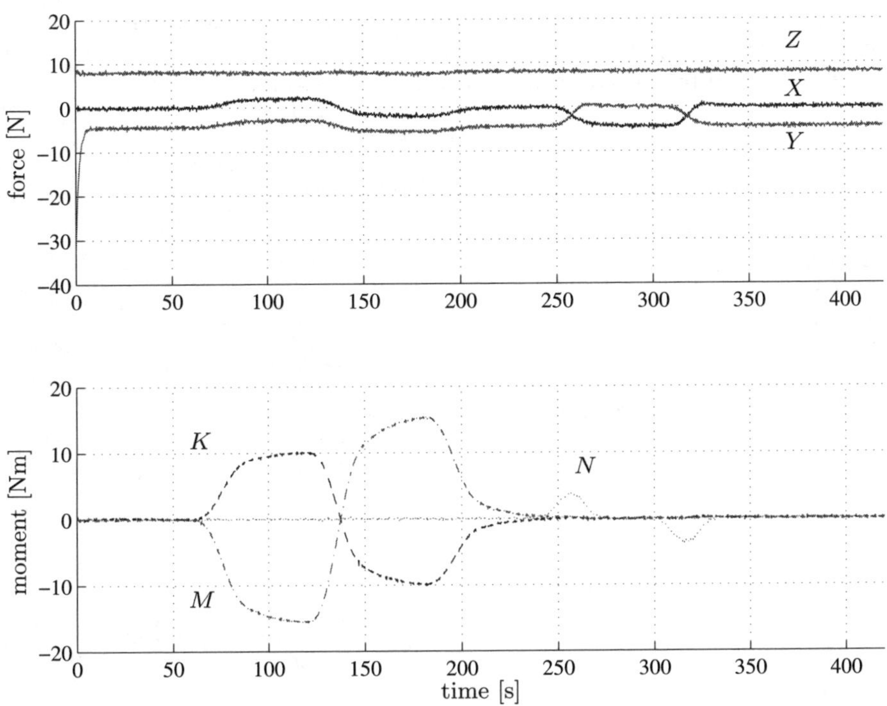


**Fig. 3.7.** Simulated force and moment for the reduced controller **C**

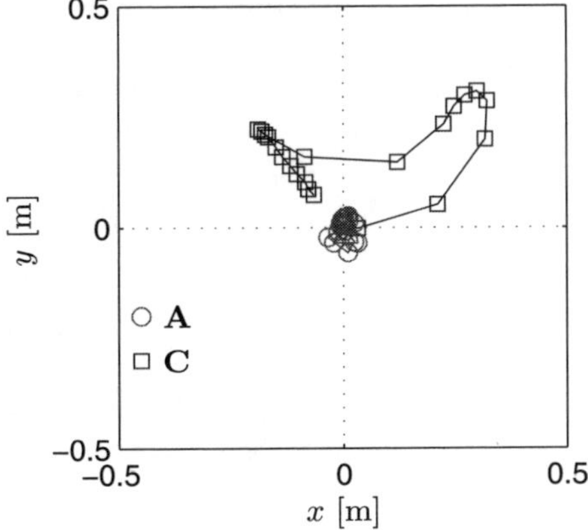


**Fig. 3.8.** Comparison of the position on the plane $xy$ for the reduced controller **A** and **C**

Figure 3.9 and 3.10 report the position, orientation and control effort for controller **D**. Even in this case, a strong coupling during the commanded yaw rotation caused by the controller itself can be observed.

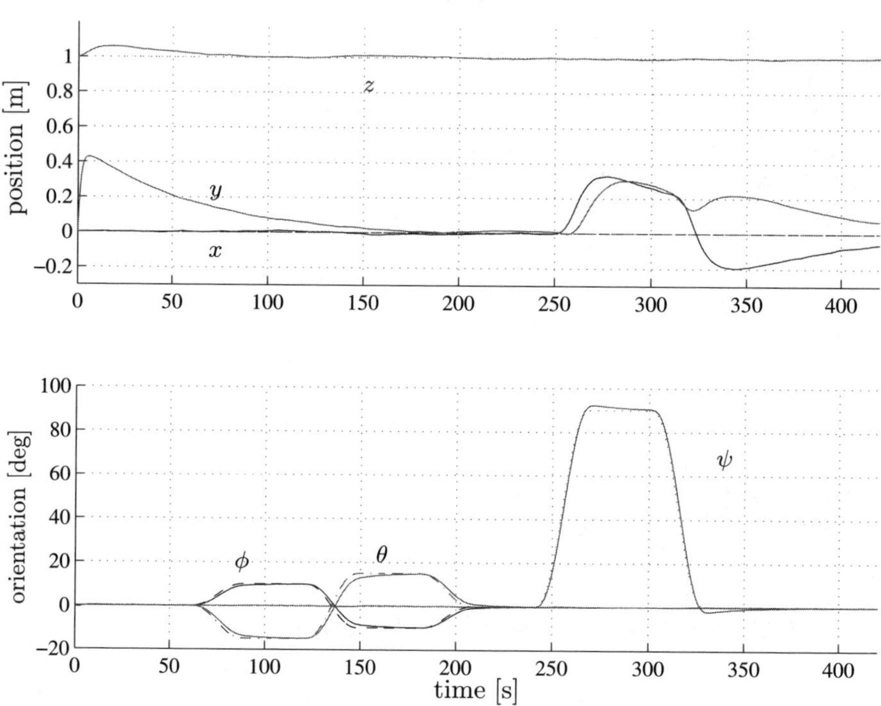


**Fig. 3.9.** Simulated position and orientation for the reduced controller **D**

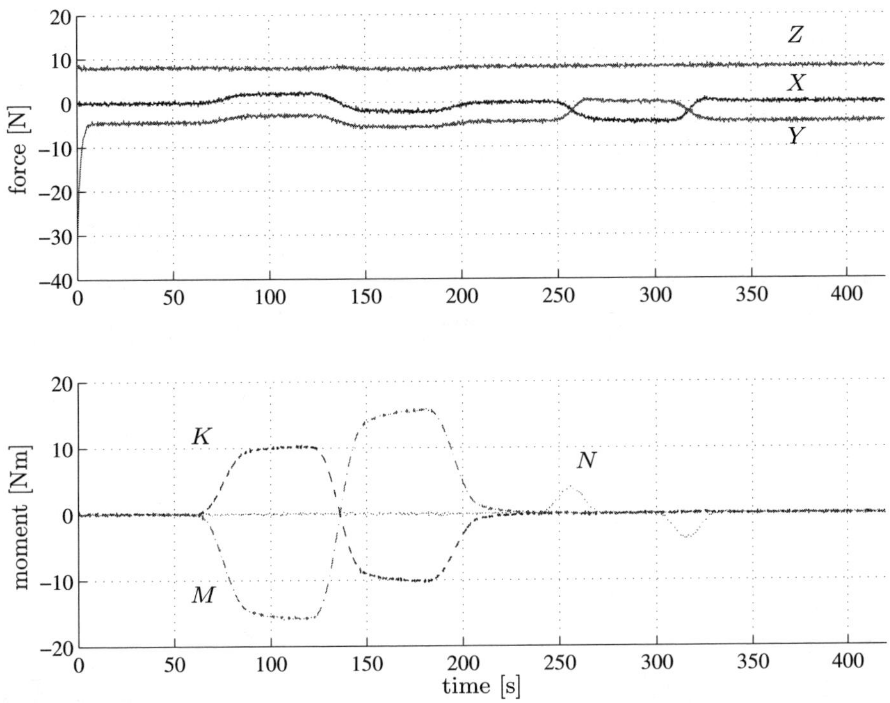


**Fig. 3.10.** Simulated force and moment for the reduced controller **D**

Figure 3.11 and 3.12 report the position, orientation and control effort for controller **E**.

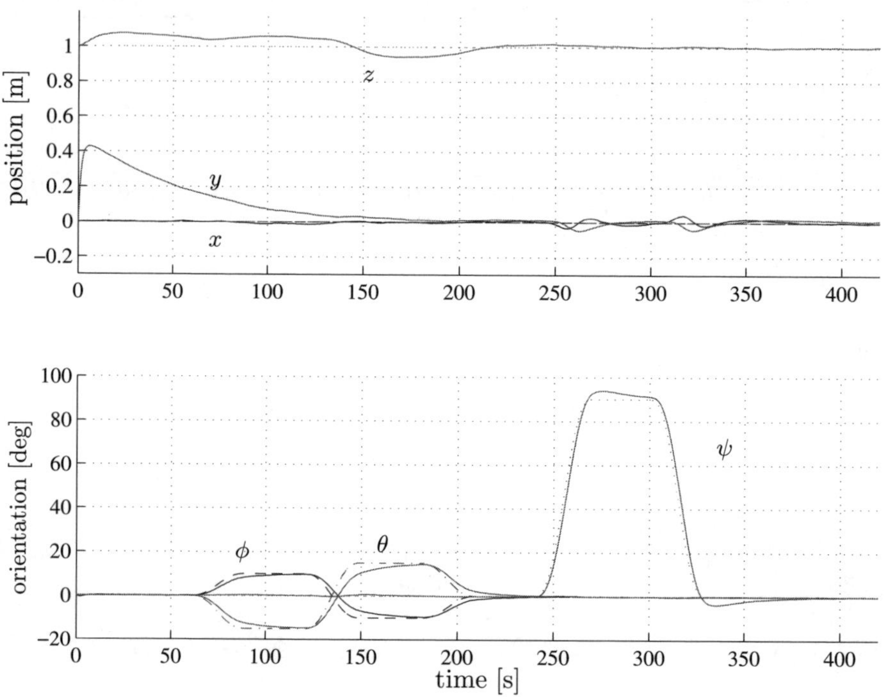


**Fig. 3.11.** Simulated position and orientation for the reduced controller **E**

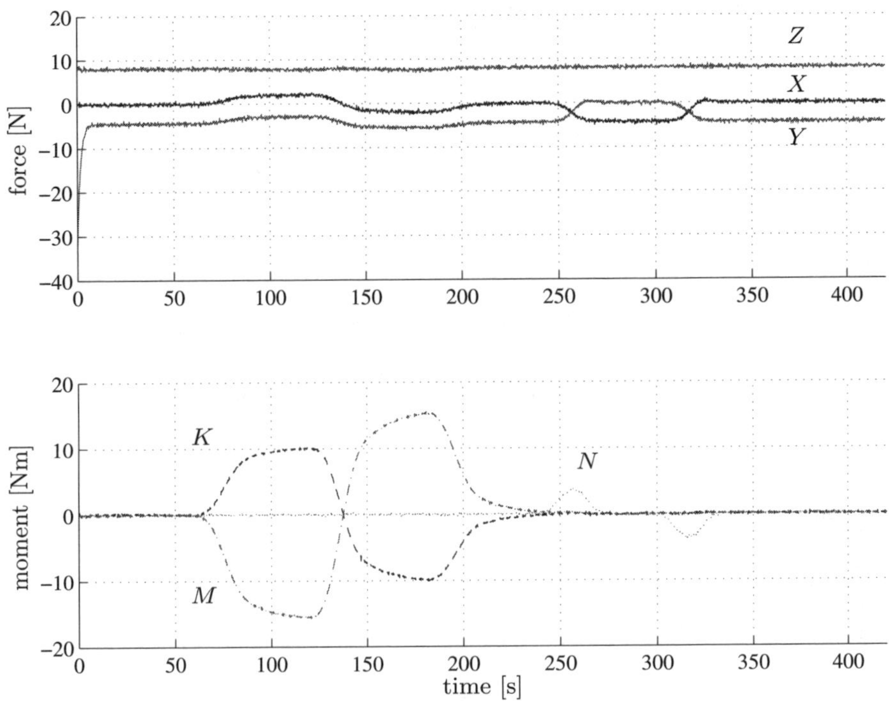


**Fig. 3.12.** Simulated force and moment for the reduced controller **E**

Figure 3.13 and 3.14 report the position, orientation and control effort for controller **F**.

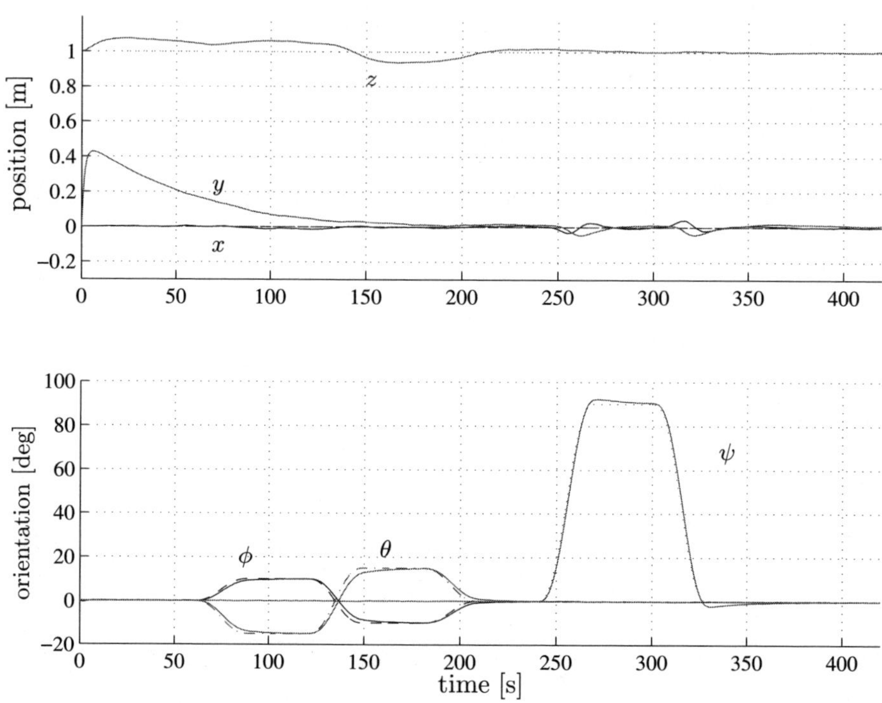


**Fig. 3.13.** Simulated position and orientation for the reduced controller **F**

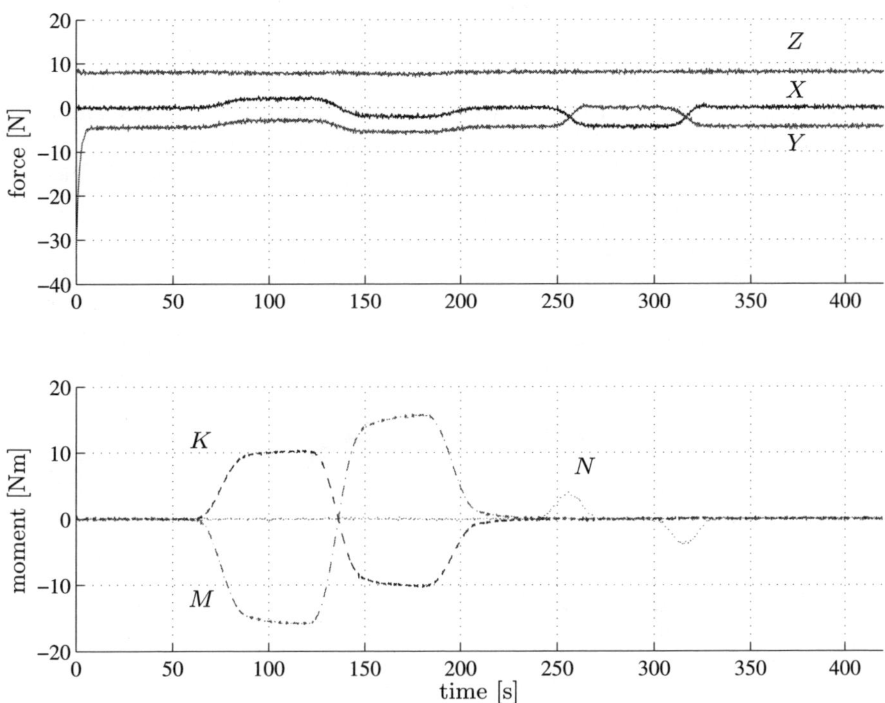


**Fig. 3.14.** Simulated force and moment for the reduced controller **F**

A further plot can be used (Figure 3.15) to understand the problem arising when compensating incorrectly the current, the controller **C** is compared with controller **E** in a normalized polar representation of the sole compensation of the ocean current during the yaw rotation. It can be noticed that, with the controller **E** (and similar) the compensation of the current rotates with the vehicle so that it still compensate with respect to a constant, earth-fixed disturbance.

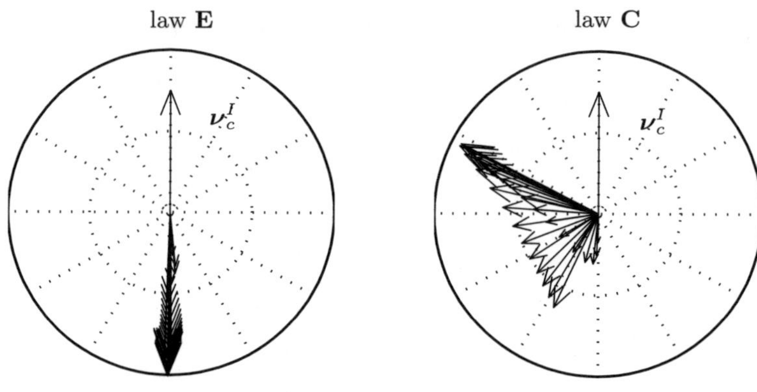


**Fig. 3.15.** Earth-fixed-frame polar representation of samples of the current compensation: controller **E** (left) and controller **C** (right). Notice that the compensation term built by the law **C** rotates together with the vehicle-fixed frame

The given simulation does not exhibit a significant error in the orientation that should be caused by the wrong restoring compensation of the controllers **A** and **F**. The numerical study has been conducted with a model largely used in the literature and tested in experimental cases; moreover, reasonable parameters' estimation has been considered; in particular only the third component of the vector $W\boldsymbol{r}_G - B\boldsymbol{r}_B$ is different from zero; finally small rotations in pitch and yaw have been commanded. Under these considerations both the controllers behave very well; the control effort, moreover, was similar to that of other controllers. It has been considered fair, thus, to report the good numerical result of these controllers despite their theoretical drawback and to avoid specific case studies where those might fail such as, e.g., with large enough $\phi$ and $\theta$ for the controller **F** and a different vector $W\boldsymbol{r}_G - B\boldsymbol{r}_B$. It might be noticed that this singularity is a representation singularity and that might arise even with the vehicle in the hovering position due to an unsatisfactory choice of the body-fixed frame or the roll-pitch-yaw convention; however, due to the common marine convention, this possibility is so uncommon that is not a real drawback.

However, to better illustrate this different behavior, a second simulation study was developed in which the vehicle is commanded to change the sole roll angle according to the time law in Figure 3.16 while keeping constant the vehicle position and the other vehicle angles at zero; for the sake of clarity, it is assumed that no ocean current is acting on the vehicle. The behavior of the reduced version of control law **E** is compared to that of the reduced version of the control law **A** in terms of the measured roll angles (Figure 3.17, left column). Remarkably, in the simple condition considered, the restoring moment to be compensated for is mainly acting around the $\boldsymbol{x}_b$ axis. In the simulation, it is possible to compute the restoring moment around this axis and compare it with the adaptive compensation of the control law **E** and the integral compensation plus the model-based compensation of the controller **A** respectively; those plots are reported in the right column of Figure 3.17. It can be observed that during the phase in which the roll angle changes from $+10°$ to $-10°$ the compensation built by the control law **A** has a lag with respect to the acting restoring moment; this is due to the integral charge/discharge time required to build a time-varying compensation term. The control law **E**, instead, performs proper compensation of the restoring moment since it accounts for the vehicle orientation in the adaptation mechanism.

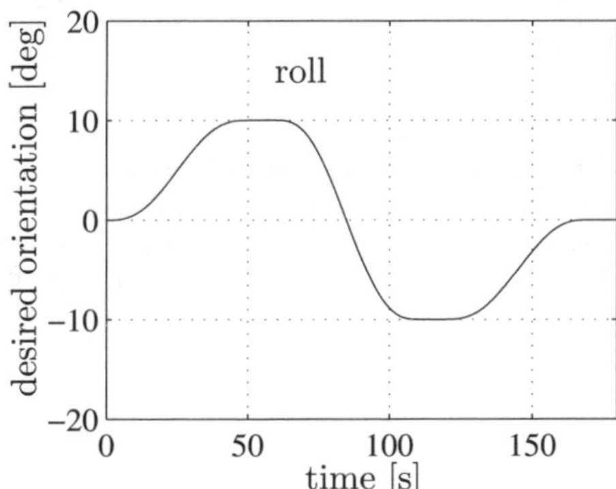


**Fig. 3.16.** Comparison of the control law **A** and the control law **E** in the second case study: Time history of the desired orientation

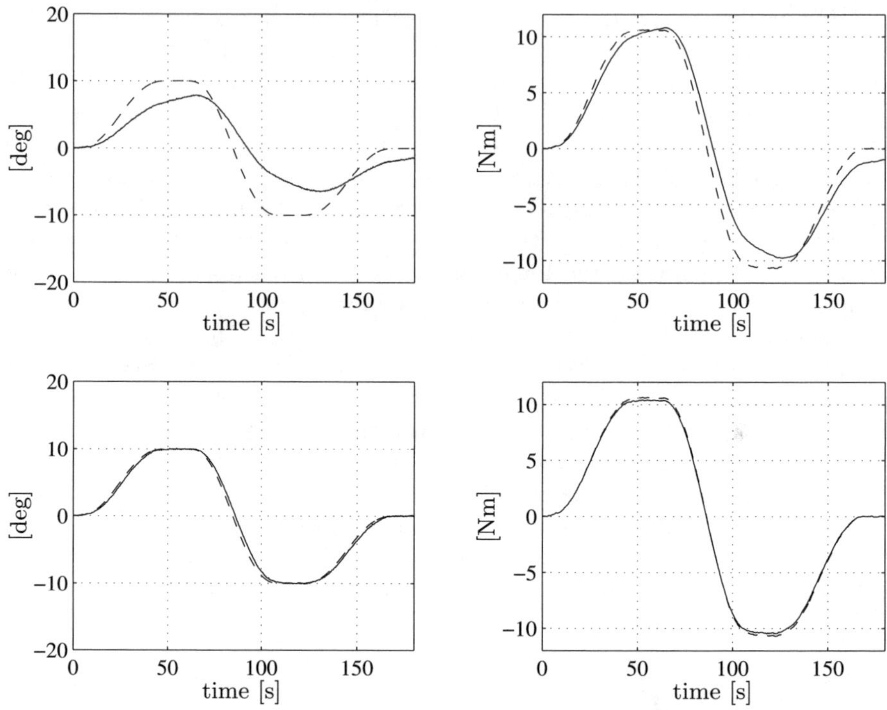


**Fig. 3.17.** Comparison of the control law **A** and the control law **E** in the second case study. Left column: Time history of measured roll angles (solid line) and the desired (dashed line); top, control law **A**, bottom, control law **E**. Right column: Time history of the restoring moment around $\boldsymbol{x}_b$ (dashed), the corresponding integral action of the control law **A** (top, solid line) and the corresponding adaptive action of the control law **E** (bottom, solid line)

### 3.9.2 Conclusions and Extension to UVMSs

In this Chapter six controllers have been compared with respect to their behavior in presence of modeling uncertainty and presence of ocean current. It is shown that, with a not proper compensation, the integral/adaptive action acts as a *disturbance* during the transients.

Numerical simulations better illustrate the theoretical results. However, an aspect to be considered for proper analysis of the simulations is that, despite any effort to chose the gains so as to ensure similar performance to the controllers has been made, this is impossible in a strict sense. Therefore, the presented results have to be read mainly by looking at the error behavior rather then focusing on strict numeric comparison. Notice, however, that the control effort is very similar for all the controllers, meaning that different behaviors are not given by a different magnitude of the inputs.

The extension of these considerations to dynamic control of UVMSs is straightforward using the virtual decomposition approach as detailed in Subsection 7.9.3.

# 4. Fault Detection/Tolerance Strategies for AUVs and ROVs

## 4.1 Introduction

Autonomous Underwater Vehicles (AUVs) and Remotely Operated Vehicles (ROVs) received increasing attention in the last years due to their significant impact in several underwater operations. Examples are the monitoring and maintenance of off-shore structures or pipelines, or the exploration of the sea bottom; see, e.g., reference [321] for a complete overview of existing AUVs with description of their possible applications and the main subsystems. The benefit in the use of unmanned vehicles is in terms of safety, due to the possibility to avoid the risk of manned missions, and economic. Generally, AUVs are required to operate over long periods of time in unstructured environments in which an undetected failure usually implies loss of the vehicle. It is clear that, even in case of failure detection, in order to terminate the mission, or simply to recover the vehicle, a fault tolerant strategy, in a wide sense, must be implemented. In fact, simple system failure can cause mission abort [154] while the adoption of a fault tolerant strategy allows to safely terminate the task as in the case of the arctic mission of Theseus [118]. In case of the use of ROVs, a skilled human operator is in charge of command the vehicle; a failure detection strategy is then of help in the human decision making process. Based on the information detected, the operator can decide in the vehicle rescue or to terminate the mission by, e.g., turning off a thruster.

Fault detection is the process of monitoring a system in order to recognize the presence of a failure; fault isolation or diagnosis is the capability to determine which specific subsystem is subject to failure. Often in literature there is a certain overlapping in the use of these terms. Fault tolerance is the capability to complete the mission also in case of failure of one or more subsystems, it is referred also as fault control, fault accommodation or control reconfiguration. In the following the terms fault detection/tolerance will be used.

The characteristics of a fault detection scheme are the capability of isolate the detected failure; the sensitivity, in terms of magnitude of the failure that can be detected and the robustness in the sense of the capability of working properly also in non-nominal conditions. The requirements of a fault tolerant scheme are the reliability, the maintainability and survivability [240]. The common concept is that, to overcome the loss of capability due to a failure, a kind of redundancy is required in the system. A general scheme is presented in Figure 4.1.

In this chapter, a survey over existing fault detection and fault tolerant schemes for underwater vehicles is presented. For these specific systems, adopting proper strategies, an hardware/software (HW/SW) sensor failure or an HW/SW thruster failure can be successfully handled in different operating conditions as it will be shown in next Sections. In some conditions, it is required that the fault detection scheme is also able to diagnostic some external not-nominal working conditions such as a multi-path phenomena affecting the echo-sounder system [61]. It is worth noticing that, for autonomous systems such as AUVs, space systems or aircraft, a fault tolerant strategy is necessary to safely recover the damaged vehicle and, obviously, there is no *panic button* in the sense that the choice of turning off the power or activate some kind of brakes is not available.

Most of the fault detection schemes are model-based [1, 3, 4, 5, 52, 61, 140, 143, 238, 306, 307] and concern the dynamic relationship between actuators and vehicle behavior or the specific input-output thruster dynamics. A model-free method is presented in [43, 173]. Higher level fault detection schemes are presented in [117, 118, 146, 324]. References [42, 177, 212, 222, 279, 308] deal with hardware/software aspects of a fault detection implementation for AUVs. Neural Network and Learning techniques have also been presented [102, 113, 142, 144, 284].

Concerning fault tolerant schemes, most of them consider a thruster redundant vehicle that, after a fault occurred in one of the thrusters, still is actuated in 6 degrees of freedom (DOFs). Based on this assumption a reallocation of the desired forces on the vehicle over the working thrusters is performed [3, 4, 5, 52, 61, 232, 233, 234, 236, 251, 306, 307]. Of interest is also the study of reconfiguration strategies if the vehicle becomes underactuated [228].

Only few papers concern the experimental results of fault detection and fault tolerant schemes [3, 4, 5, 42, 52, 61, 117, 118, 212, 232, 233, 251, 306, 307]; for all the above references it is worth noticing that the successfully results has been achieved with the implementation of simple algorithms.

In Section 4.2 a small list of failures occurred during wet operations is reported; Section 4.3 and 4.4 report the description of fault detection and tolerant strategies for underwater vehicles. Since the implementation of such strategies in a real environment is not trivial Section 4.5 describes in more detail some successfully experiments. Finally, the conclusions are drawn in Section 4.6.

## 4.2 Experienced Failures

In this section, a small list of possible ROVs/AUVs' failures is reported.

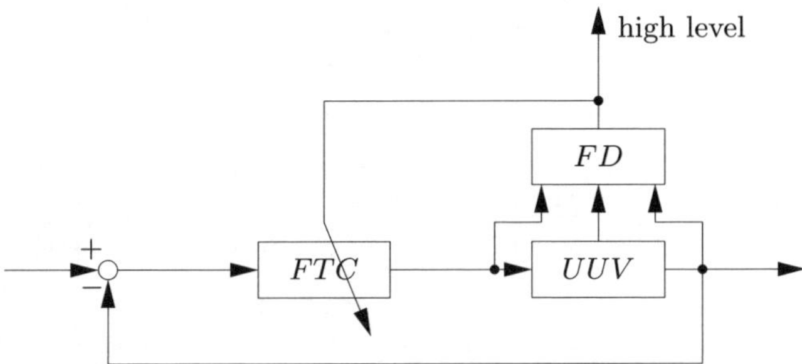


**Fig. 4.1.** General fault detection/tolerant control scheme for an Unmanned Underwater Vehicle (UUV). The Fault Detection (FD) block is in charge of detecting the failure, send a message to the higher level supervisor and, eventually, modifying the Fault Tolerant Controller (FTC)

**sensor failure** The underwater vehicles are currently equipped with several sensors in order to provide information about their localization and velocity. The problem is not easy, it does not exist a single, reliable sensor that gives the required position/velocity measurement or information about the environment, e.g., about the presence of obstacles. For this reason the use of sensor fusion by, e.g., a Kalman filtering approach, is a common technique to provide to the controller the required variables. This structural redundancy can be used to provide fault detection capabilities to the system. In detail, a failure can occur in one of the following sensors:

- IMU (Inertial Measurement Unit): it provides information about the vehicle's linear acceleration and angular velocity;
- Depth Sensor: by measuring the water pressure gives the vehicle's depth;
- Altitude and frontal sonars: they are used to detect the presence of obstacles and the distance from the sea bottom;
- Ground Speed Sonar: it measures the linear velocity of the vehicle with respect to the ground;
- Currentmeter: it measures the relative velocity between vehicle and water;
- GPS (Global Positioning System): it is used to reset the drift error of the IMU and localize exactly the vehicle; it works only at the surface;
- Compass: it gives the vehicle yaw;
- Baseline Acoustic: with the help of one or more transmitters it allows exact localization of the vehicle in a specific range of underwater environment;
- Vision system: it can be used to track structures such as pipelines.

For each of the above sensors the failure can consist in an output zeroing if, e.g., there is an electrical trouble or in a loss of meaning. It can be considered as sensor failure also an external disturbance such as a multi-path reading of the sonar that can be interpreted as a sensor fault and correspondingly detected.

**thruster blocking** It occurs when a solid body is between the propeller blades. It can be checked by monitoring the current required by the thruster. It has been observed, e.g., during the Antarctic mission of Romeo [61]: in that occurrence it was caused by a block of ice.

**flooded thruster** A thruster flooded with water has been observed during a Romeo's mission [61]. The consequence has been an electrical dispersion causing an increasing blade rotation velocity and thus a thruster force higher then the desired one.

**fin stuck or lost** This failure can causes a loss of steering capability as discussed by means of simple numerical simulations in [143].

**rotor failure** A possible consequence of different failures of the thrusters is the zeroing of the blade rotation. The thruster in question, thus, simply stops working. This has been intentionally experienced during experiments with ODIN [232, 233, 306, 307], RAUVER [140] and Roby 2 [4] and during another Romeo's mission [61].

**hardware-software failure** A crash in the hardware or software implemented on the vehicle can be experienced. In this case, redundancy techniques can be implemented to handle such situations [42].

## 4.3 Fault Detection Schemes

In [3, 4] a model-based fault detection scheme is presented to isolate actuators' failures in the horizontal motion. Each thruster is modeled as in [127]. The algorithm is based on a bank of Extended Kalman Filters (EKFs) the outputs of which are checked in order to detect behaviors not coherent with the dynamic model. In case of two horizontal thrusters and horizontal motion 3 EKFs are designed to simulate the 3 behaviors: nominal behavior, left thruster fault, right thruster fault. The cross-checking of the output allows efficient detection as it has been extensively validated experimentally (details are given in Section 4.5). A sketch of this scheme is given in Figure 4.3 where $\boldsymbol{u}$ is the vector of thruster inputs and the vehicle yaw $\psi$ is measured by means of a compass. In [5], the same approach is investigated with the use of a sliding-mode observer instead of the EKF. The effectiveness of this approach is also discussed by means of experiments.

The work in [52, 61] focuses on the thruster failure detection by monitoring the motor current and the propeller's revolution rate. The non-linear nominal characteristic has been experimentally identified, thus, if the measured couple current-propeller's rate is out of a specific bound, then a fault is experienced. Based on a mapping of the i-o axis the possible cause is also specified with a message to the remote human operator. The two failures corresponding to a thruster flooding or to a rotor failure, in fact, fall in different axis regions and can be isolated. Interesting experiments are given in Section 4.5.

**Fig. 4.2.** Romeo (courtesy of M. Caccia, National Research Council-ISSIA, Italy)

The fault in a thruster is also monitored in [306, 307] by the use of a hall-effect sensor mounted on all the thrusters. The input is the desired voltage as computed by the controller and the TCM, the output is the voltage as measured by the hall-effect sensor; the mismatching between the measured and the predicted voltage is considered as a fault. The paper also considers fault tolerance for sensor and actuator faults and experiments, as shown in following sections.

The vehicle Theseus [117, 118] is equipped with a Fault Manager subsystem. This provides some kind of high-level failure detection in the sense that the mission is divided in a number of phases (each phase is a series of manoeuvres between way points); in case of failure of a phase there is a corresponding behavior to be activated. See Section 4.5 for detail about a practical intervention of the Fault Manager. A hierarchical control system developed for future implementation on Theseus is described in [324], this is based on the layered control concepts [58].

References [42, 212] present an architecture for AUVs that integrates fault detection capabilities of the subsystems. The hardware and software architecture, named AUVC (Autonomous Underwater Vehicle Controller) imple-

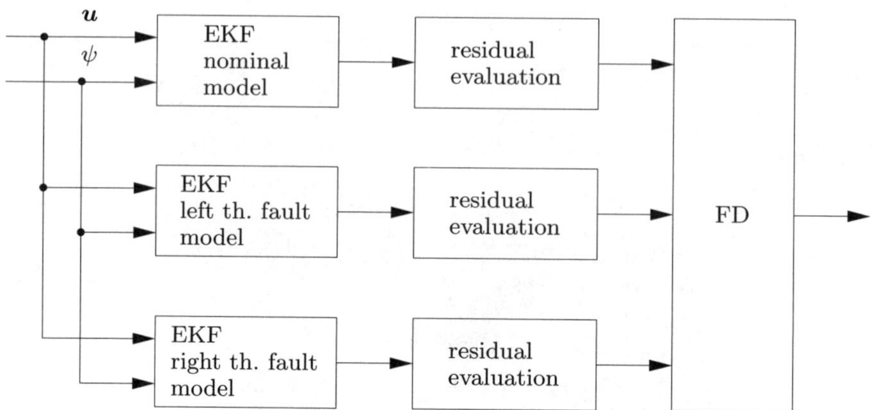


**Fig. 4.3.** Fault detection strategy for one of the horizontal thruster failure proposed by A. Alessandri, M. Caccia and G. Verrugio

ments a fault detection strategy based on five rule-based systems that monitor all the subsystems. The five systems concern the Navigation, the Power/Propulsion, the Direction Control and the Communication; they are coordinated by a Global Diagnoser that avoid contradictory actions. A specific attention has been given at the hardware reliability: as a matter of fact the AUVC is distributed on a redundant network of 18 loosely coupled processors. AUVC has been also used to test the approach proposed in [222], a redundancy management technique based on CLIPS expert system shell to identify faults affecting depth and heading control. In [308] an architecture developed for the vehicle ARICS with fault detection/tolerant capabilities is presented. A software developed for ROVs to help the remote operator that integrates some elementary fault detection algorithms is presented in [279]. The paper in [177] describes the first results on the development of a long endurance AUV that is currently ongoing at the MBARI (Monterey Bay Aquarium Research Institute, California, USA). The fault detection approach is mainly ported from the MIT (Massachusetts Institute of technology, Massachusetts, USA) vehicle Odyssey (I and II) [46] and it is based on the Layered Control Architecture. The software architecture is based on C++, QNX-based modules, which offers multi-tasking capabilities suited for fault-tolerant operations. A single thread suppression and restart can be implemented to recover from software failure. Short-duration operations in open sea and long-duration operations in the lab proved the effectiveness of this approach.

In [143] a model-based observer is used to generate residual between the sensor measured behavior and the predicted one. The model also takes into account the presence of waves in case of operations near the surface. When the residual is larger than a given threshold a Fuzzy Inference System is in charge of isolate the source of this mismatching (see sketch in Figure 4.4). A planar simulation is provided in case of low speed under wave action and a stuck fin.

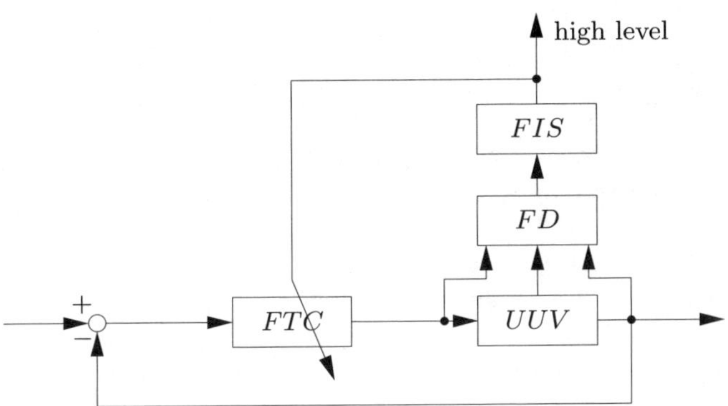


**Fig. 4.4.** Fuzzy fault detection/tolerant control scheme proposed by A.J. Healey: the FIS (Fuzzy Inference System) block is in charge of isolating faults observed by the FD (Fault Detection) block between fin stroke, servo error, residual and wave activity detectors

A model-free fault detection method is proposed in [43, 173]: this is based on the Hotelling $T^2$ statistic and it is a data-driven approach. The validation is based on a 6-DOFs simulation affected by stern plane jams and rudder jams.

A model-based, integrated heterogeneous knowledge approach is proposed in [140]. A multi-dimensional correlation analysis allows to increase the confidence in the detected fault and to detect also indirectly sensed subsystems. Some preliminary results with the vehicle RAUVER are also given.

In [1] a model-based fault detection scheme for thrusters and sensors is proposed. It has been designed based on the identified model of the 6-thruster ROV Linotip and it is composed of a bank of single-output Luenberger observers. Its effectiveness is verified by simulations. Another model-based fault detection scheme verified by simulations is provided in [238]. A robust approach in [199].

A neuro-symbolic hybrid system is used in [102] to perform fault diagnosis on AUVs with learning capability. The method is simulated on the planar motion of the VORTEX mathematical model. Another learning technique is proposed in [113] and verified by means of 6-DOFs simulations. In [284], a neural network mathematical model is used to set-up a self-diagnosis scheme of the AUV. A software for health monitoring of AUVs' missions with learning capabilities is also described in [151].

The work in [39] studies a systematic, quantitative approach in order to maximize the mission and return success probabilities. The failed sensor is de-activated and the information obtained by a backup sensor able to recover the vehicle. No dynamic simulations are provided.

In [100], the wavelet theory is used to detect the fault in the vehicle's navigation angle fault.

Finally, [258] developed a software tool to test intelligent controllers for AUVs. This is done by using learning techniques from the artificial intelligence theory.

## 4.4 Fault Tolerant Schemes

Most of the fault tolerant controllers developed for thruster-driven underwater vehicles are based on a suitable inversion of the TCM in eq. (2.48). It is self-evident that, if the matrix is low rectangular it is still possible to turn off the broken thruster and to control the vehicle in all the 6 DOFs. When the vehicle becomes underactuated, or when it is driven by control surfaces, the problem is mathematically more complex. In this case only few solutions to specific set-up have been developed.

The work in [52, 61] reports a fault tolerant approach for ROMEO, a thruster redundant ROV with 8 thrusters. The strategy, experimentally verified, simply consists in deleting the column corresponding to the broken thruster from the TCM. The mapping from the vehicle force/moment to the thrusters' forces thus, does not concern the failed component. A similar approach is used in [306, 307] by exploiting the thruster redundancy of ODIN, an AUV developed at the Autonomous Systems Laboratory (ASL) of the University of Hawaii, HI, USA. The proposed approach is sketched in Figure 4.5 where the subscript $d$ denotes the desired trajectory, $\boldsymbol{V}_m$ is the motor input voltage and $\Omega$ the propeller angular velocity of the thrusters.

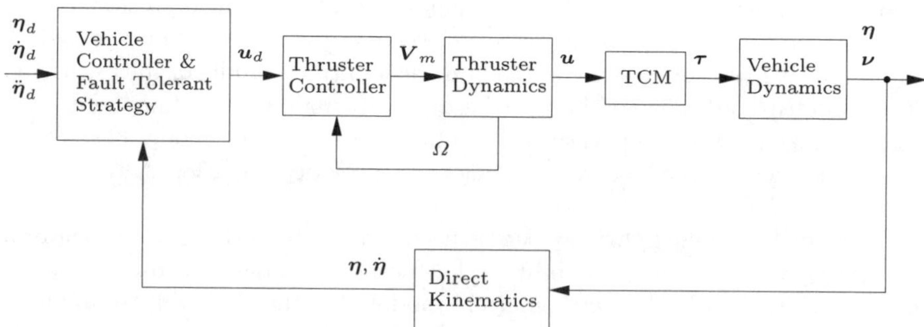


**Fig. 4.5.** Fault tolerance strategy proposed by K.S. Yang, J. Yuh and S.K. Choi

In [232, 233, 234, 236, 251], a task-space-based, fault tolerant control for vehicles with redundant actuation is proposed. The control law is model based and it handles the thruster redundancy by a pseudo-inverse approach of the TCM that guarantees the minimization of the actuator quadratic norm. The thruster dynamics, with the model described in [147], is also taken into account.

The work [306, 307] presents a fault detection-tolerant scheme for sensor faults. In detail, the depth of the vehicle is measured by using a pressure sensor and a bottom sonar sensor. A third, virtual, sensor is added: this is basically an ARX (AutoRegressive eXogeneous) model of the vehicle depth dynamics. By comparing the measured values with the predicted ones, the residual is calculated and the failed sensor eventually disconnected for the remaining portion of the mission. A sketch of the scheme is shown in Figure 4.6, in nominal working conditions, the 3 residual $R_i$ are close to the null value. It is worth noticing that this approach requires exact knowledge of the ocean depth.

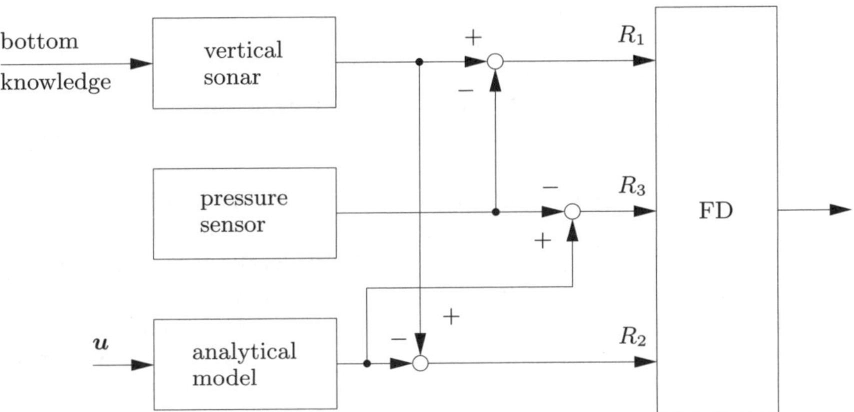


**Fig. 4.6.** Fault tolerance strategy for sensor fault proposed by K.C. Yang, Y. Yuh and S.K. Choi

In [228], the case of an underactuated AUV controlled by control surfaces is considered. The vehicle tries to move in the unactuated DOFs by using elementary motions in the actuated DOFs. The method has been tested on the vehicle ARCS and showed that, in this form, it is not applicable. This study provides information on structural changes to be adopted in the vehicle in order to develop a vehicle suitable for implementing this method.

One of the first works of reconfiguration control for AUVs is given in [283], where, however, only a superficial description of a possible fault tolerant strategy is provided. Recently, [288] proposed a reconfiguration strategy to accommodate actuator faults: this is based on a mixed $H_2/H_\infty$ problem. Simulation results are provided.

## 4.5 Experiments

Roby 2 is a ROV developed at the National Research Council-ISSIA, Italy. It has been object of several wet tests aiming at validating fault detection approaches. The horizontal motion is obtained by the use of two fore thrusters that control the surge velocity as well as the vehicle heading; the depth is regulated by means of two vertical thrusters. In [3, 4, 5] experiments of different fault detection schemes have been carried out by causing, on purpose, an actuator failure: one of the thrusters has been simply turned off. The experiments in [3, 4] have been carried out in a pool, in [5] the experiments also concern a comparison between EKFs and sliding-mode observers. The latter is a result of a bilateral project with the Naval Postgraduate School, Monterey, CA.

The Italian National Research Council (CNR-ISSIA) also developed the ROV ROMEO and tested, in an antarctic mission, both fault detection and tolerant schemes [52, 61]. In particular the case of flooded and blocked thrusters occurred. In both cases the fault has been detected and the information could be reported to the human operator in order to activate the reconfiguration procedure. Figure 4.7 shows the expected and measured motor currents in case of flooded thruster: it can be observed a persistent mismatching between the output of the model and the measured values.

The vehicle Theseus manufactured by ISE Research Ltd with the Canadian Department of National Defence successfully handled a failure during an Arctic mission of cable laying [117, 118]. In details, the vehicle did not terminate a homing step, probably due to poor acoustic conditions and the Fault Manager activated a safe behavior: stop under the ice and wait for further instructions. This allowed to re-establish acoustic telemetry and surface tracking and safely recover the vehicle. Notice that his fault wasn't intentionally caused [118].

The vehicle ODIN, an AUV developed at the Autonomous Systems Laboratory (ASL) of the University of Hawaii, HI, USA, has been used for several experiments. In [306, 307] the fault detection and tolerant schemes are experimentally validated. The thruster fault has been tested by zeroing the output voltage by means of software, the fault detection scheme identified the trouble and correctly reconfigured the force allocation by properly modifying of the TCM. The fault tolerant scheme with respect to depth sensor fault has also been tested by zeroing the sensor reading and verifying that the algorithm, after a programmed time of 1 s, correctly switched on the other sensor. While the theory has been developed for a 6-DOFs vehicle, the experiments results only present the vehicle depth.

The same vehicle has been used to validate the fault tolerant approach developed in [232, 233, 251] in 6-DOFs experiments. Different experiments have been carried out by zeroing the voltage on one or two thrusters simultaneously that, however, did not cause the vehicle becoming under-actuated. The implemented control law is based on an identified reduced ODIN model

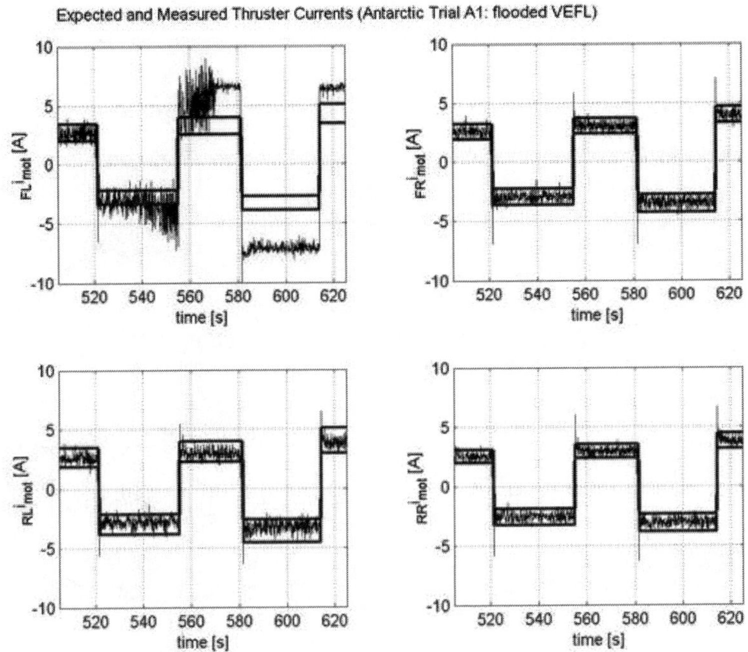


**Fig. 4.7.** Expected and measured motor currents for the vehicle Romeo in case of flooded thruster (courtesy of M. Caccia, National Research Council-ISSIA)

and does not make use of thruster model neither it needs the vehicle acceleration as required by the theory; the block diagram, thus, is simply given by Figure 4.8. Details on the control law are given in the referenced papers, the basic formulation of the controller is given by:

$$\boldsymbol{u}_v = \boldsymbol{E}^{\dagger} \left[ (\ddot{\boldsymbol{\eta}}_d - \boldsymbol{\beta}) + \boldsymbol{K}_v \dot{\tilde{\boldsymbol{\eta}}} + \boldsymbol{K}_p \tilde{\boldsymbol{\eta}} \right] \tag{4.1}$$

where $\boldsymbol{K}_v$ and $\boldsymbol{K}_v$ are control gains, the vector $\tilde{\boldsymbol{\eta}}$ is the position/orientation error, $\boldsymbol{\beta}$ represents the compensation of the nonlinear terms of the equation of motions. The matrix $\boldsymbol{E}$ takes into account the TCM matrix, the inertia matrix and the Jacobian matrix that converts body-fixed to inertial-fixed velocities. Generalization about control of a desired task is given in [233]. The experiments validated the proposed approach; in Figure 4.9 the voltages are shown: it can be recognized that thrusters 2 and 6 (one horizontal and one vertical) are turned off at $t = 260$ s and $t = 300$ s, respectively, this causes an augmentation of the chattering of the remaining thrusters that, however, still can perform the desired task (Figure 4.10).

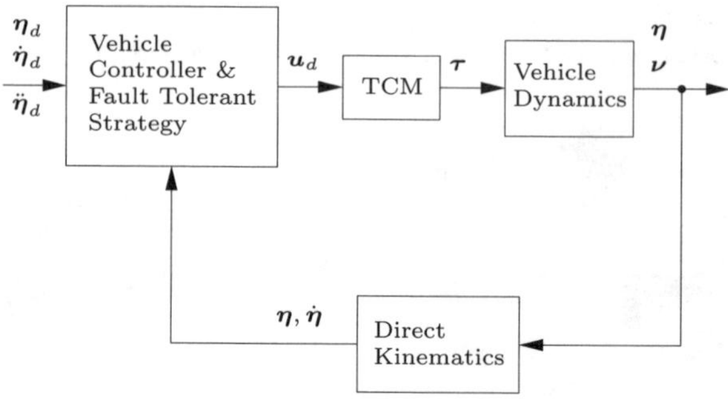


**Fig. 4.8.** Sketch of the fault tolerance strategy implemented by N. Sarkar, T.K. Podder and G. Antonelli

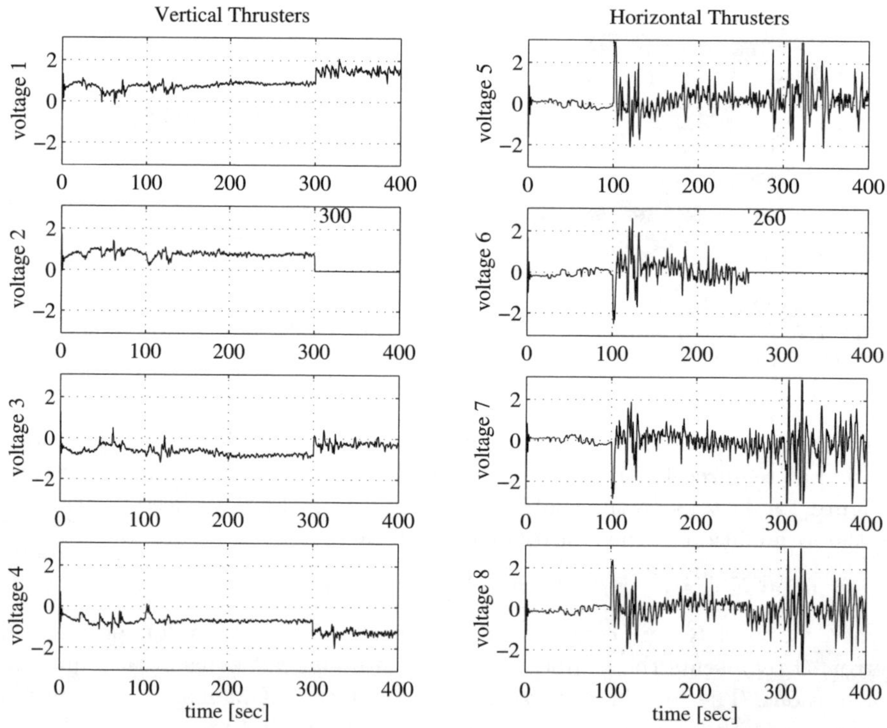


**Fig. 4.9.** Voltage profile - vertical thrusters (on the left) and horizontal thrusters (on the right) for the N. Sarkar and T.K. Podder algorithm

The AUVC described in [42, 212] has been tested on a six-processor version on the Large Diameter Unmanned Underwater Vehicle of the Naval Undersea Warfare Center.

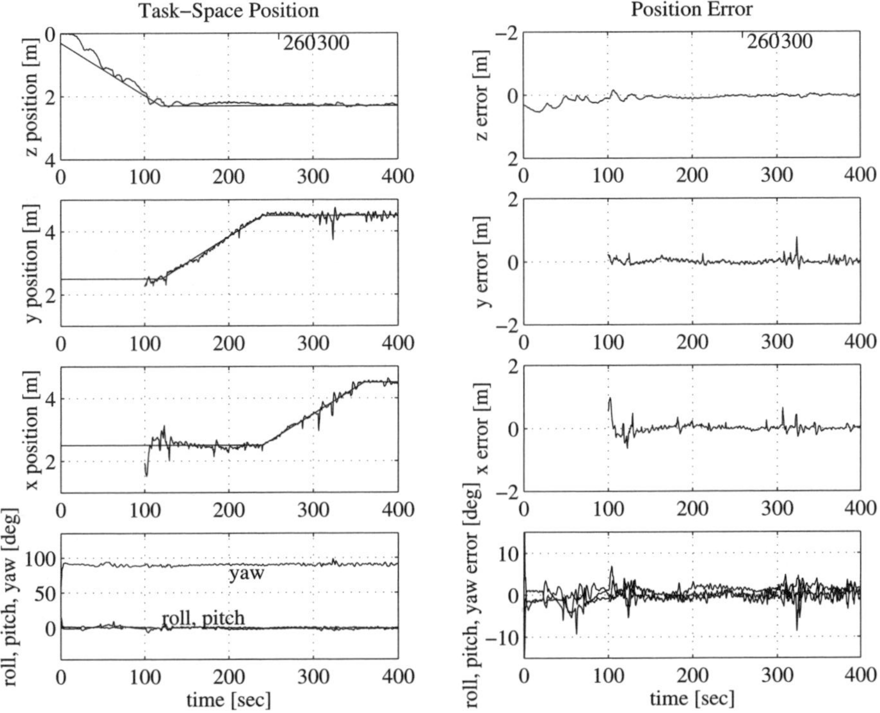


**Fig. 4.10.** Task-space trajectories (left column) and their errors (right column) for the N. Sarkar and T.K. Podder algorithm

## 4.6 Conclusions

An overview over existing fault detection and fault tolerant schemes for underwater vehicles has been presented. The case of failures for autonomous missions in un-structured environment is, obviously, a dramatic occurrence to handle. In this sense, the underwater community would benefit from research studies with a strong practical orientation rather than theoretical-only approaches. Failures in a redundant sensor seem to be a solved problem; however, a particular attention needs to be paid to the tuning of the detection gains, real-data experiments for off-line tuning seems to be a reliable way to select those gains. As far as thrusters are of concern, experiments have shown that current AUVs can be controlled at low velocity with 6 thrusters, also if the original symmetric allocation is lost, without a strong performance deterioration. Some possible research areas concern the case of a thruster-driven AUV that becomes under-actuated and the case of vehicles controlled by means of control surfaces that failed; in both cases some practical approaches might be useful for the underwater community.

# 5. Experiments of Dynamic Control of a 6-DOF AUV

## 5.1 Introduction

In this Chapter some experimental results on dynamic control of a 6-DOF AUV are given; practical aspects of the implementation are also discussed. The experiments have been conducted in the pool of the University of Hawaii using ODIN, an AUV developed at the Autonomous Systems Laboratory (ASL) [81].

**Implemented control law.** The control law is briefly rewritten:

$$\boldsymbol{u}_v = \boldsymbol{B}_v^{\dagger}[\boldsymbol{K}_D \boldsymbol{s}_v' + \boldsymbol{\Phi}_v(\mathcal{Q}, \boldsymbol{\nu}, \dot{\boldsymbol{\nu}}_a)\hat{\boldsymbol{\theta}}_v] \tag{5.1}$$

$$\dot{\hat{\boldsymbol{\theta}}}_v = \boldsymbol{K}_\theta^{-1} \boldsymbol{\Phi}_v^{\mathrm{T}}(\mathcal{Q}, \boldsymbol{\nu}, \dot{\boldsymbol{\nu}}_a)\boldsymbol{s}_v \tag{5.2}$$

where $\boldsymbol{B}_v^{\dagger}$ is the pseudoinverse of matrix $\boldsymbol{B}_v$ (see the Appendix), $\boldsymbol{K}_\theta > \boldsymbol{O}$ and $\boldsymbol{\Phi}_v$ is the vehicle regressor. The vectors $\boldsymbol{s}_v' \in \mathbb{R}^6$ and $\boldsymbol{s}_v \in \mathbb{R}^6$ are defined as follows

$$\begin{aligned} \boldsymbol{s}_v' &= \begin{bmatrix} \tilde{\boldsymbol{\nu}}_1 \\ \tilde{\boldsymbol{\nu}}_2 \end{bmatrix} + \left(\boldsymbol{\Lambda} + \boldsymbol{K}_D^{-1}\boldsymbol{K}_P\right) \begin{bmatrix} \boldsymbol{R}_I^B \tilde{\boldsymbol{\eta}}_1 \\ \tilde{\boldsymbol{\varepsilon}} \end{bmatrix} \\ &= \tilde{\boldsymbol{\nu}} + \left(\boldsymbol{\Lambda} + \boldsymbol{K}_D^{-1}\boldsymbol{K}_P\right)\tilde{\boldsymbol{y}}, \end{aligned} \tag{5.3}$$

$$\boldsymbol{s}_v = \tilde{\boldsymbol{\nu}} + \boldsymbol{\Lambda}\tilde{\boldsymbol{y}}\,, \tag{5.4}$$

with $\tilde{\boldsymbol{\eta}}_1 = [\,x_d - x \quad y_d - y \quad z_d - z\,]^{\mathrm{T}}$, $\tilde{\boldsymbol{\nu}}_1 = \boldsymbol{\nu}_{1,d} - \boldsymbol{\nu}_1$, where the subscript $d$ denotes desired values for the relevant variables. The matrix $\boldsymbol{\Lambda} \in \mathbb{R}^{6\times 6}$ is defined as $\boldsymbol{\Lambda} = \text{blockdiag}\{\lambda_p \boldsymbol{I}_3, \lambda_o \boldsymbol{I}_3\}$, $\boldsymbol{\Lambda} > \boldsymbol{O}$. The matrix $\boldsymbol{K}_P \in \mathbb{R}^{6\times 6}$ is defined as $\boldsymbol{K}_P = \text{blockdiag}\{k_p \boldsymbol{I}_3, k_o \boldsymbol{I}_3\}$, $\boldsymbol{K}_P > \boldsymbol{O}$. Finally, it is

$$\boldsymbol{\nu}_a = \boldsymbol{\nu}_d + \boldsymbol{\Lambda}\tilde{\boldsymbol{y}}$$

and $\boldsymbol{K}_D > \boldsymbol{O}$.

## 5.2 Experimental Set-Up

ODIN is an autonomous underwater vehicle developed at the Autonomous Systems Laboratory of the University of Hawaii. A picture of the vehicle is shown in Figure 5.1. It has a near-spherical shape with horizontal diameter of 0.63 m and vertical diameter of 0.61 m, made of anodizied Aluminum (AL 6061-T6). Its dry weight is about 125 kg. It is, thus, slightly positive buoyant. The processor is a Motorola 68040/33 MHz working with VxWorks 5.2 operating system. The power supply is furnished by 24 Lead Gel batteries, 20 for the thrusters and 4 for the CPU, which provide about 2 hours of autonomous operations. The actuating system is made of 8 marine propellers built at ASL; they are actuated by brushless motors. Each motor weighs about 1 kg and can provide a maximum thrust force of about 27 N. The sensory system is composed of: a pressure sensor for depth measuring, with an accuracy of 3 cm; 8 sonars for position reconstruction and navigation, each with a range $0.1 \div 14.4$ m; an Inertial System for attitude and velocity measures.

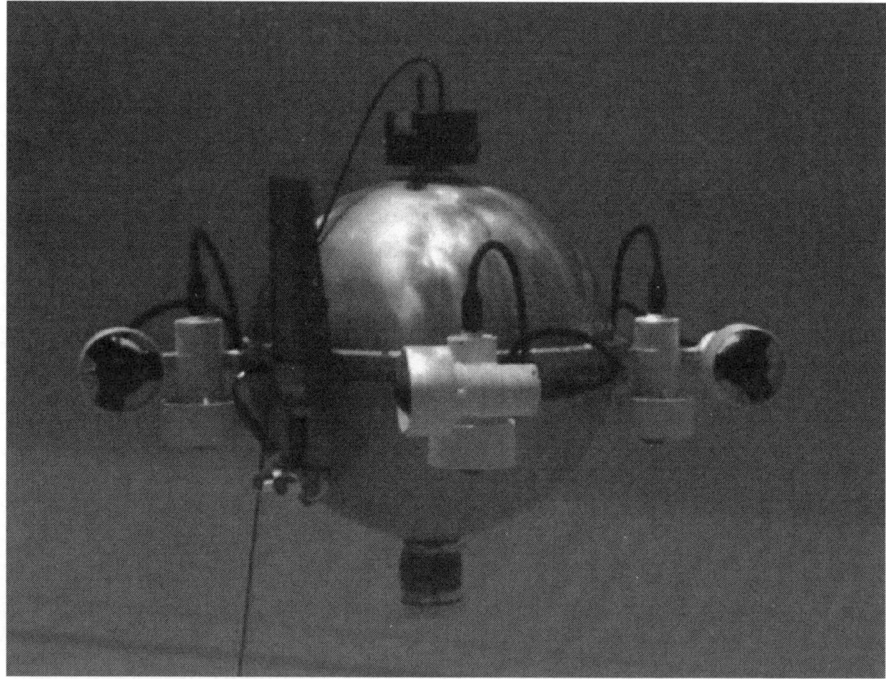

**Fig. 5.1.** ODIN, the AUV used to experimentally test adaptive and fault tolerant control strategies

## 5.3 Experiments of Dynamic Control

Despite the closed environment in which the experiments have been conducted, the pool of the University of Hawaii, it is necessary to take into account the presence of a current as an irrotational, constant disturbance [127]. The modeling aspects of including the current in the dynamic model have been discussed in Section2.4.3. Since the measure of the current is not available in ODIN, it has been taken into account as a disturbance $\boldsymbol{\tau}_{v,C}$ acting at the force/moment level on the vehicle-fixed frame. Moreover, since the number of dynamic parameters could be very large ($n_{\theta,v} > 100$, see [127]) it was implemented a reduced version of the regressor matrix in order to adapt only with respect to the restoring force/moments and to the current. In other words, the control law implemented was the reduced controller of the one discussed in Section 3.2.

The experiment was conceived in the following way: the vehicle had to follow a desired trajectory with trapezoidal profile. Since the sonars need to be under the surface of the water to work properly the first movement planned was in the $z$ direction (see Figure 5.2 for the relevant frames). The vehicle planned to move 2 m in the $y$ direction and 2 m in the $x$ direction.

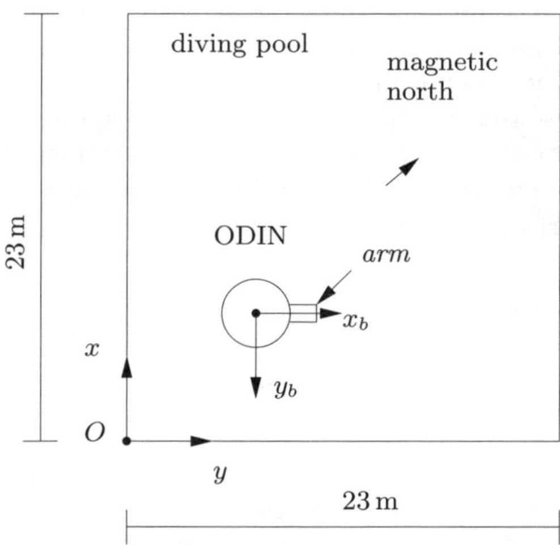


**Fig. 5.2.** Sketch of the pool used for experiments with relevant frames

The control law has been designed using quaternions, however the specifications of the desired trajectory and the output results are given in Euler angles because of their immediate comprehension. Notice that the transformation from Euler angles to quaternions is free from representation singularities. The attitude must be kept constant at the value of

$$\boldsymbol{\eta}_{d,2} = [0 \quad 0 \quad 90]^{\mathrm{T}} \quad \text{deg}.$$

Since the vehicle is not perfectly balanced, at rest, i.e., with the thrusters off, its position is $\phi \approx 5°$ and $\theta \approx 15°$ with the yaw depending on the current. The desired orientation, thus, is a set-point for the control task.

The following control gains have been used:

$$\begin{aligned}
\boldsymbol{\Lambda} &= \mathrm{diag}\{0.5, 0.5, 0.15, 0.8, 0.8, 0.8\}\,, \\
\boldsymbol{K}_D &= \mathrm{diag}\{12, 12, 5, 2, 2, 8\}\,, \\
k_p &= 1\,, \\
k_o &= 1\,, \\
\boldsymbol{K}_\theta &= \mathrm{blockdiag}\{0.01, 0.3\boldsymbol{I}_3\}
\end{aligned}$$

In Figure 5.3 the time history of the position of the vehicle in the inertial frame and the attitude of the vehicle and the orientation error in terms of the Euler angles is shown. Notice that, due to technical characteristics of the horizontal sonars [215], the $x$ and $y$ position data for the first 100 s are not available. The vehicle is not controlled in those directions and subjected to the pool's current. At $t = 100$ s, it recovers the desired position and starts tracking the trajectory. In the first seconds the vehicle does not track the desired depth. This is based on the assumption that all the dynamic parameters are unknown; at the beginning, thus, the action of the adaptation has to be waited. From this plots it is also possible to appreciate the different noise characteristics of the two position sensors: the sonar for the horizontal plane and the pressure sensor for the depth. It can be noticed that the desired attitude is kept for the overall length of the experiment and that a small coupling can be seen when, at $t = 100$ s the vehicle recovers the desired position in the horizontal plane.

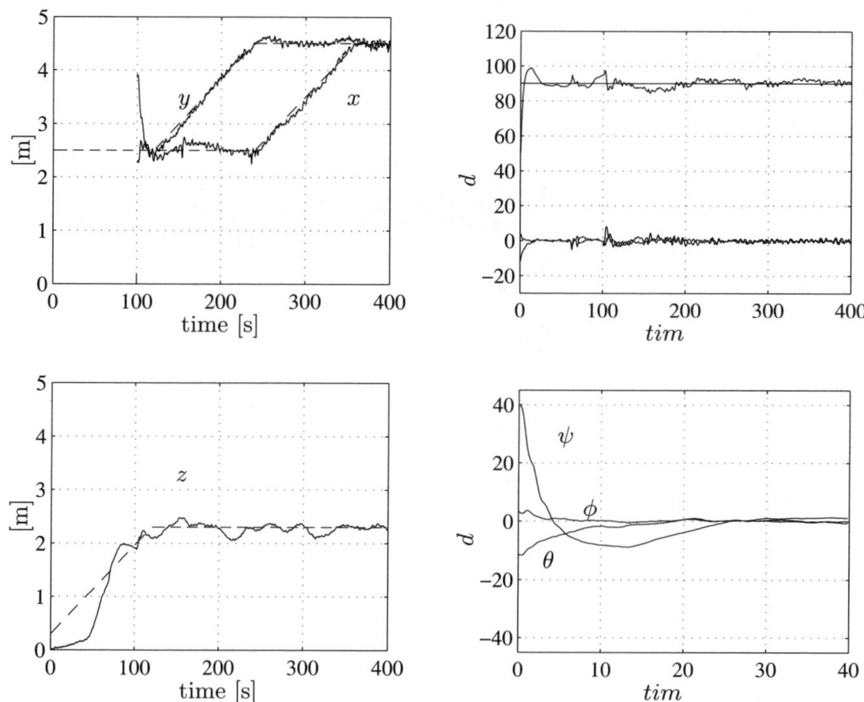


**Fig. 5.3.** Experiments of adaptive control on ODIN. Left: time history of the vehicle positions. Top: $x$ and $y$ positions. Bottom: $z$ position. Right: Top: time history of the vehicle attitude in terms of Euler angles. Bottom: vehicle attitude errors in terms of Euler angles (particular)

In Figure 5.4 the control actions are shown. The *peak* in $\tau_x$ is due to the big error *seen* by the controller at $t = 100\,\text{s}$.

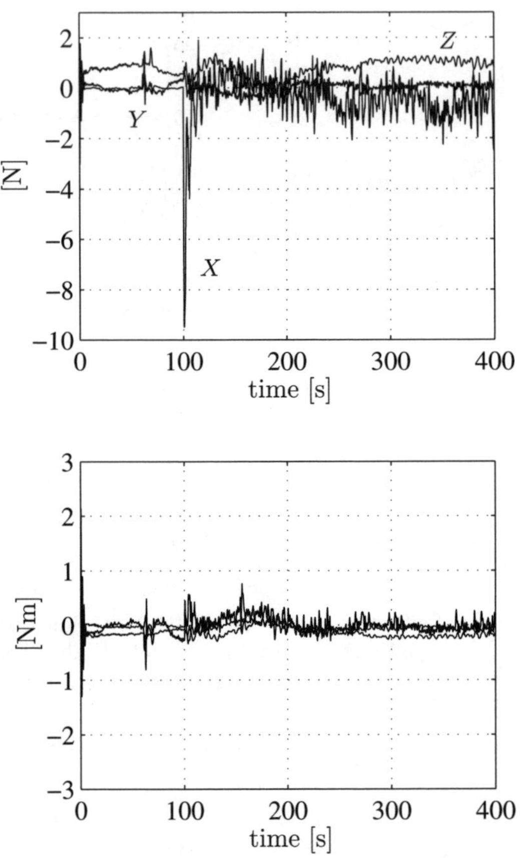


**Fig. 5.4.** Experiments of adaptive control on ODIN. Top: control forces. Bottom: control moments

In Figure 5.5 the path in the horizontal plane is shown. It can be noted that the second segment is followed with a smaller error than the first one, due to the adaptation action.

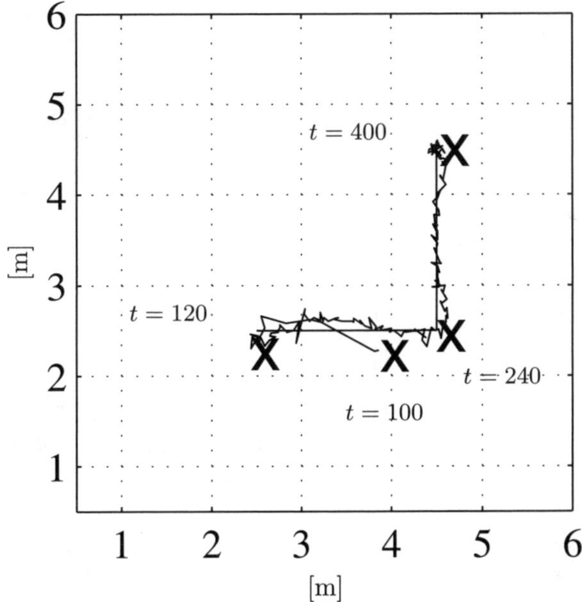


**Fig. 5.5.** Experiments of adaptive control on ODIN. Path on the $xy$ plane

In Figure 5.6 the 2-norm of the position error ($x, y$ and $z$ components) is shown. It can be noted that the mean error at steady state, in the last 40 s, is about 7 cm. These data need to be be related to the low accuracy and to the big noise that affect the sonar's data.

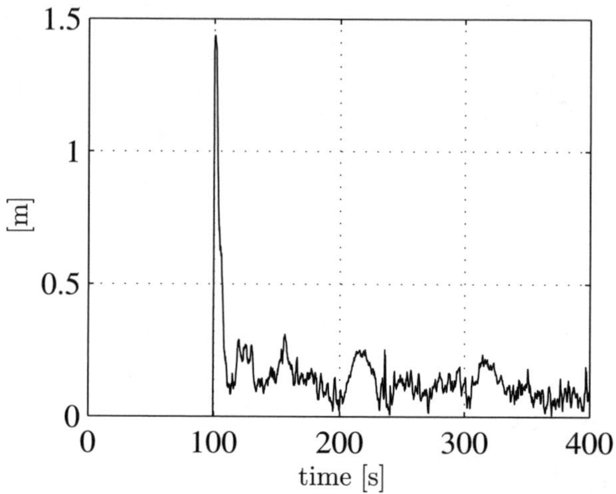


**Fig. 5.6.** Experiments of adaptive control on ODIN. 2-norm of the 3D position errors in the inertial frame

## 5.4 Experiments of Fault Tolerance to Thrusters' Fault

The desired trajectory is the same shown in the previous Section. The experiments were run several times simulating different thruster's fault. The fault was simulated via software just imposing zero voltage to the relevant thrusters. It is worth noticing that, while the desired trajectory is always the same, some minor differences arise among the different experiments due to different factors: the presence of strong noise on the sonar that affected the $xy$ movement (see [215] for details), the presence of a pool current. Meanwhile the controller guarantees a good tracking performance tolerant to the occurrence of thruster's fault and it is robust with respect to the described disturbances.

In next plots the movement of the vehicle without fault and with 2 faults are reported. The two faults arise as follow: at $t = 260\,\mathrm{s}$, one horizontal thruster is off, at $t = 300\,\mathrm{s}$ also the corresponding vertical thruster is off. We chose to test a fault of the same side thrusters because this appears to be the worst situation. Notice that the control law is different from the control law tested in the previous Section. For details, see [232, 233, 234, 251].

In Figure 5.7 the behavior of the vehicle along $xy$ without and with fault is reported. It can be noted that, as for the previous experiments, the vehicle is controlled only after 100 s in order for the sonar to work properly and wait for the transient of the position filter [216]. From the right plot in Figure 5.7 we can see that the first fault, at $t = 260\,\mathrm{s}$, it does not affect the tracking error while for the second, at $t = 300\,\mathrm{s}$, it causes only a small perturbation that is fully recovered by the controller after a transient. Some comments are required for the sonar based error: during the experiment with fault it has been possible to see a small perturbation in the $xy$ plane but, according to the data, this movement has been of 60 cm in 0.2 s, this is by far a wrong data caused by the sonar filter.

In Figure 5.8 the behavior of the vehicle along $z$ without and with fault is reported. It can be noted that the vehicle tracks the trajectory with the same error with and without thrusters' faults. Since the control law was designed in 6 DOFs, the difference in the behavior between the horizontal plane, where a small perturbation has been observed, and the vertical plane is caused mainly by the different characteristics of the sensors rather than by the controller itself. In other words, we can expect this nice behavior also in the horizontal plane if a more effective sensorial system would be available.

Another comment need to be done about the depth data. In Figure 5.3 the plot of the depth for a different control law is shown. It could appear that the tracking performance is better in the last experiment shown in Figure 5.8. However it has to be underlined that the experiments are expensive in terms of time and of people involved, the latter are simply successive to the first, the gains, thus, are better tuned.

In Figure 5.9 the attitude without and with fault is reported. It can be noted that, in case of fault, only after when the two faults arise simultaneously

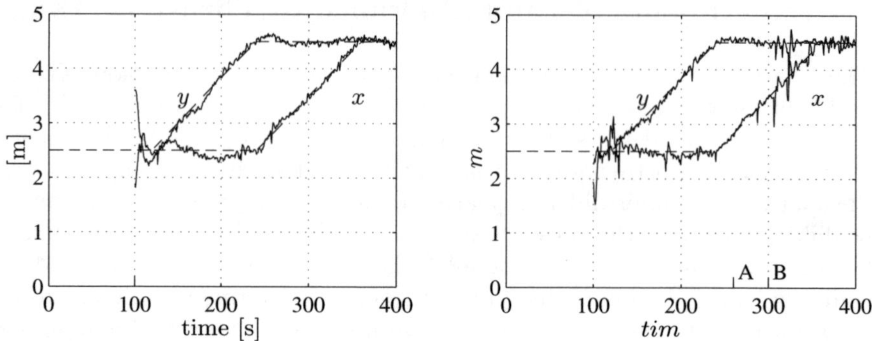


**Fig. 5.7.** Experiments of fault tolerant control on ODIN. Left: vehicle movement along $xy$ without thrusters' faults. Right: vehicle movement along $xy$ with thrusters' faults. A, first fault at $t = 260\,\mathrm{s}$ at one horizontal thruster. B, second fault at $t = 300\,\mathrm{s}$ at one vertical thruster

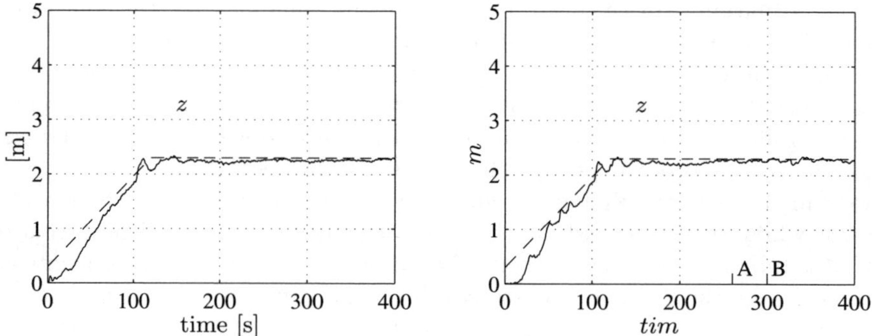


**Fig. 5.8.** Experiments of fault tolerant control on ODIN. Left: vehicle movement along $z$ without thrusters' faults. Right: vehicle movement along $z$ with thrusters' faults. A, first fault at $t = 260\,\mathrm{s}$ at one horizontal thruster. B, second fault at $t = 300\,\mathrm{s}$ at one vertical thruster

there is a significant transient in the yaw angle that is quick recovered. Comments similar to that done for the linear error in the horizontal plane could be done.

In Figure 5.10 the voltages of the vertical and horizontal thrusters are reported. We can see that when the thrusters are off, the desired force/moment on the vehicle are redistributed on the working thrusters. From these plots we can also appreciate the different noise on the control caused by the different sensors; due to the null roll and pitch angles, in fact, the vertical thrusters work mainly to track the vertical direction using the depth sensor while the horizontal are mainly use to move the vehicle on the plane.

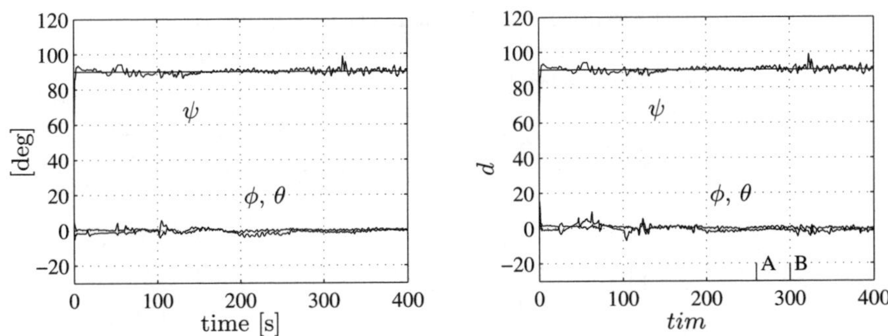


**Fig. 5.9.** Experiments of fault tolerant control on ODIN. Left: roll, pitch and yaw without thrusters' faults. Right: roll, pitch and yaw with thrusters' faults. A, first fault at $t = 260\,\mathrm{s}$ at one horizontal thruster. B, second fault at $t = 300\,\mathrm{s}$ at one vertical thruster

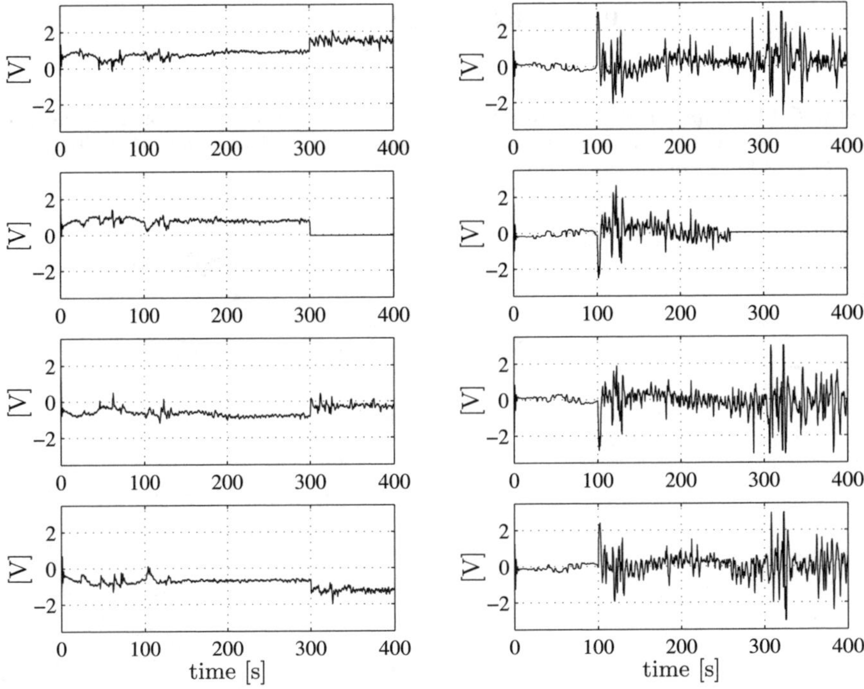


**Fig. 5.10.** Experiments of fault tolerant control on ODIN. Left: voltages at vertical thrusters. Right: voltages at horizontal thrusters

In Figure 5.11 the 2-norm of the linear error is reported. It can be noted that the first fault does not affect the error while for the second there is a small transient that is recovered.

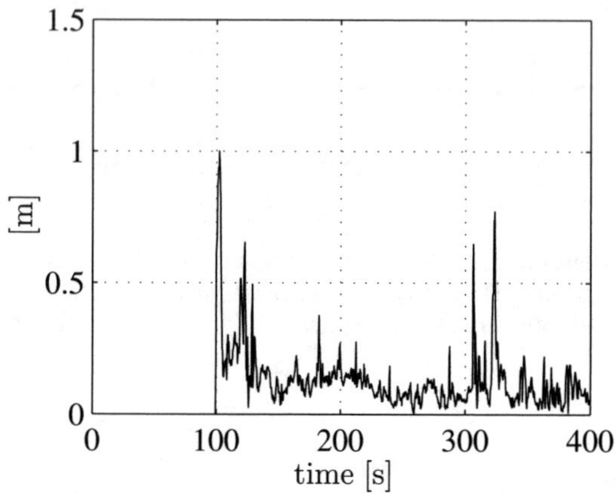


**Fig. 5.11.** Experiments of fault tolerant control on ODIN. 2-Norm of the position errors. A, first fault at $t = 260\,\mathrm{s}$ at one horizontal thruster. B, second fault at $t = 300\,\mathrm{s}$ at one vertical thruster

# 6. Kinematic Control of UVMSs

*"... mais de toutes les sciences la plus absurde, à mon avis et celle qui est la plus capable d'étouffer toute espèce de génie, c'est la géométrie. Cette science ridicule a pour objet des surfaces, des lignes et des points qui n'existent pas dans la nature. On fait passer en esprit cent mille lignes courbes entre un cercle et une ligne droite qui le touche, quoique, dans la réalité, on n'y puisse pas passer un fétu. La géométrie, en vérité, n'est qu'une mauvaise plaisanterie."*
*Voltaire, "Jeannot et Colin" 1764.*

## 6.1 Introduction

A robotic system is kinematically redundant when it possesses more degrees of freedom than those required to execute a given task. A generic manipulation task is usually given in terms of position/orientation trajectories for the end effector. In this sense, an Underwater Vehicle-Manipulator System is always kinematically redundant due to the DOFs provided by the vehicle itself. However, it is not always efficient to use vehicle thrusters to move the manipulator end effector because of the difficulty of controlling the vehicle in hovering. Moreover, due to the different inertia between vehicle and manipulator, movement of the latter is energetically more efficient. On the other hand, reconfiguration of the whole system is required when the manipulator is working at the boundaries of its workspace or close to a kinematic singularity; motion of the sole manipulator, thus, is not always possible or efficient. Also, off-line trajectory planning is not always possible in unstructured environments as in case of underwater autonomous missions.

When a manipulation task has to be performed with an UVMS, the system is usually kept in a confined space (e.g., underwater structure maintenance). The vehicle is then used to ensure station keeping. However, motion of the vehicle can be required for specific purposes, e.g., inspection of a pipeline, reconfiguration of the system, real-time motion coordination while performing end-effector trajectory tracking.

According to the above, a redundancy resolution technique might be useful to achieve system coordination in such a way as to guarantee end-effector tracking accuracy and, at the same time, additional control objectives, e.g., energy savings or increase of system manipulability. To this purpose the task priority redundancy resolution technique [195, 210] is well suited in that it allows the specification of a primary task which is fulfilled with higher priority with respect to a secondary task.

Control of end-effector position/orientation can be obtained also with dynamic control by suitably expressing the mathematical model [255, 256, 285]. This approach, successfully implemented for industrial robots, seems not to be suitable for UVMSs for two main reasons: first, in underwater environment the dynamic parameters are usually poorly known; second, the redundancy of the system is not exploited. Some, approaches, moreover, are specifically designed for a 6-DOFs manipulator only.

By limiting our attention to UVMSs, few papers have addressed the problem of inverse kinematics resolution. Reference [237] proposes a local motion planner solved in parallel by a distributed search; this provides an iterative algorithm for an approximate solution. In [20], a task priority approach has been proposed aimed at fulfilling secondary tasks such as reduction of fuel consumption, improvement of system manipulability, and obstacle avoidance. This approach has been further integrated with a fuzzy approach in [24, 25, 26, 29] and it will be deeply analyzed in this Chapter. In [249, 250], a second-order inverse kinematics approach is developed to reduce the total hydrodynamic drag forces of the system. Simulations results are performed on a 6-link vehicle carrying a 3-link manipulator. The same authors also developed a dynamic-based algorithm in [235] that generates the joint trajectories by taking into account the natural frequencies of the two subsystems: vehicle and manipulator; the task-space trajectory is represented by Fourier series and suitably projected on the subsystems. Reference [252] reports an adaptive dynamic controller that uses, as reference trajectory, the output of a first-order inverse kinematics algorithm aimed at satisfying joint limits. In [163], the Authors develop two cost functions devoted at increase the manipulability and respect the joint limits to be used in a task priority approach. In [160, 161], a genetic algorithm-based motion planner is proposed; dividing the workspace in cells, the presence of obstacles and the minimization of the drag forces are taken into consideration. Finally, in [73] a distributed kinematic control was developed for coordination of a multi-manipulator system mounted under a free-flying base such as, e.g., an underwater vehicle; the case of a possible under-actuated vehicle is explicitly taken into account.

## 6.2 Kinematic Control

A manipulation task is usually given in terms of position and orientation trajectory of the end effector. The objective of kinematic control is to find suitable vehicle/joint trajectories $\boldsymbol{\eta}(t)$, $\boldsymbol{q}(t)$ that correspond to a desired end-effector trajectory $\boldsymbol{\eta}_{ee,d}(t)$. The output of the inverse kinematics algorithm $\boldsymbol{\eta}_r(t)$, $\boldsymbol{q}_r(t)$ provides the reference values to the control law of the

UVMS. This control law will be in charge of computing the driving forces aimed at tracking the reference trajectory for the system while counteracting dynamic effects, external disturbances, and modeling errors.

Equation (2.64)

$$\boldsymbol{\eta}_{ee} = \boldsymbol{k}(\boldsymbol{\eta}, \boldsymbol{q}) \tag{2.64}$$

is invertible only for specific kinematic structures with fixed base. Moreover, the complexity of the relation and the number of solutions, i.e., different joint configurations that correspond to the same end-effector posture, increases with the degrees of freedom. As an example, a simple two-link planar manipulator with fixed base admits two different solutions for a given end-effector position, while up to 16 solutions can be found in the case of 6-DOFs structures. At differential level, however, the relation between joints and end-effector velocities is much more tractable and a theory of kinematic control has been established aimed at solving inverse kinematics of generic kinematic structures.

Equation (2.68)

$$\dot{\boldsymbol{x}}_E = \begin{bmatrix} \dot{\boldsymbol{\eta}}_{ee1} \\ \boldsymbol{R}_n^I \boldsymbol{\nu}_{ee2} \end{bmatrix} = \boldsymbol{J}(\boldsymbol{R}_B^I, \boldsymbol{q})\boldsymbol{\zeta} \tag{2.68}$$

maps the $(6+n)$-dimensional vehicle/joint velocities into the $m$-dimensional end-effector task velocities. If the UVMS has more degrees of freedom than those required to execute a given task, i.e., if $(6+n) > m$, the system is redundant with respect to the specific task and the equations (2.64)–(2.68) admit infinite solutions. Kinematic redundancy can be exploited to achieve additional task, beside the given end-effector task. In the following, the typical case $(6+n) \geq m$ will be considered.

The configurations at which $\boldsymbol{J}$ is rank deficient, i.e., $\text{rank}(\boldsymbol{J}) < m$, are termed kinematic singularities. Kinematic singularities are of great interest for several reasons; at a singularity, in fact,

- The mobility of the structure is reduced. If the manipulator is not redundant, this implies that it is not possible to give an arbitrary motion to the end effector;
- Infinite solutions to the inverse kinematics problem might exist;
- *Close* to a kinematic singularity at small task velocities can correspond large joint velocities.

Notice that, in case of UVMS, the Jacobian has always full rank due to the mobility of the vehicle, i.e., a rigid body with 6-DOFs. However, as it will be shown in next Sections, movement of the vehicle has to be avoided when unnecessary.

### Pseudoinverse

The simplest way to invert the mapping (2.68) is to use the pseudoinverse of the Jacobian matrix [301]

$$\boldsymbol{\zeta}_r = \boldsymbol{J}^\dagger(\boldsymbol{\eta}, \boldsymbol{q})\dot{\boldsymbol{x}}_{E,d}\,, \tag{6.1}$$

where $\dot{\boldsymbol{x}}_{E,d}$ is the end-effector task defined in (2.68) and

$$\boldsymbol{J}^\dagger(\boldsymbol{\eta}, \boldsymbol{q}) = \boldsymbol{J}^{\mathrm{T}}(\boldsymbol{\eta}, \boldsymbol{q})\left(\boldsymbol{J}(\boldsymbol{\eta}, \boldsymbol{q})\boldsymbol{J}^{\mathrm{T}}(\boldsymbol{\eta}, \boldsymbol{q})\right)^{-1}.$$

This solution corresponds to the minimization of the vehicle/joint velocities in a least-square sense [254], i.e., of the function:

$$E = \frac{1}{2}\boldsymbol{\zeta}^{\mathrm{T}}\boldsymbol{\zeta}\,.$$

Notice that subscript $r$ in $\boldsymbol{\zeta}_r$ stands for *reference value*, meaning that those velocities are the desired values for the low-level motion control of the manipulator (see also Figure 6.1, where a closed-loop inverse kinematics, detailed in next Subsections, is sketched). It is possible to minimize a weighted norm of the vehicle/joint velocities

$$E = \frac{1}{2}\boldsymbol{\zeta}^{\mathrm{T}}\boldsymbol{W}\boldsymbol{\zeta}$$

leading to the *weighted* pseudoinverse:

$$\boldsymbol{J}^\dagger_{\boldsymbol{W}} = \boldsymbol{W}^{-1}\boldsymbol{J}^{\mathrm{T}}\left(\boldsymbol{J}\boldsymbol{W}^{-1}\boldsymbol{J}^{\mathrm{T}}\right)^{-1}. \tag{6.2}$$

With this approach, however, the problem of handling kinematic singularities is not addressed and their avoidance cannot be guaranteed.

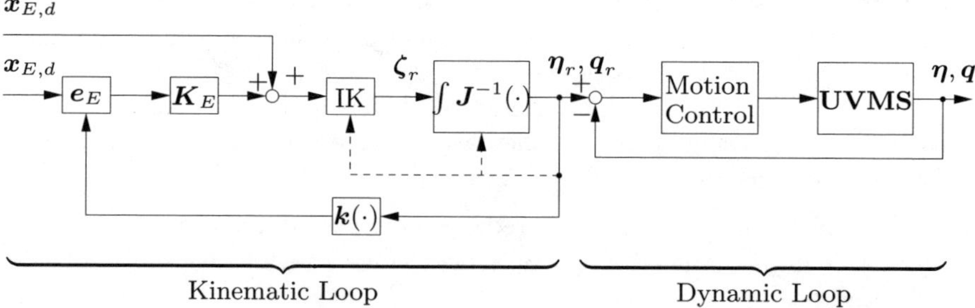


**Fig. 6.1.** Kinematic and dynamic loops. The block labeled $\boldsymbol{e}_E$ is defined by equation (6.8)

### Augmented Jacobian

Another approach to redundancy resolution is the augmented Jacobian [107]. In this case, a constraint task is added to the end-effector task as to obtain a square Jacobian matrix which can be inverted.

The main drawback of this technique is that new singularities might arise in configurations in which the end-effector Jacobian $\boldsymbol{J}$ is still full rank. Those singularities, named *algorithmic singularities*, occur when the additional task does cause conflict with the end-effector task.

A similar approach, with the same drawback, is the *extended Jacobian* approach.

### Task Priority Redundancy Resolution

By solving (2.68) in terms of a minimization problem of the quadratic cost function $\boldsymbol{\zeta}^{\mathrm{T}}\boldsymbol{\zeta}$ the general solution [191] is given:

$$\boldsymbol{\zeta}_r = \boldsymbol{J}^{\dagger}(\boldsymbol{\eta},\boldsymbol{q})\dot{\boldsymbol{x}}_{E,d} + \left(\boldsymbol{I}_N - \boldsymbol{J}^{\dagger}(\boldsymbol{\eta},\boldsymbol{q})\boldsymbol{J}(\boldsymbol{\eta},\boldsymbol{q})\right)\boldsymbol{\zeta}_a\,, \tag{6.3}$$

where $N = 6+n$ and $\boldsymbol{\zeta}_a \in \mathbb{R}^{6+n}$ is an arbitrary vehicle/joint velocity vector.

It can be recognized that the operator $\left(\boldsymbol{I}_N - \boldsymbol{J}^{\dagger}(\boldsymbol{\eta},\boldsymbol{q})\boldsymbol{J}(\boldsymbol{\eta},\boldsymbol{q})\right)$ projects a generic joint velocity vector in the null space of the Jacobian matrix. This corresponds to generating an internal motion of the manipulator arm that does not affect the end-effector motion.

Solution (6.3) can be seen in terms of projection of a secondary task, described by $\boldsymbol{\zeta}_a$, in the null space of the higher priority primary task, i.e., the end-effector task. A first possibility is to choose the vector $\boldsymbol{\zeta}_a$ as the gradient of a scalar objective function $H(\boldsymbol{q})$ in order to achieve a local minimum [191]:

$$\boldsymbol{\zeta}_a = -k_H \nabla H(\boldsymbol{q})\,, \tag{6.4}$$

where $k_H$ is a scalar gain factor. Another possibility is to chose a primary task $\boldsymbol{x}_{p,d} \in \mathbb{R}^m$ and a correspondent Jacobian matrix $\boldsymbol{J}_p(\boldsymbol{q}) \in \mathbb{R}^{m\times(6+n)}$

$$\dot{\boldsymbol{x}}_{p,d} = \boldsymbol{J}_p(\boldsymbol{q})\boldsymbol{\zeta}\,.$$

and to design a secondary task $\boldsymbol{x}_{s,d} \in \mathbb{R}^r$ and a correspondent Jacobian matrix $\boldsymbol{J}_s(\boldsymbol{q}) \in \mathbb{R}^{r\times(6+n)}$:

$$\dot{\boldsymbol{x}}_{s,d} = \boldsymbol{J}_s(\boldsymbol{q})\boldsymbol{\zeta}\,.$$

for which the vector of joint velocity is then given by [195, 210]:

$$\boldsymbol{\zeta}_r = \boldsymbol{J}_p^{\dagger}\dot{\boldsymbol{x}}_{p,d} + \left(\boldsymbol{J}_s\left(\boldsymbol{I}_N - \boldsymbol{J}_p^{\dagger}\boldsymbol{J}_p\right)\right)^{\dagger}\left(\dot{\boldsymbol{x}}_{s,d} - \boldsymbol{J}_s\boldsymbol{J}_p^{\dagger}\dot{\boldsymbol{x}}_{p,d}\right)\,. \tag{6.5}$$

However, for this solution too, the problem of the algorithmic singularities still remains unsolved. In this case, it is possible to experience an algorithmic singularity when $\boldsymbol{J}_s$ and $\boldsymbol{J}_p$ are full rank but the matrix $\boldsymbol{J}_s\left(\boldsymbol{I}_N - \boldsymbol{J}_p^{\dagger}\boldsymbol{J}_p\right)$ looses rank. Extension of the approach to several tasks for highly redundant systems can be achieved by generalization of (6.5), as described in [263].

### Singularity-Robust Task Priority Redundancy Resolution

A robust solution to the occurrence of the algorithmic singularities is based on the following mapping [78]:

$$\boldsymbol{\zeta}_r = \boldsymbol{J}_p^{\dagger}(\boldsymbol{\eta},\boldsymbol{q})\dot{\boldsymbol{x}}_{p,d} + \left(\boldsymbol{I}_N - \boldsymbol{J}_p^{\dagger}(\boldsymbol{\eta},\boldsymbol{q})\boldsymbol{J}_p(\boldsymbol{\eta},\boldsymbol{q})\right)\boldsymbol{J}_s^{\dagger}(\boldsymbol{\eta},\boldsymbol{q})\dot{\boldsymbol{x}}_{s,d}\,. \tag{6.6}$$

This algorithm has a clear geometrical interpretation: the two tasks are separately inverted by the use of the pseudoinverse of the corresponding Jacobian; the vehicle/joint velocities associated with the secondary task are further projected in the null space of the primary task $\boldsymbol{J}_p$. Similarly to [263], extension to several tasks for highly redundant systems can be easily achieved by recursive application of (6.6).

### Damped Least-Squares Inverse Kinematics Algorithms

The problem of inverting ill-conditioned matrices that might occur with all the above algorithms can be avoided by resorting to the damped least-square inverse given by [209]:

$$\boldsymbol{J}^{\#}(\boldsymbol{\eta},\boldsymbol{q}) = \boldsymbol{J}^{\mathrm{T}}(\boldsymbol{\eta},\boldsymbol{q})\left(\boldsymbol{J}(\boldsymbol{\eta},\boldsymbol{q})\boldsymbol{J}^{\mathrm{T}}(\boldsymbol{\eta},\boldsymbol{q}) + \lambda^2\boldsymbol{I}_m\right)^{-1},$$

where $\lambda \in \mathbb{R}$ is a damping factor.

In this case, the introduction of a damping factor allows solving the problem from the numerical point of view but, on the other hand, it introduces a reconstruction error in all the velocity components. Better solutions can be found with variable damping factors or damped least-squares with numerical filtering [196, 209].

### Closed-Loop Inverse Kinematic Algorithms

The numerical implementation of the above algorithms would lead to a numerical drift when obtaining vehicle/joint positions by integrating the vehicle/joint velocities. A closed loop version of the above equations can then be adopted. By considering as primary task the end-effector position/orientation, (6.6), as an example, would become:

$$\begin{aligned}\boldsymbol{\zeta}_r = {} & \boldsymbol{J}^{\dagger}(\boldsymbol{\eta},\boldsymbol{q})\left(\dot{\boldsymbol{x}}_{E,d} + \boldsymbol{K}_E\boldsymbol{e}_E\right) + \\ & + \left(\boldsymbol{I}_N - \boldsymbol{J}^{\dagger}(\boldsymbol{\eta},\boldsymbol{q})\boldsymbol{J}(\boldsymbol{\eta},\boldsymbol{q})\right)\boldsymbol{J}_s^{\dagger}(\boldsymbol{\eta},\boldsymbol{q})\left(\dot{\boldsymbol{x}}_{s,d} + \boldsymbol{K}_s\boldsymbol{e}_s\right),\end{aligned} \tag{6.7}$$

where $\boldsymbol{e}_E$ and $\boldsymbol{e}_s$ are the numerical reconstruction errors and $\boldsymbol{K}_E \in \mathbb{R}^{m\times m}$ and $\boldsymbol{K}_s \in \mathbb{R}^{r\times r}$ are design matrix gains to be chosen so as to ensure convergence to zero of the corresponding errors.

If the task considered is position control, its reconstruction error is simply given by the difference between the desired and the reconstructed values. In case of the orientation, however, care in the definition of such error is required to ensure convergence to the desired value. In this work, the quaternion attitude representation is used [246]; the vector $\boldsymbol{e}_E$ for the task defined in (2.68) is then given by [60, 80]:

$$\boldsymbol{e}_E = \begin{bmatrix} \boldsymbol{\eta}_{\mathrm{ee1},d} - \boldsymbol{\eta}_{\mathrm{ee1},r} \\ \eta_r \boldsymbol{\varepsilon}_d - \eta_d \boldsymbol{\varepsilon}_r - \boldsymbol{S}(\boldsymbol{\varepsilon}_d)\boldsymbol{\varepsilon}_r \end{bmatrix}, \tag{6.8}$$

where $\mathcal{Q}_d = \{\eta_d, \boldsymbol{\varepsilon}_d\}$ and $\mathcal{Q}_r = \{\eta_r, \boldsymbol{\varepsilon}_r\}$ are the desired and reference attitudes expressed by quaternions, respectively, and $\boldsymbol{S}(\cdot)$ is the matrix operator performing the cross product.

The obtained $\boldsymbol{\zeta}_r$ can then be used to compute the position and orientation of the vehicle $\boldsymbol{\eta}_r$ and the manipulator configuration $\boldsymbol{q}_r$:

$$\begin{aligned} \begin{bmatrix} \boldsymbol{\eta}_r(t) \\ \boldsymbol{q}_r(t) \end{bmatrix} &= \int_0^t \begin{bmatrix} \dot{\boldsymbol{\eta}}_r(\sigma) \\ \dot{\boldsymbol{q}}_r(\sigma) \end{bmatrix} d\sigma + \begin{bmatrix} \boldsymbol{\eta}(0) \\ \boldsymbol{q}(0) \end{bmatrix} \\ &= \int_0^t \boldsymbol{J}_k^{-1}(\sigma)\boldsymbol{\zeta}_r(\sigma) d\sigma + \begin{bmatrix} \boldsymbol{\eta}(0) \\ \boldsymbol{q}(0) \end{bmatrix}. \end{aligned} \tag{6.9}$$

As customary in kinematic control approaches, the output of the above inverse kinematics algorithm provides the reference values to the dynamic control law of the vehicle-manipulator system (see Figure 6.1). This dynamic control law will be in charge of computing the driving forces, i.e., the vehicle thrusters and the manipulator torques. The kinematic control algorithm is independent from the dynamic control law as long as the latter is a vehicle/joint space-based control, i.e., it requires as input the reference vehicle-joint position and velocity. In the literature number of such control laws have been proposed that are suitable to be used within the proposed kinematic control approach; a literature survey is presented in the next Chapter. For the seek of simplicity, in the Figure, only the primary task is shown.

Remarkably, all those inverse kinematics approaches are suitable for real-time implementation. Of course, depending on the specific algorithm, a different computational load is required [78].

### Transpose of the Jacobian

A simple algorithm, conceptually similar to the closed loop approach, is given by the use of the transpose of the Jacobian. In this case, the joint velocities are given by [254]:

$$\boldsymbol{\zeta}_r = \boldsymbol{J}^{\mathrm{T}} \boldsymbol{K}_E \boldsymbol{e}_E,$$

where a direct relationship between the reference joint velocities and the end-effector reconstruction error is obtained.

## 6.3 The Drag Minimization Algorithm

In 1999 N. Sarkar and T.K. Podder ([249, 250]) suggest to use the system redundancy in order to minimize the total hydrodynamic drag. Roughly speaking the proposed kinematic control generates a coordinate vehicle/manipulator motion so that the resulting trajectory incrementally reduce the total drag encountered by the system.

The second-order version of (6.3) is given by

$$\dot{\boldsymbol{\zeta}}_r = \boldsymbol{J}^{\dagger}(\boldsymbol{\eta}, \boldsymbol{q}) \left( \ddot{\boldsymbol{x}}_{E,d} - \dot{\boldsymbol{J}}(\boldsymbol{\eta}, \boldsymbol{q})\boldsymbol{\zeta} \right) + \left( \boldsymbol{I}_N - \boldsymbol{J}^{\dagger}(\boldsymbol{\eta}, \boldsymbol{q})\boldsymbol{J}(\boldsymbol{\eta}, \boldsymbol{q}) \right) \dot{\boldsymbol{\zeta}}_a , \tag{6.10}$$

where, again, the vector $\dot{\boldsymbol{\zeta}}_a$ is arbitrary and can be used to optimize some performance criteria that can be chosen, similarly to eq. (6.4), as

$$\dot{\boldsymbol{\zeta}}_a = -k_H \nabla H(\boldsymbol{q}) . \tag{6.11}$$

The Authors propose the following scalar objective function

$$H(\boldsymbol{q}) = \boldsymbol{D}^{\mathrm{T}}(\boldsymbol{q}, \boldsymbol{\zeta}) \boldsymbol{W} \boldsymbol{D}(\boldsymbol{q}, \boldsymbol{\zeta}) \tag{6.12}$$

where $\boldsymbol{D}(\boldsymbol{q}, \boldsymbol{\zeta})$ is the damping matrix and $\boldsymbol{W} \in \mathrm{I\!R}^{N \times N}$ is a positive definite weight matrix. By properly selecting the weight matrix it is possible to shape the influence of the drag of the individual components on the total system's drag. A possible choice for $\boldsymbol{W}$ is a diagonal matrix [250].

The method is tested in detailed simulations where the drag coefficients are supposed to be known. As noticed by the Authors, in practical situations, the drag coefficient need to be identified and this can not be an easy task; it must be noted, however, that theoretical drag coefficient are available in the literature for the most common shapes. As a first approximation it is possible to model the vehicle as an ellipsoid and the manipulator arms as cylinders. The proposed method, thus, even being approximated, provides information of wider interest.

It is worth noticing that drag minimization has been the objective of several approaches, even not based on kinematics control, such as, e.g., [160, 161]; in [164], the Authors propose to utilize a genetic algorithm to be trained over a periodic motion in order to estimate the trajectory's parameters that minimize the directional drag force in the task space.

## 6.4 The Joint Limits Constraints

N. Sarkar, J. Yuh and T.K. Podder, in 1999 ([252]) take into account the problem of handling the manipulators' joints limits.

The Authors propose to suitably modifying the weight matrix $\boldsymbol{W}$ in Eq. (6.2). In detail, $\boldsymbol{W}$ is chosen as a diagonal matrix the entries of which are related to a proper function. The approach might also take into account the vehicle position, but, for seek of simplicity, let us consider only the joints positions by defining

$$H(\boldsymbol{q}) = \sum_{i=1}^{n} \frac{1}{c_i} \frac{q_{i,\max} - q_{i,\min}}{(q_{i,\max} - q_i)(q_i - q_{i,\min})}$$

with $c_i > 0$ and the subscript max and min that obviously denotes the two joint limits. This function ([252]) inherits the concepts developed in [44].

Its partial derivative with respect to the joint positions is given by

$$\frac{\partial H(\boldsymbol{q})}{\partial q_i} = \frac{1}{c_i} \frac{(q_{i,\max} - q_{i,\min})(2q_i - q_{i,\max} - q_{i,\min})}{(q_{i,\max} - q_i)^2 (q_i - q_{i,\min})^2} .$$

The elements of the weight matrix are then defined as

$$W_{i,i} = 1 + \left\| \frac{\partial H(\boldsymbol{q})}{\partial q_i} \right\| ,$$

in fact, it can be easily observed that the element goes to 1 when the joint is in the center of its allowed range and goes to infinity when the joint is approaching its limit.

As a further improvement it is possible to relate the weight also to the direction of the joint by defining

$$W_{i,i} = \begin{cases} 1 + \left\| \frac{\partial H(\boldsymbol{q})}{\partial q_i} \right\| & \text{if } \Delta \left\| \frac{\partial H(\boldsymbol{q})}{\partial q_i} \right\| > 0 \\ 1 & \text{if } \Delta \left\| \frac{\partial H(\boldsymbol{q})}{\partial q_i} \right\| \leq 0 \end{cases}$$

In Section 6.6, the joint limits are part of a number of tasks handled with a fuzzy approach. In [163], B.H. Jun, P.M. Lee and J. Lee propose a first order task priority approach with the optimization of specific cost functions developed for the ROV named KORDI. The joints constraints are taken into account also.

## 6.5 Singularity-Robust Task Priority

To achieve an effective coordinated motion of the vehicle and manipulator while exploiting the redundant degrees of freedom available, G. Antonelli and S. Chiaverini, in [20], resort to the singularity-robust task priority redundancy resolution technique. The velocity vector $\boldsymbol{\zeta}_r$ is then computed as shown in (6.7).

In the case of a UVMS, the primary task vector will usually include the end-effector task vector, while the secondary task vector might include the vehicle position coordinates. This choice is aimed at achieving station keeping of the vehicle as long as the end-effector task can be fulfilled with the sole manipulator arm. It is worth noticing that this approach is conceptually similar to the macro-micro manipulator approach [107]; the main difference is that the latter requires dynamic compensation of the whole system while the former is based on a kinematic control approach. This is advantageous for underwater applications in which uncertainty on dynamic parameters is experienced.

### Simulations

Let consider a 9-DOF UVMS constituted by the Naval Postgraduate School AUV Phoenix [145] with a 3-DOF planar manipulator arm. For the sake of clarity, in this first group of simulations, the attention we was restricted to planar tasks described in the plane of the manipulator, that is mounted horizontally. Therefore, let us consider six degrees of freedom in the system which is characterized by the three vehicle coordinates $x$, $y$, $\psi$, and the three end-effector coordinates $x_E$, $y_E$, $\psi_E$ all expressed in a earth-fixed frame; the three vehicle coordinates $z$,$\theta$,$\phi$ are assumed to be constant. A sketch of the system as seen from the bottom is reported in Figure 6.2, where the earth-fixed, body-fixed, and end-effector reference frames are also shown.

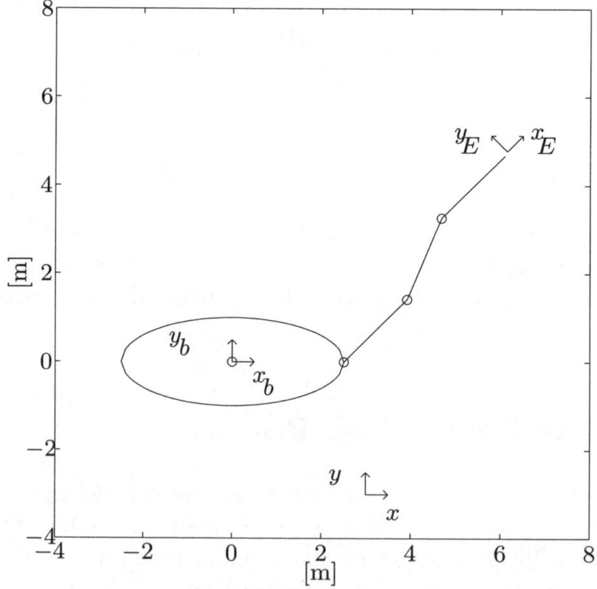


**Fig. 6.2.** Sketch of the simulated UVMS seen from the bottom

A station keeping task is considered as first case study. During station keeping, the thrusters must react to the ocean current the strength of which exhibits a quadratic dependence on the relative velocity [127]. However, if the model of the NPS AUV [145] is considered it can be easily recognized that, the drag in the $\boldsymbol{x}_b$ direction is really smaller with respect to the drag in the $\boldsymbol{y}_b$. This suggests to attempt keeping the fore aft direction of the vehicle aligned with the ocean current in order to reduce energy consumption.

To implement the proposed approach, it is proposed to consider as primary task both the end-effector position+orientation and the vehicle orientation, i.e.,

$$\boldsymbol{x}_p = [\, x_E \quad y_E \quad \psi_E \quad \psi \,]^{\mathrm{T}} ,$$

and as secondary task the vehicle position, i.e.,

$$\boldsymbol{x}_s = [\, x \quad y \,]^{\mathrm{T}} .$$

In a first simulation the end-effector has to maintain its position and orientation while the vehicle will change its orientation to minimize the effect of the ocean current; it is desired to keep the vehicle position constant, if possible. Let the initial configuration of the vehicle be

$$\begin{aligned} x &= 0 \quad \mathrm{m} , \\ y &= 0 \quad \mathrm{m} , \\ \psi &= 0 \quad \mathrm{rad} , \end{aligned}$$

and the manipulator joint angles be

$$\boldsymbol{q} = [\, 1.47 \quad -1 \quad 0.3 \,]^{\mathrm{T}} \quad \mathrm{rad} ,$$

corresponding to the end-effector location

$$\begin{aligned} x_E &= 5.92 \quad \mathrm{m} , \\ y_E &= 4.29 \quad \mathrm{m} , \\ \psi_E &= 0.77 \quad \mathrm{rad} . \end{aligned}$$

The desired values of the end-effector variables and vehicle position are coincident with their initial value. The desired final value of the vehicle orientation is 0.78 rad as given, e.g., by a current sensor; the time history of the desired value $\psi_d$ is computed according to a quintic polynomial interpolating law with null initial and final velocities and accelerations and a duration of 10 s. The algorithm's gains are

$$\begin{aligned} \boldsymbol{K}_p &= \mathrm{diag}\{10, 10, 10, 1\} , \\ \boldsymbol{K}_s &= \mathrm{diag}\{2, 2\} . \end{aligned}$$

The simulation results are reported in Figure 6.3 and 6.4. It can be recognized that the task is successfully executed, in that the end-effector location and vehicle position are held while the vehicle body is re-oriented to align with the ocean current. Remarkably, the obtained vehicle reference trajectory is smooth.

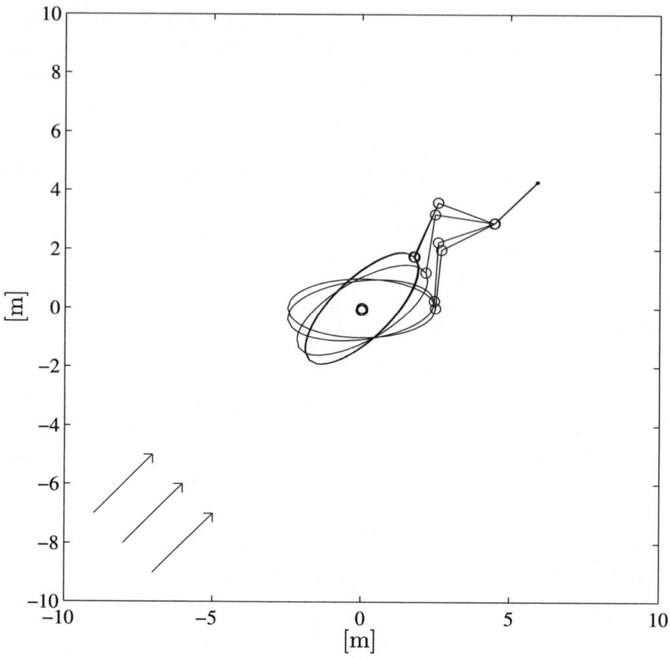


**Fig. 6.3.** Re-orientation of the vehicle body with fixed end-effector location

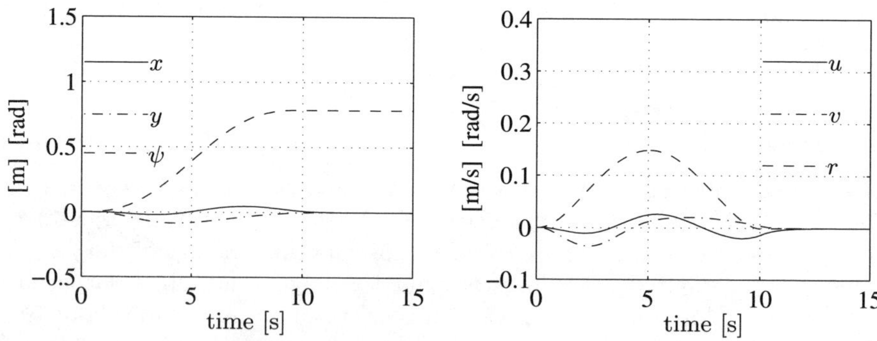


**Fig. 6.4.** Time history of vehicle position and velocity variables when the vehicle reconfigure itself while keeping a fixed end-effector position/orientation

A second simulation, starting from the same initial system configuration, considers an end-effector trajectory that cannot be tracked by sole manipulator motion. Therefore, the vehicle must be moved to allow the manipulator end-effector to track its reference trajectory. Also in this simulation, alignment of the vehicle fore aft direction with the ocean current is pursued.

The desired end-effector trajectory is a straight-line motion starting from the same initial location as in the previous simulation and lasting at the final location

$$
\begin{aligned}
x_E &= 8.00 \quad \text{m}\,,\\
y_E &= 9.00 \quad \text{m}\,,\\
\psi_E &= 0.78 \quad \text{rad}\,.
\end{aligned}
$$

The path is followed according to a quintic polynomial interpolating law with null initial and final velocities and acceleration and a duration of 10 s. The other task variables and gains are the same as in the previous simulation; remarkably, the desired values of the vehicle position variables are coincident with their initial value also in this case.

The simulation results are reported in Figure 6.5 and 6.6. It can be recognized that the primary task is successfully executed, in that the end-effector location and vehicle orientation achieve their target. On the other hand, the vehicle moves from its initial position despite the secondary task demands for station keeping. Remarkably, the obtained vehicle reference trajectory is smooth.

To show generality of the proposed approach a second case study has been developed. A drawback of the previous case study might be that the manipulator arm is almost completely stretched out when the end-effector trajectory requires large displacements going far from the vehicle body. Nevertheless, this is related to our choice to keep the position of the vehicle constant and to align the fore aft direction with the ocean current. To overcome this drawback, a different choice of the tasks to be fulfilled is necessary. In particular, the task of vehicle re-orientation might be replaced with the task of keeping the manipulator arm in dexterous configurations. To this aim, it would be possible to use a task variable expressing a manipulability measure of the manipulator arm [312]. In this simple case, it is clear that arm singularities occur when $q_2 = 0$; therefore, the use of $q_2$ as manipulability task variable would reduce the computational burden of the algorithm.

To implement the proposed approach in this second case study, both the end-effector position+orientation and the second manipulator joint variable are thus considered as primary task, i.e.

$$
\boldsymbol{x}_p = \begin{bmatrix} x_E & y_E & \psi_E & q_2 \end{bmatrix}^{\mathrm{T}}\,,
$$

and as secondary task the vehicle position, i.e.

$$
\boldsymbol{x}_s = \begin{bmatrix} x & y \end{bmatrix}^{\mathrm{T}}\,.
$$

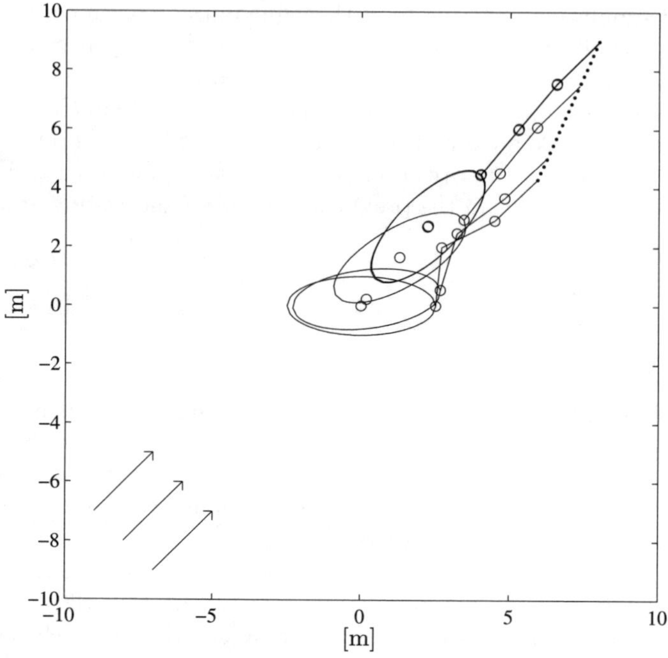

Fig. 6.5. Re-orientation of the vehicle body with given end-effector trajectory

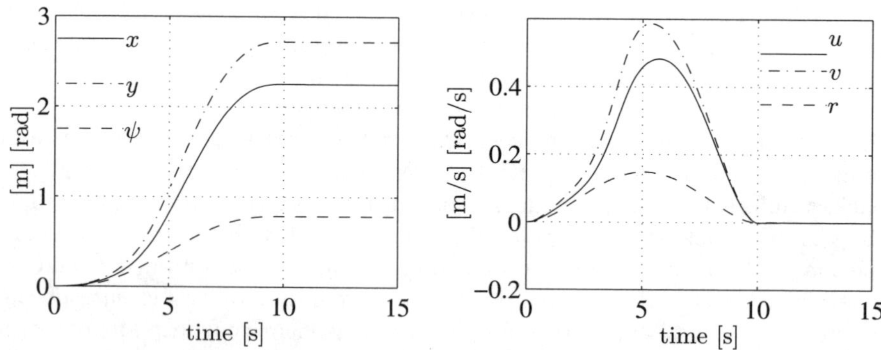


Fig. 6.6. Time history of vehicle position and velocity variables with given end-effector trajectory

Starting from the same initial system configuration as before, in the simulation the same end-effector trajectory has been assigned while manipulator joint 2 is driven far from zero; it is desired to keep the vehicle position constant, if possible.

The desired final value of $q_2$ is $-0.78$ rad; the time history of the desired value $q_{2,d}$ is computed according to a quintic polynomial interpolating law with null initial and final velocities and acceleration and a duration of 10 s. The algorithm's gains are

$$\boldsymbol{K}_p = \text{diag}\{10, 10, 10, 1\}\,, \tag{6.13}$$

$$\boldsymbol{K}_s = \text{diag}\{2, 2\}\,. \tag{6.14}$$

The simulation results are reported in Figure 6.7 and 6.8.

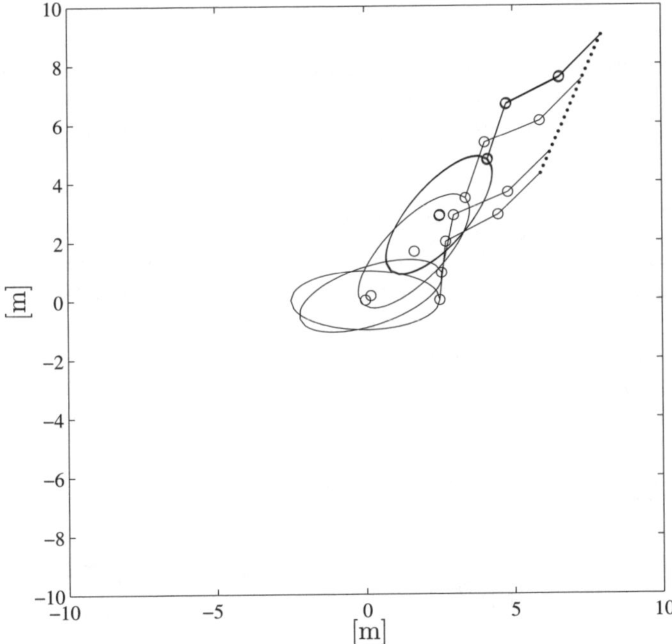


**Fig. 6.7.** Tracking of a given end-effector trajectory with manipulator dexterity

It can be recognized that the primary task is successfully executed, in that the end-effector location and manipulator joint 2 achieve their target. On the other hand, the vehicle moves despite the secondary task demands for station keeping. Remarkably, the obtained vehicle reference trajectory is smooth.

To underline the energetic difference in the station keeping task considered in the first case study when executed with and without vehicle re-orientation,

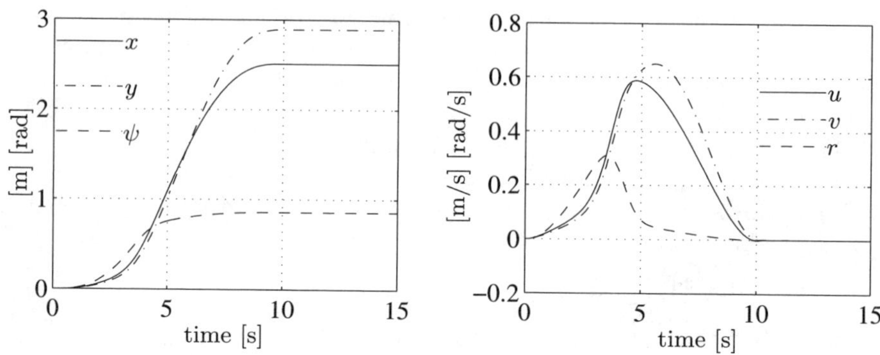


**Fig. 6.8.** Time history of vehicle position and velocity variables with given end-effector trajectory and manipulator dexterity

a third case study has been developed. Two simulations have been performed considering the full-dimensional dynamic model of the NPS AUV under a sliding mode control law [21] and the following constant and irrotational current

$$\boldsymbol{\nu}_c^I = [0.1\sqrt{2} \quad 0.1\sqrt{2} \quad 0 \quad 0 \quad 0 \quad 0]^{\mathrm{T}} .$$

In the first simulation the vehicle stays still and the generalized control forces are required to only compensate the current effect, since the NPS AUV is neutrally buoyant and $\theta = \phi = 0$. In the second simulation the vehicle moves according to the results of the first simulation in the first case study; thus, the generalized control forces are used to both move the vehicle and to compensate the current effect.

Figure 6.9 reports the time histories of the 2-norm of the control forces and moments acting on the vehicle as obtained in the two simulations.

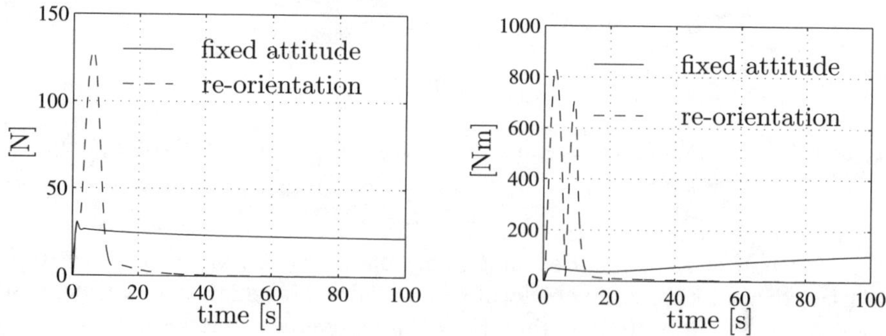


**Fig. 6.9.** Time history of the 2-norm of the vehicle forces (left) and moments (right)

It is easy to recognize that during the reconfiguration the proposed solution is more energy-consuming than the fixed-attitude solution; nevertheless, after the re-orientation has been achieved, the energy consumption required by the proposed technique is negligible. Therefore, the proposed solution becomes the more attractive the longer is the duration of the manipulation task.

For the sake of argument, Table 6.1 reports the time integral of the 2-norms of force and moment obtained in the two simulations over a 100 s task duration.

**Table 6.1.** Time integral of the force and moment 2-norms: a) without re-orientation; b) with re-orientation

| | a | b |
|---|---|---|
| $\int \|\|f\|\|$ | 2300 | 800 |
| $\int \|\|m\|\|$ | 9500 | 5800 |

## 6.6 Fuzzy Inverse Kinematics

Because of the different inertia characteristics of the vehicle and of the manipulator, it would be preferable to perform fast motions of small amplitude by means of the manipulator while leaving to the vehicle the execution of slow gross motions. This might be achieved by adopting the weighted pseudoinverse of Eq. (6.2) with the $(6+n) \times (6+n)$ matrix $\boldsymbol{W}^{-1}$

$$\boldsymbol{W}^{-1}(\beta) = \begin{bmatrix} (1-\beta)\boldsymbol{I}_6 & \boldsymbol{O}_{6\times n} \\ \boldsymbol{O}_{n\times 6} & \beta\boldsymbol{I}_n \end{bmatrix}, \tag{6.15}$$

where $\beta$ is a weight factor belonging to the interval $[0,1]$ such that $\beta = 0$ corresponds to sole vehicle motion and $\beta = 1$ to sole manipulator motion.

During the task execution, setting a constant value of $\beta$ would mean to fix the motion distribution between the vehicle and the manipulator. Nevertheless, the use of a fixed weight factor inside the interval $[0,1]$ has a drawback: it causes motion of the manipulator also if the desired end-effector posture is out of reach; on the other hand, it causes motion of the vehicle also if the manipulator alone could perform the task.

Another problem is the need to handle a large number of variables; UVMSs, in fact, are complex systems and several variables must be monitored during the motion, e.g., the manipulator manipulability, the joint range limits to avoid mechanical breaks, the vehicle roll and pitch angles for correct tuning of the proximity sensors, the yaw angle to exploit the vehicle shape in presence of ocean current, etc. As it can be easily understood, it is quite difficult to handle all these terms without a kinematic control approach. Nevertheless, the existing techniques do not allow to find a flexible and reliable solution.

To overcome this drawback a fuzzy theory approach has therefore been considered at two different levels. First, it is required to manage the distribution of motion between the vehicle and the manipulator; second, it is required to consider multiple secondary tasks that are activated only when the corresponding variable is outside (inside) a desired range. This can be done using different weight factors adjusted on-line according to the Mamdani fuzzy inference system ([103]) shown in Figure 6.10.

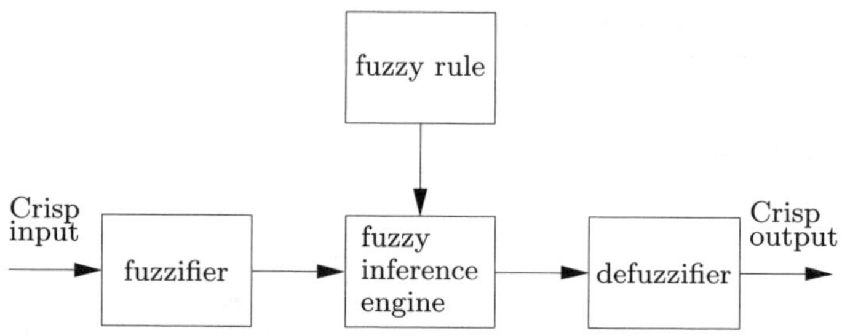


**Fig. 6.10.** Mamdani fuzzy inference system

In detail, the crisp outputs are the scalar $\beta$ of (6.15) that distributes the desired end-effector motion between the vehicle and the manipulator and a vector of coefficients $\alpha_i$ that are used in the task priority equation as follows

$$\boldsymbol{\zeta} = \boldsymbol{J}_{\boldsymbol{W}}^{\dagger} \left( \dot{\boldsymbol{x}}_{E,d} + \boldsymbol{K}_E \boldsymbol{e}_E \right) + \left( \boldsymbol{I} - \boldsymbol{J}_{\boldsymbol{W}}^{\dagger} \boldsymbol{J}_{\boldsymbol{W}} \right) \left( \sum_i \alpha_i \boldsymbol{J}_{s,i}^{\dagger} \boldsymbol{w}_{s,i} \right) , \quad (6.16)$$

where $\boldsymbol{w}_{s,i}$ are suitably defined secondary task variables and $\boldsymbol{J}_{s,i}$ are the corresponding Jacobians. Both $\beta$ and $\alpha_i$'s are tuned according to the state of the system and to given behavioral rules. The inputs of the fuzzy inference system depend on the variables of interest in the specific mission. As an example, the end-effector error, the ocean current measure, the system's dexterity, the force sensor readings, can be easily taken into account by setting up a suitable set of fuzzy rules.

To avoid the exponential growth of the fuzzy rules to be implemented as the number of tasks is increased, the secondary tasks are suitably organized in a hierarchy. Also, the rules have to guarantee that only one $\alpha_i$ is high at a time to avoid conflict between the secondary tasks. An example of application of the approach is described in the Simulation Section.

### Simulations

The proposed fuzzy technique has been verified in full-DOFs case studies. An UVMS has been considered constituted by a vehicle with the size of the NPS Phoenix [145] and a manipulator mounted on the bottom of the vehicle. The kinematics of the manipulator considered is that of the SMART-3S manufactured by COMAU. Its Denavit-Hartenberg parameters are given in Table 6.2. The overall system, thus, has 12 DOFs. Figure 6.11 shows the configuration in which all the joint positions are zero according to the used convention.

**Table 6.2.** D-H parameters [m,rad] of the manipulator mounted on the underwater vehicle

| | $a$ | $d$ | $\theta$ | $\alpha$ |
|---|---|---|---|---|
| link 1 | 0.150 | 0 | $q_1$ | $-\pi/2$ |
| link 2 | 0.610 | 0 | $q_2$ | 0 |
| link 3 | 0.110 | 0 | $q_3$ | $-\pi/2$ |
| link 4 | 0 | 0.610 | $q_4$ | $\pi/2$ |
| link 5 | 0 | −0.113 | $q_5$ | $-\pi/2$ |
| link 6 | 0 | 0.103 | $q_6$ | 0 |

The simulations are aimed at proving the effectiveness of the fuzzy kinematic control approach; for seek of clarity, thus, only the kinematic loop performance is shown (see Figure 6.1). The real vehicle/joint position will be affected by a larger error since the tracking error too has to be taken into account. It is worth noticing that, as long as the law level dynamic controller is suitably designed, this tracking error is bounded. Moreover, it does not affects the kinematic loop performance.

The primary task is to track a position/orientation trajectory of the end effector. The system starts from the initial configuration:

$$\begin{aligned} \boldsymbol{\eta} &= [0 \quad 0 \quad 0 \quad 0 \quad 0 \quad 0]^{\mathrm{T}} && \text{m,deg} \\ \boldsymbol{q} &= [0 \quad -30 \quad -110 \quad 0 \quad -40 \quad 90]^{\mathrm{T}} && \text{deg} \end{aligned}$$

that corresponds to the end-effector position/orientation

$$\begin{aligned} \boldsymbol{\eta}_{ee1} &= [0.99 \quad -0.11 \quad 2.99]^{\mathrm{T}} && \text{m} \\ \boldsymbol{\eta}_{ee2} &= [0 \quad 0 \quad -90]^{\mathrm{T}} && \text{deg}\,. \end{aligned}$$

The end effector has to track a segment of $-30\,\text{cm}$ along $z$, stop there, and track a segment of $1\,\text{m}$ along $x$. Both segments have to be executed with a quintic polynomial time law in $12\,\text{s}$. During the translation, the end effector orientation has to be kept constant. The initial configuration and the desired

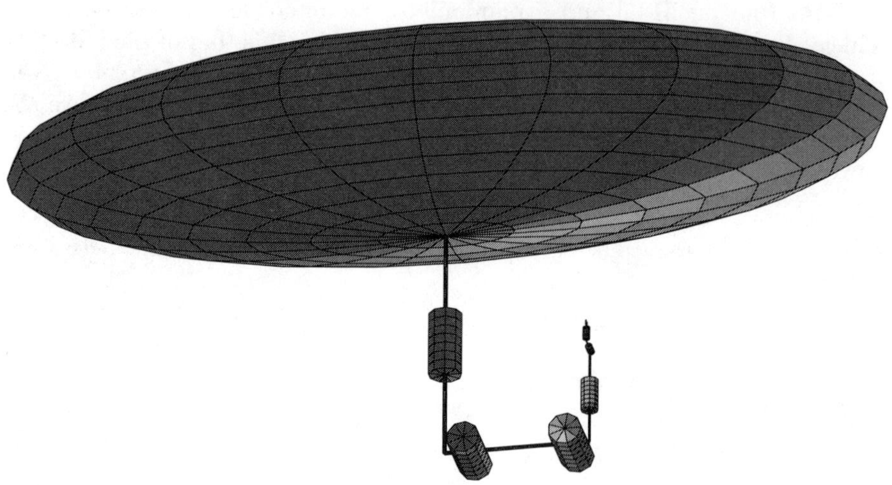

**Fig. 6.11.** UVMS in the configuration with null joint positions

path are shown in Figure 6.12. The duration of the simulation is 50 s. The algorithm is implemented at a sampling frequency of 20 Hz. Notice that the desired path cannot be tracked by the manipulator alone since it goes outside of its workspace. It is then necessary, somehow, to move the vehicle as well. Finally, the task is to be executed in real-time; no off-line knowledge of the task is available.

Different simulations will be shown:

Case Study n. 1. Simple pseudoinversion of (2.68);
Case Study n. 2. Pseudoinversion of (2.68) by the use of a weighted pseudoinverse;
Case Study n. 3. Singularity robust task priority algorithm;
Case Study n. 4. Integration of the former algorithm with the proposed fuzzy technique in presence of several secondary tasks.

For all the simulations a CLIK algorithm is considered, moreover, the orientation error is represented by the use of quaternions. The desired orientation, however, is still assigned in terms of Euler angles, since the transformation from Euler angles to quaternions is free from representation singularities.

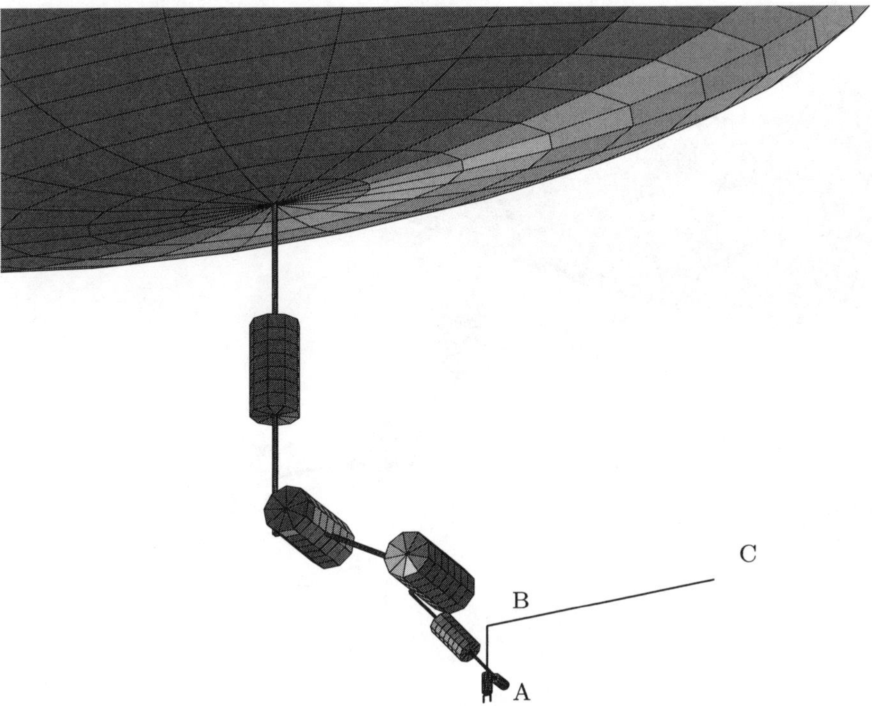


**Fig. 6.12.** Initial configuration of the UVMS for all the case studies. Desired end-effector position: at the start time (A); after the first movement of 12 s (B); at the final time (C)

## Case Study n. 1

The first simulation has been run by the use of a simple pseudoinversion of (2.68) with CLIK gain:

$$\boldsymbol{K}_E = \text{blockdiag}\{1.6\boldsymbol{I}_3, \boldsymbol{I}_3\}\,.$$

This solution is not satisfactory in presence of such a large number of degrees of freedom. The system, in fact, is not taking into account the different nature of the degrees of freedom (vehicle and manipulator) leading to evident drawbacks: large movement of the vehicle in position and orientation, final configuration not suitable for sensor tuning, possible occurrence of kinematic singularities or joint mechanical limits. As an example, Figure 6.13 reports the sketch of the final configuration where the bottom sonar would not work properly since the pitch is $\approx 18\,\text{deg}$.

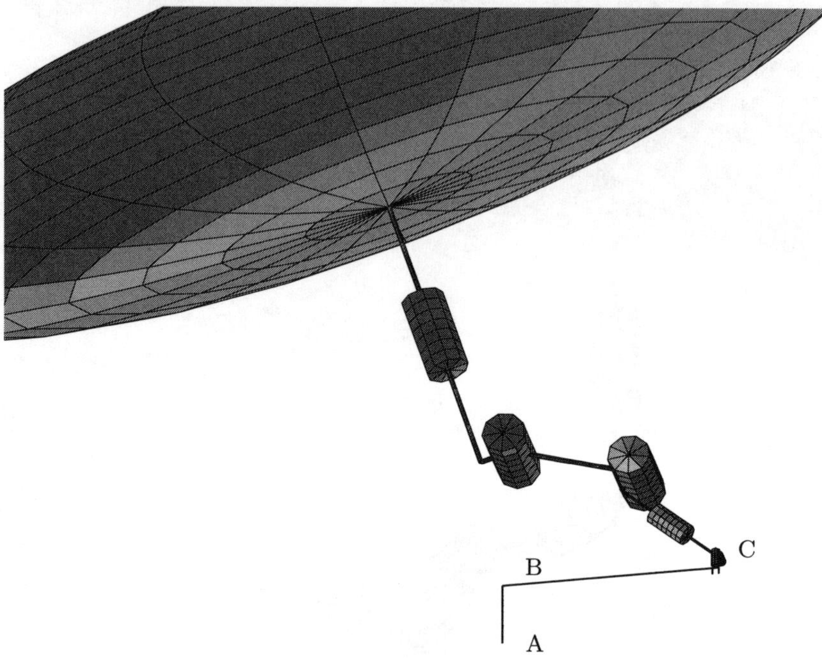


**Fig. 6.13.** Final configuration of the UVMS for the first case study. The redundancy is not exploited and the possible occurrence of undesired configurations is not avoided

## Case Study n. 2

In the second case study a weighted pseudoinverse is added in order to redistribute the motion between vehicle and manipulator including a cost factor that can be considered, e.g., proportional to the ratio of their inertias. The following matrix of gain has been used:

$$\boldsymbol{W}^{-1} = \text{blockdiag}\{0.01\boldsymbol{I}_6, \boldsymbol{I}_6\}.$$

Despite the much different costs of the two movements, the vehicle is still required to move in order to contribute to the end effector motion. However it would be preferable to move the vehicle only when absolutely necessary, leading to sole movement of the manipulator in ordinary working conditions. As an example, the vehicle attitude, in terms of Euler angles, for the last simulation is shown in Figure 6.14. The vehicle still has a pitch of about 20 deg.

## Case Study n. 3

The drawback shown by the algorithm as presented in the Case Study n. 2 can be easily avoided by resorting to a singularity-robust task priority redundancy

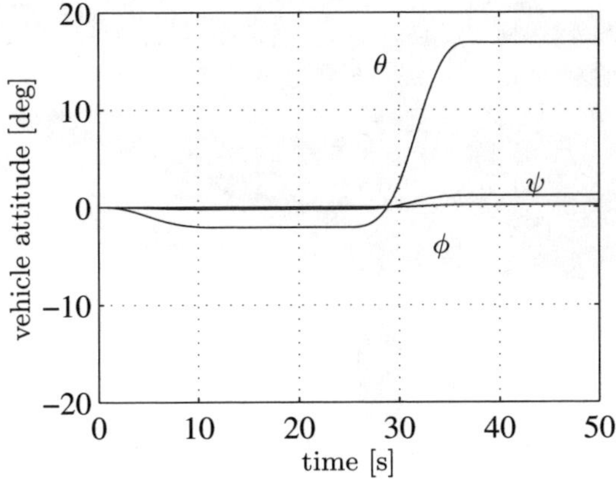


**Fig. 6.14.** Case study n. 2. Vehicle attitude in terms of Euler angles. Despite the weight factor, the vehicle can reach non-dexterous configurations

resolution [78]. The same task is now simulated with the introduction of the secondary task:

$$\boldsymbol{x}_s = \begin{bmatrix} \phi \\ \theta \end{bmatrix}$$

with $\boldsymbol{x}_{s,d} = [0 \quad 0]^{\mathrm{T}}$, meaning that the vehicle has to maintain an horizontal configuration all along the task execution. Its Jacobian is simply given by:

$$\boldsymbol{J}_s = \begin{bmatrix} 0 & 0 & 0 & 1 & 0 & 0 & 0 & 0 & 0 & 0 & 0 & 0 \\ 0 & 0 & 0 & 0 & 1 & 0 & 0 & 0 & 0 & 0 & 0 & 0 \end{bmatrix}.$$

Notice that, for this simple matrix, it is $\boldsymbol{J}_s^{\dagger} = \boldsymbol{J}_s^{\mathrm{T}}$. Obviously, the vehicle position and yaw are not limited.

Figure 6.15 reports the sketch of the final configuration; it can be observed that the pitch is now close to zero, as can be seen also from Figure 6.16.

### Case Study n. 4

In this simulation, the implementation of the proposed kinematic control approach is presented. In an UVMS several variables are of interest in order to achieve a successful mission:

- avoidance of kinematic singularities;
- keeping the joints far from the mechanical limits;
- keeping the vehicle with small roll and pitch;
- avoidance of obstacles;
- alignment of the vehicle fore-aft direction with the ocean current.

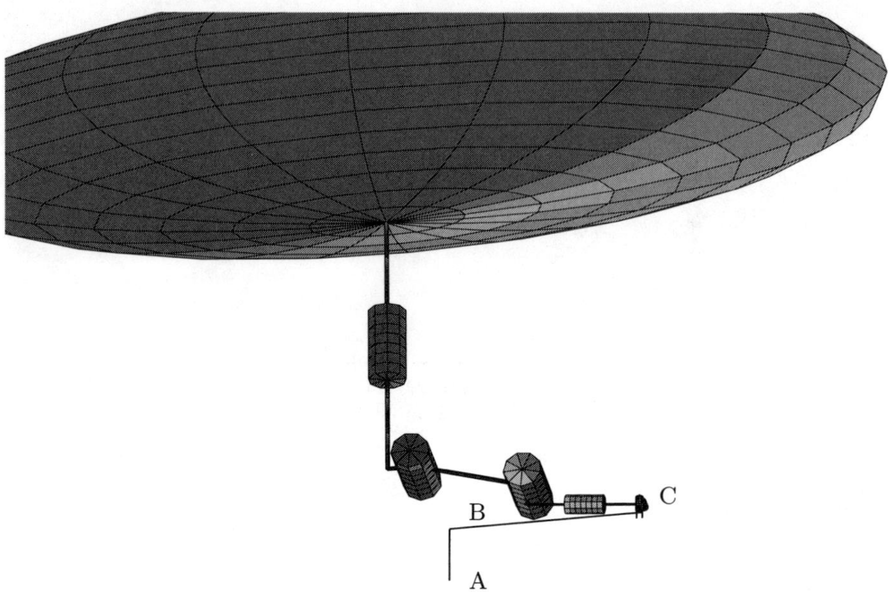


**Fig. 6.15.** Case study n. 3. Final configuration; the roll and pitch angles are now kept close to zero by exploiting the redundancy with the singularity-robust task priority algorithm

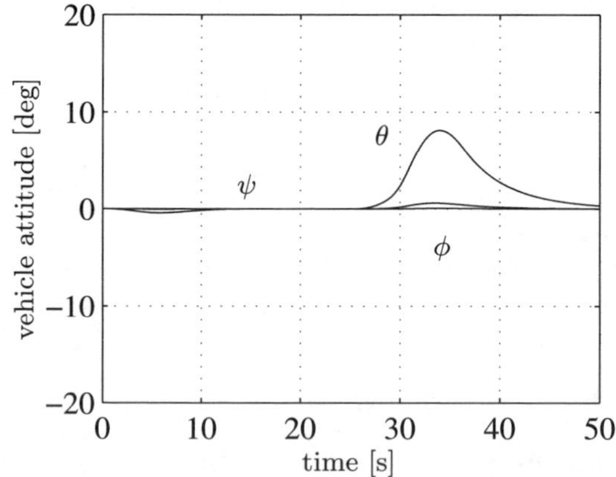


**Fig. 6.16.** Case study n. 3. Vehicle attitude in terms of Euler angles

It is underlined that some of the above items are critical; the alignment with the current, however, can be significant in order to reduce power consumption [20]. As an example, in the previous case study, the use of a weight factor requires now a larger movement of the manipulator. This can lead to the occurrence of kinematic singularities. In fact, if the trajectory is assigned in real-time, it is possible that the manipulator is asked to move to the border of its workspace, where the possibility to experience a kinematic singularity is high. Also, when the manipulator is outstretched, mechanical joint limits can be encountered. In the simulations, the following joint limits have been assumed:

$$\begin{aligned} \boldsymbol{q}_{\text{min}} &= [-100\ -210\ -210\ -150\ -80\ -170]^{\text{T}}\ \text{deg} \\ \boldsymbol{q}_{\text{max}} &= [100\quad 30\quad 10\quad 150\quad 80\quad 170]^{\text{T}}\ \text{deg}\,. \end{aligned}$$

In Figure 6.17, the minimum distance to such limits for the previous case study is shown. It can be observed that the manipulator hits a mechanical limit at $t \approx 35\,\text{s}$.

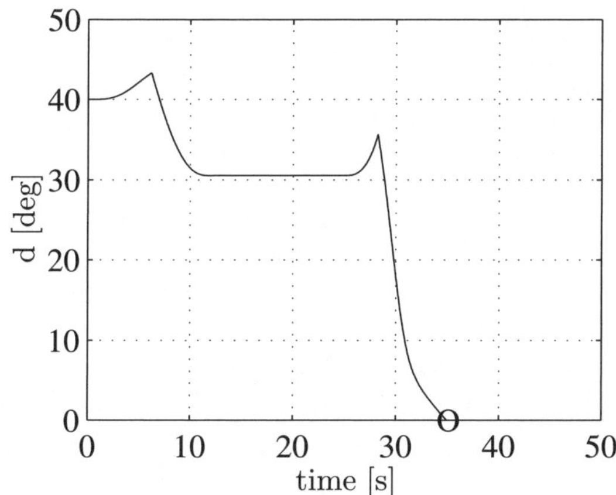


**Fig. 6.17.** Case study n. 3. Minimum distance of the 6 joints from their mechanical limits. It can be observed that large movements of the manipulator may cause hitting of the joint limits

In order to match all the constraints of such problem, a solution as been proposed based on the use of the singularity-robust task priority merged with fuzzy techniques.

Let consider the following tasks:

- **End-effector position/orientation**. The primary task is given, as for the previous simulations by the end-effector position and orientation. The corresponding Jacobian $\boldsymbol{J}_p$ is $\boldsymbol{J}$ given in (2.68);

- **Manipulability**. Since the fuzzy approach tries to move the manipulator alone, its manipulability has to be checked. A computationally limited measure of the manipulability can be obtained by checking the minimum singular value of the Jacobian [195, 77]. Since the manipulability function is strongly non-linear it is possible to adopt the following approach: when close to a singular configuration, the system tries to reconfigure itself in a dexterous configuration. The task, thus, is a nominal manipulator configuration whose Jacobian is given by:

$$\boldsymbol{J}_{s1} = [\,\boldsymbol{O}_{6\times 6} \quad \boldsymbol{I}_6\,] \,;$$

- **Mechanical limits**. Due to the mechanical structure, each joint has a limited allowed range. In case of a real-time trajectory, avoidance of such limits is crucial. For this reason the minimum distance from a mechanical limit is considered as another secondary task. Notice that the Jacobian is the same of to the previous task:

$$\boldsymbol{J}_{s2} = [\,\boldsymbol{O}_{6\times 6} \quad \boldsymbol{I}_6\,] \,;$$

- **Vehicle attitude**. As for the previous cases, the vehicle attitude (roll and pitch angles) has to be kept null when possible. The Jacobian is then given by

$$\boldsymbol{J}_{s3} = \begin{bmatrix} 0 & 0 & 0 & 1 & 0 & 0 & 0 & 0 & 0 & 0 & 0 & 0 \\ 0 & 0 & 0 & 0 & 1 & 0 & 0 & 0 & 0 & 0 & 0 & 0 \end{bmatrix} .$$

Due to the simple structure of the matrices, the pseudoinversion of the secondary tasks is trivial: $\boldsymbol{J}^{\dagger}_{s1} = \boldsymbol{J}^{\mathrm{T}}_{s1}$, $\boldsymbol{J}^{\dagger}_{s2} = \boldsymbol{J}^{\mathrm{T}}_{s2}$, $\boldsymbol{J}^{\dagger}_{s3} = \boldsymbol{J}^{\mathrm{T}}_{s3}$.

As shown in Section 6.6, the above tasks are activated by fuzzy variables. The fuzzy inference system has 3 inputs, namely: a measure of the robot manipulability, a measure of the distance from the joints limits and a measure of the vehicle attitude. Hence, 3 linguistic variables can be defined that can take the values:

$$\begin{aligned} \textit{manipulability} &= \{\text{singular, not singular}\} \\ \textit{joint limits} &= \{\text{close, not close}\} \\ \textit{vehicle attitude} &= \{\text{small, not small}\} \end{aligned}$$

The output is given by the linguistic variables $\beta$ and the 3 $\alpha_i$'s. The latter can take the following values:

$$\alpha_i = \{\text{high, low}\}.$$

The linguistic variable $\beta$, named *motion* can take the following values:

$$\textit{motion} = \{\text{vehicle, manipulator}\} .$$

As an example, the membership function of the linguistic variable *joint limits* is reported in Figure 6.18.

The FIS outputs are considered at two different levels. The variable *motion* ($\beta$ in (6.15)) can be considered at a higher level with respect to the

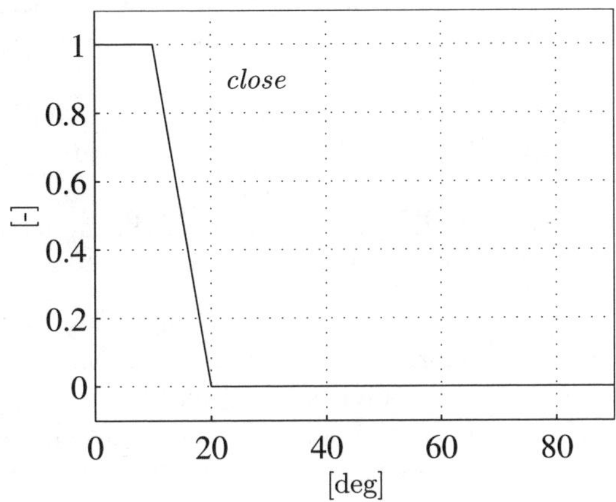


**Fig. 6.18.** Membership function of the linguistic variable: *joint limits* used in the fuzzy singularity-robust task priority technique

variables $\alpha_i$'s. Roughly speaking, the motion has to be normally assigned to the manipulator, i.e., if all the $\alpha_i$'s are null. Hence, the value of $\beta$ is related to the value of the $\alpha_i$'s and the first set of rules to be designed concerns the 3 linguistic variables $\alpha_1$, $\alpha_2$ and $\alpha_3$. A complete and consistent set of fuzzy rules for 3 linguistic variables each of them defined in two fuzzy sets requires $2^3$ rules for every output leading to 32 total rules. The rules have been arranged in a hierarchical structure that gives higher priority to the kinematic singularity of the manipulator and lower to the vehicle attitude. Hence, the following 8 rules have been used:

1. `if (manipulator is singular) then (`$\alpha_1$` is high);`
2. `if (manipulator is not singular) then (`$\alpha_1$` is low);`
3. `if (manipulator is singular) then (`$\alpha_2$` is low);`
4. `if (joint limits is not close) then (`$\alpha_2$` is low);`
5. `if (manipulator is not singular) and (joint limits is close)`
   `then (`$\alpha_2$` is high);`
6. `if (manipulator is singular) or (joint limits is close)`
   `then (`$\alpha_3$` is low);`
7. `if (vehicle attitude is small) then (`$\alpha_3$` is low);`
8. `if (manipulator is not singular)`
   `and (joint limits is not close)`
   `and (vehicle attitude is not small) then (`$\alpha_3$` is high).`

In detail, the rules are developed as follows:

- The first two rules concern the primary task. In this case, the manipulator singularity is of concern and the variable $\alpha_1$ is activated when the manipulator is close to a singularity.

- A second task ($\alpha_2$), with lower priority with respect to the first, has to be added. Rule n. 3, thus, is aimed at avoiding activation of this task when the primary ($\alpha_1$) is high.
- Rule n. 4 and n. 5 are aimed at activating $\alpha_2$. Notice that the activation of $\alpha_2$ is in `and` with the condition that does not activate the higher priority task ($\alpha_1$).
- Repeat for the third task, in order of priority, the same rules as done for the second task by taking into account that two tasks are now of higher priority.

Table 6.3 is aimed at clarifying the rules development with respect to two tasks, 1 and 2 in which the first is of higher priority with respect to the second. The fuzzy sets are very simple, i.e., an input $u_i$ high requires the activation of this task by imposing $\alpha_i$ high. It can be recognized, thus, that $\alpha_2$ respects its lower priority.

**Table 6.3.** Examples of the fuzzy set rules for two tasks: $u_1$ is the input of a generic task of higher priority with respect to $u_2$ corresponding to a secondary task. $\alpha_1$ and $\alpha_2$ are the corresponding output

| | $u_1$ = low | $u_1$ = high |
|---|---|---|
| $u_2$ = low | $\alpha_1$ = low<br>$\alpha_2$ = low | $\alpha_1$ = high<br>$\alpha_2$ = low |
| $u_2$ = high | $\alpha_1$ = low<br>$\alpha_2$ = high | $\alpha_1$ = high<br>$\alpha_2$ = low |

It is worth noticing that the rules presented could be grouped, e.g., rules n. 1 and n. 3. The list presented, however, keeps the logical structure used to develop the rules and should be clearer to the reader. Obviously, in the simulation the rules have been compacted. With this logical approach the rules are complete, consistent and continuous [103].

The `and`–`or` operations have been calculated by resorting to the *min–max* operations respectively, the implication–aggregation operations too have been calculated by resorting to the *min–max* operations respectively, the values of $\alpha_i \in [0, 1]$ are obtained by defuzzification using the centroid technique and a normalization. Finally, the value of $\beta \in ]0, 1]$ is given by $\beta = 1 - \max_i(\alpha_i)$. Notice that the extremities of the range in which $\beta$ is defined do not involve a singular configuration since, if $\beta = 1$ the manipulator alone is moving and it is not close to a kinematic singularity. On the other hand, it is preferable to have a certain degree of mobility of the manipulator avoiding $\beta = 0$; this to guarantee that the manipulator reconfigures itself in a dexterous posture.

A simulation has been run with the proposed kinematic control leading to satisfactory results. Figures 6.19–6.24 show some plots of interest. In detail,

Figure 6.19 shows the vehicle position and attitude, Figure 6.20 the joint positions, Figure 6.21 reports the variables considered as secondary tasks and the corresponding FIS outputs. It can be observed that, in the execution of the segment A-B, the vehicle is not requested to move since the manipulator is working in a safe posture; this can be observed from Figure 6.21 where it can be noticed that the $\alpha_i$'s are null in the first part of the simulation. When $t \approx 30\,\mathrm{s}$ joint 5 is approaching its mechanical limit and the corresponding $\alpha_2$ is increasing, requesting the vehicle to contribute to the end-effector motion while the manipulator reconfigures itself; thus, by always keeping a null end-effector position/orientation error, the occurrence of an hit is avoided. The same can be observed for the pitch of the vehicle; since it is at the lower hierarchical level in the FIS, its value is recovered to the null value only when the other $\alpha_i$'s are null.

Figure 6.22 shows a sketch of the initial and final configuration of the system. Figure 6.23 shows the system velocities. It can be remarked that the proposed algorithm outputs smooth trajectories. As for all the simulations shown, the end-effector position/orientation error is practically null (Figure 6.24) due to the use of a CLIK algorithm.

In order to show handling of the fuzzy rules under the proposed approach while avoiding exponential growth of their number, we finally add as 4th task, specification of the vehicle yaw. Our aim is to align the vehicle fore-aft direction with the current in order to get energetic benefit from the low drag of such configuration. To limit the number of rules to be implemented we assign to this task the last priority among the secondary tasks. In this case, considering two fuzzy sets also for this last variable ($yaw$ = {aligned, not aligned}), only the following 3 rules have to be added to the previous 8, leading to 11 rules in total instead of 64:

```
9. if (manipulator is singular) or (joint limits is close) or
   (vehicle attitude is not small) then (α4 is low);
10. if (yaw is aligned) then (α4 is low);
11. if (manipulator is not singular)
    and (joint limits is not close)
    and (vehicle attitude is small)
    and (yaw is not aligned) then (α4 is high).
```

The 3 rules have the following aim: rule n. 9 is aimed at giving the lower priority to this specific task; rule n. 10 is aimed at guaranteeing that the output is always low when the corresponding input is inside the safe range; finally, rule n. 11 activates $\alpha_4$ only for the given specific combination of inputs.

The integration of fuzzy technique with established inverse kinematic techniques exhibits promising results, the fuzzy theory can give an added value in handling complex situations as missions in remotely, unknown, hazardous underwater environments. In a certain way, a fuzzy approach could be considered to implement an higher level supervisor that is in charge of distributing the motion between vehicle and manipulator while taking into

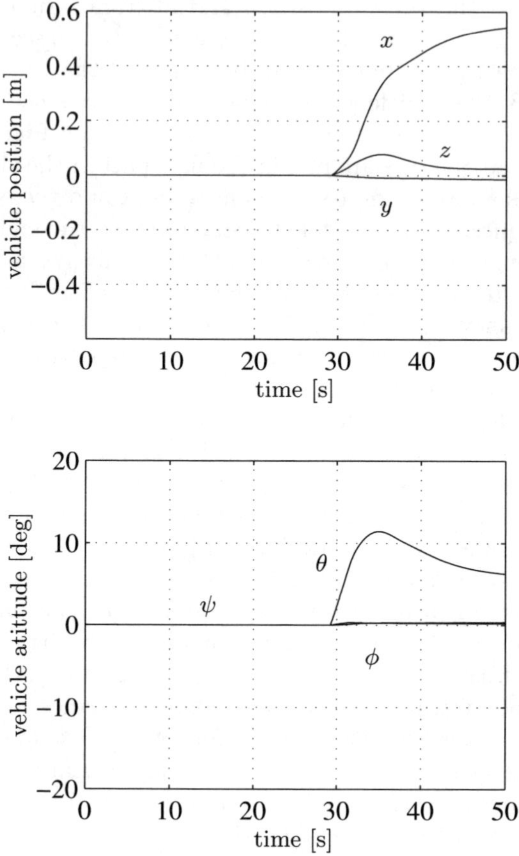


**Fig. 6.19.** Case study n. 4. Vehicle position (top) and attitude in terms of Euler angles (bottom). The movement of the vehicle is not required in the execution of the first segment (A-B, first 12 s) when the manipulator is working in dexterous configuration

account the big amount of constraints of UVMSs: joint's limits, vehicle's roll and pitch, robot's manipulability, obstacle avoidance, etc.

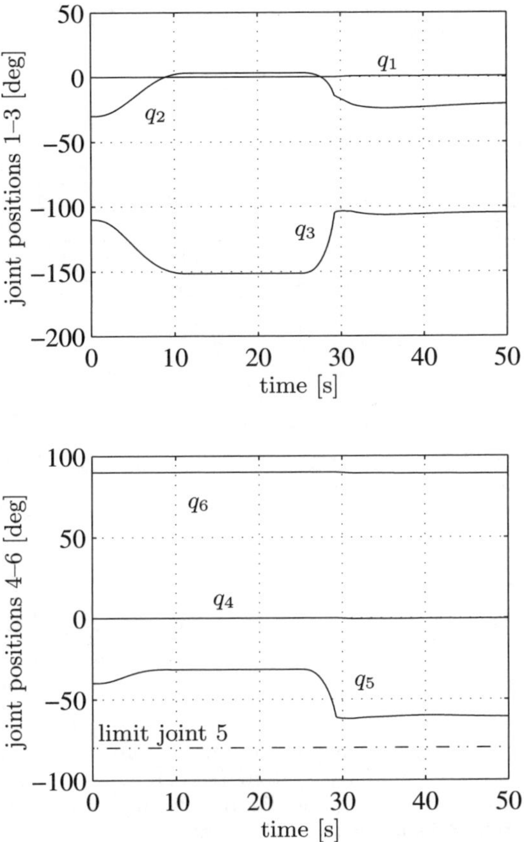


**Fig. 6.20.** Case study n. 4. Joint positions. The mechanical limit of joint 5 is highlighted; it can be observed that the system reconfigures itself in order to avoid working close to the mechanical limit

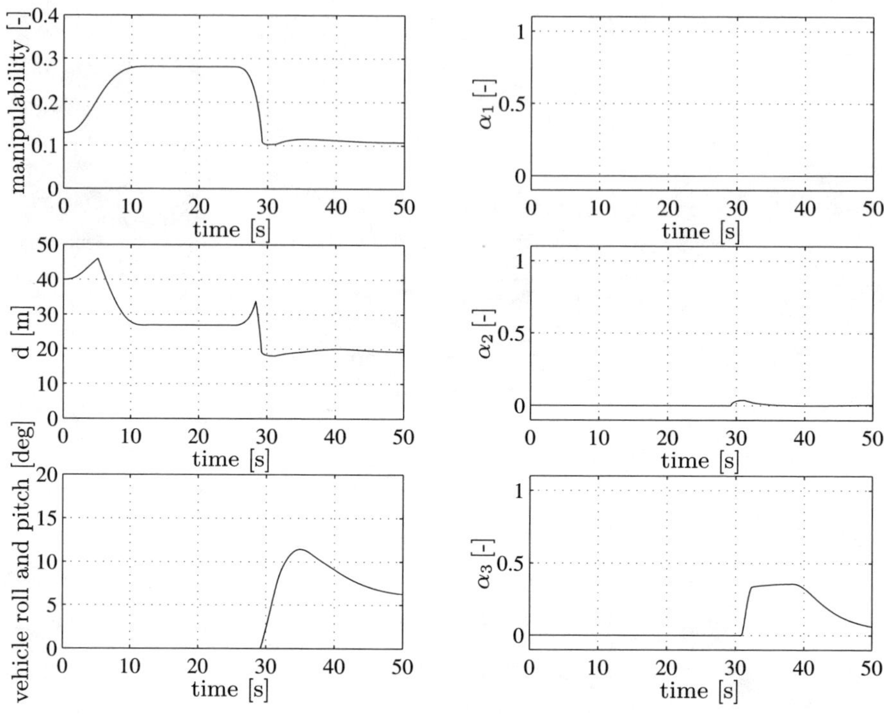


**Fig. 6.21.** Case study n. 4. Variables of interest for the secondary task (left) and output of the fuzzy inference system (right). For this specific mission, the manipulability task is not excited, the distance from the mechanical limit and the vehicle roll and pitch tasks are kept in their safe range

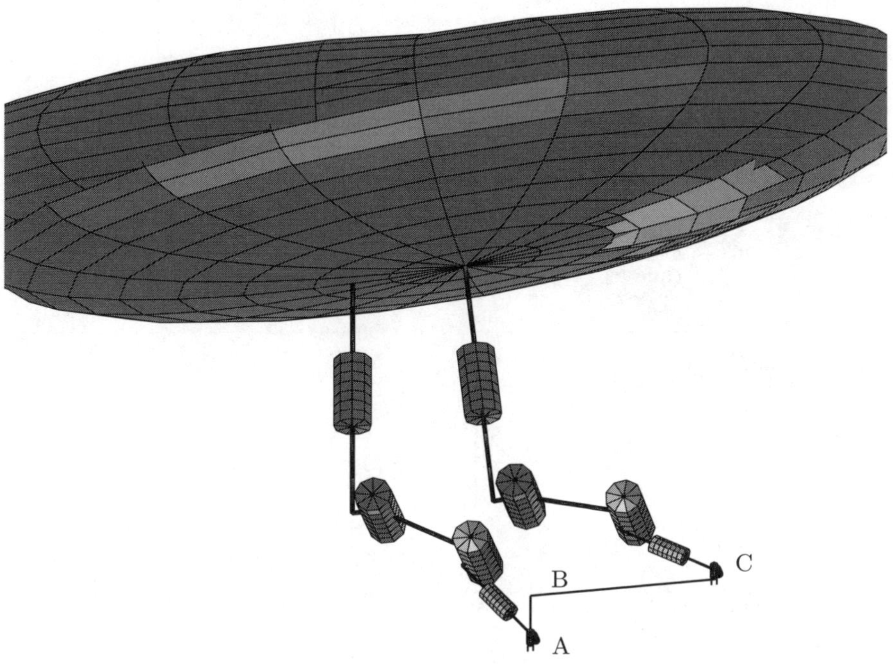


**Fig. 6.22.** Case study n. 4. Final configuration. The proposed kinematic control allows handling several variables of interest

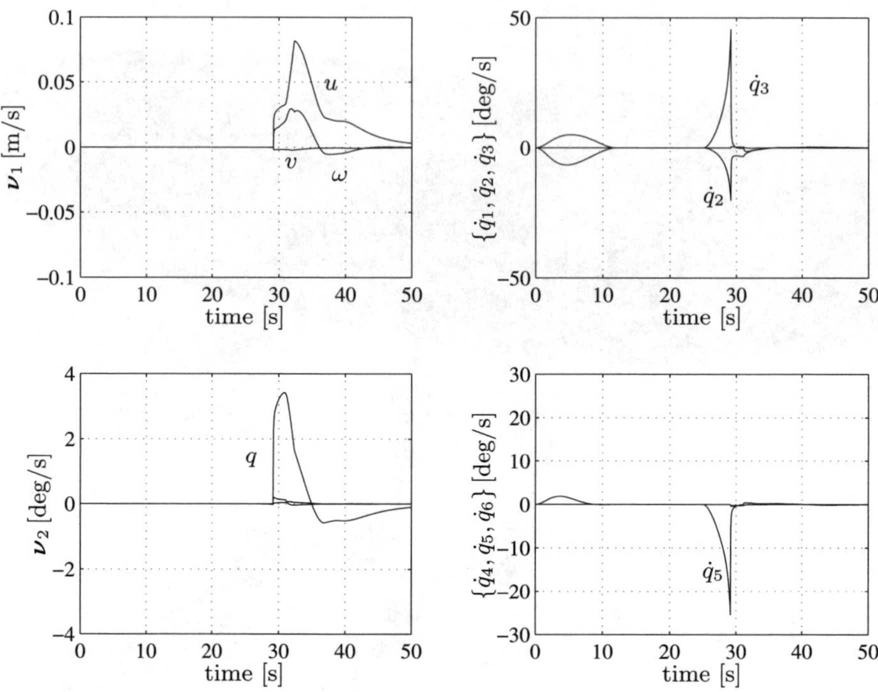


**Fig. 6.23.** Case study n. 4. System velocities

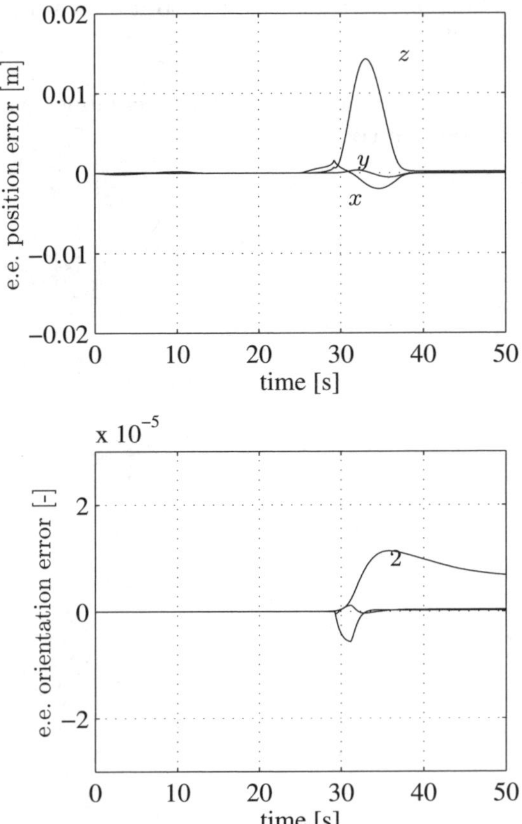


**Fig. 6.24.** Case study n. 4. End-effector position/orientation errors

## 6.7 Conclusions

With a view to implementing autonomous missions of robotic systems, kinematic control plays an important role. The manipulation task is naturally defined in the operational space. A mapping between the task space and the vehicle/joint space is then necessary to achieve the desired task.

If this mapping is implicitly performed via a model based dynamic control the natural redundancy of the system is not exploited, e.g., is not possible to take into account additional constraints. Moreover dynamic compensation of underwater robotic systems is difficult to obtain. On the other hand, off-line planning of the vehicle/joint positions is not advisable, since the mission has to be accomplished in an unstructured, generally unknown, environment. For these reasons real-time kinematic control seems to be the right approach to motion control of UVMSs.

The use of techniques well known in robotics such as the task priority approach seems to offer good results. They allow to reliably exploit the redundancy and do not require compensation of the system's dynamics. As it will be shown in Chap. 8, kinematic control techniques can be successfully integrated with interaction control schemes.

# 7. Dynamic Control of UVMSs

## 7.1 Introduction

In Chap. 2 the equations of motion of Underwater Vehicle-Manipulator Systems (UVMSs) have been presented. Their expression in matrix form (2.71), is formally close to the equations of motion of ground fixed manipulators for which a wide control literature exists. This has suggested a suitable translation/implementation of existing control algorithms. However, some differences, crucial from the control aspect, need to be underlined. UVMSs are complex systems characterized by several strong constraints:

- Uncertainty in the model knowledge, mainly due to the poor knowledge of the hydrodynamic effects;
- Complexity of the mathematical model;
- Kinematic redundancy of the system;
- Difficulty to control the vehicle in hovering, mainly due to the poor thrusters performance;
- Dynamic coupling between vehicle and manipulator;
- Low bandwidth of the sensor's readings.

In [197, 198] a discrete adaptive control strategy for coordinated control of UVMSs is presented. Numerical simulations, on a planar task, show that the use of a centralized controller, better than two separate controllers, one for the vehicle and one for the manipulator, guarantees performance improvement.

Reference [124] shows an adaptive macro-micro control for UVMSs. Inverse kinematics is obtained by inversion of the Jacobian matrix; hence, a manipulator with 6 degrees of freedom is required. A stability analysis in Lyapunov sense is provided.

An adaptive control law for an underwater manipulator is proposed in [59]. A self-tuning PID controller is developed and tested in simulation on a 2-link manipulator with fixed base. In [190] too, an underwater manipulator with fixed base is considered. A description of a telerobotic control system is provided in [181].

The use of multiple manipulators to be used as stabilizing paddles is investigated by means of simulations in [168]. Those concern a vehicle carrying a 6-DOF manipulator plus a pair of 2-link manipulators counteracting the interaction force between vehicle and manipulator as paddles.

In [105, 106] some dynamic considerations are given to underline the existence of a dynamic coupling between vehicle and manipulator. From this analysis, based on a specific structure of an UVMS, a Sliding Mode approach with a feedforward compensation term is presented. Numerical simulation results show that the knowledge of the dynamics allows improvement of the tracking performance.

In [86, 87], during the manipulation motion, the vehicle is assumed to be inactive and modeled as a passive joint. A robust controller, with a disturbance observer-based action is used and its effectiveness verified in a 1-DOF vehicle carrying a 3-DOF manipulator.

Reference [186] presents a sliding mode controller that benefits from the compensation of a multilayer neural network. Simulation with a 2-DOF, ground-fixed, underwater manipulator is provided.

In [131, 245] the possibility to mount a force/torque sensor at the base of the manipulator is considered in order to compensate for the vehicle/manipulator dynamic interaction. In case of absence of the sensor, [245] also proposes a disturbance observer. The possibility to use information coming from an other sensor is interesting; however, the practical implementation of such algorithms is not trivial.

Reference [169] presents an iterative learning control experimentally validated on a 3-DOF manipulator with fixed base. This is first moved in air and then in water in order to *learn* the hydrodynamic dynamic contribution and then use it in a feedforward compensation.

Reference [174], after having reported some interesting dynamic considerations about the interaction between the vehicle and the manipulator, propose a two-time scale control. The vehicle, characterized by low bandwidth actuators that can not compensate for the high manipulator bandwidth, is controlled by a simple P-type action, while the manipulator is controlled by a feedback linearizing controller.

In [162] the model of a planar motion of the AUV Twin-Burger equipped with a 2-link manipulator is developed. A Resolved Acceleration Control is then applied and simulated on the 5-DOFs.

Reference [281] proposes an adaptive action mainly based on the transpose of the Jacobian. The approach is validated on simulations involving a 2-DOF model of ODIN carrying a 2-DOF planar manipulator.

A problem slightly different is approached in [200], where the 4-DOF manipulator *Sherpa*, mounted under the ROV Victor 6000 developed at the Ifremer, is considered. This manipulator has been originally deigned to be controlled in open-loop by a remote operator via a master/slave configuration using a joystick: it is, thus, not-provided with proprioceptive sensors. The Authors propose closed-loop system based on an eye-to-hand visual servoing approach to control its displacement.

## 7.2 Feedforward Decoupling Control

In 1996 T. McLain, S. Rock and S. Lee, in [205, 206], present a control law for UVMSs with some interesting experimental results conducted at the Monterey Bay Aquarium Research Institute (MBARI). A 1-link manipulator is mounted on the vehicle OTTER (see Figure 7.1) controlled in all the 6-DOFs by mean of 8 thrusters. A coordinated control is then implemented to improve the tracking error of the end effector.

**Fig. 7.1.** OTTER vehicle developed in the Aerospace Robotics Laboratory at Stanford University (courtesy of T. McLain, Brigham Young University)

In order to explain the control implemented let rewrite the equations of motion for the sole vehicle in matrix form as

$$\boldsymbol{M}_v\dot{\boldsymbol{\nu}} + \boldsymbol{C}_v(\boldsymbol{\nu})\boldsymbol{\nu} + \boldsymbol{D}_{RB}(\boldsymbol{\nu})\boldsymbol{\nu} + \boldsymbol{g}_{RB}(\boldsymbol{R}_B^I) = \boldsymbol{\tau}_v - \boldsymbol{\tau}_m(\boldsymbol{R}_B^I, \boldsymbol{q}, \boldsymbol{\zeta}, \dot{\boldsymbol{\zeta}}) \quad (7.1)$$

where $\boldsymbol{\tau}_m(\boldsymbol{R}_B^I, \boldsymbol{q}, \boldsymbol{\zeta}, \dot{\boldsymbol{\zeta}}) \in \mathbb{R}^6$ represents the coupling effect caused by the presence of the manipulator. The control of the vehicle and of the arm is achieved, independently, by classical control technique. The coordination action is obtained by adding to the vehicle thrusters a feedforward compensation term that is an estimate of $\boldsymbol{\tau}_m$:

$$\boldsymbol{\tau}_v = \boldsymbol{\tau}_{control} + \hat{\boldsymbol{\tau}}_m(\boldsymbol{R}_B^I, \boldsymbol{q}, \boldsymbol{\zeta}, \dot{\boldsymbol{\zeta}}) \tag{7.2}$$

where $\boldsymbol{\tau}_{control} \in \mathbb{R}^6$ is the control action output by the sole vehicle controller.

In [206] there are experimental results conducted with the vehicle OTTER that is about 2.5 m long, 0.95 m wide, and 0.45 m tall and weights about 145 kg in air. The arm used has a 7.1 cm diameter and is 1 m long. The control benefits from a variety of commercial sensors: an acoustic short-baseline for the horizontal position, a pressure transducer for the depth, a dual-axis inclinometer for the roll and pitch angles, a flux-gate compass for the yaw, and solid-state gyros for the angular velocities. The control loop has been implemented at a frequency of 100 Hz, i.e., the frequency of all the sensors expect the baseline acoustic that worked at 2.5 Hz.

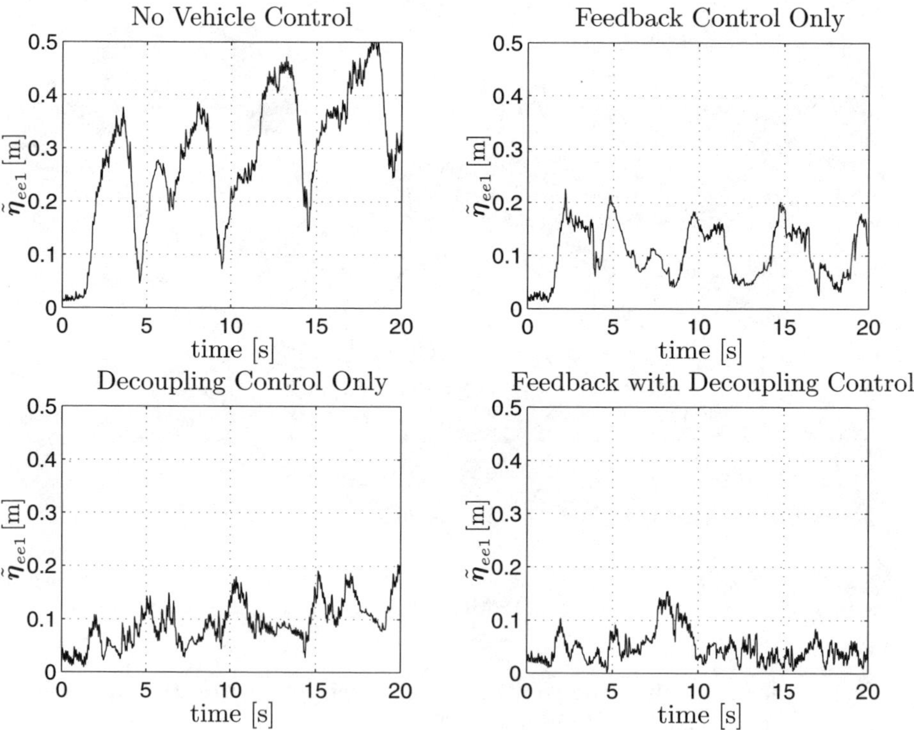


**Fig. 7.2.** Feedforward Decoupling control. End-effector errors of a periodic motion. Top-left: without vehicle control; top-right: with the sole decoupling action; bottom-left: with the sole vehicle feedback control; bottom-right: the proposed approach (courtesy of T. McLain, Brigham Young University)

The experiment is conducted under some assumptions: the vehicle does not influence the manipulator dynamic due to its small movements; the desired joint acceleration are used instead of the measured/filtered ones; there is no current and the lift forces are small compared to the in-line forces. For this specific experiment, moreover, the control of the vehicle is obtained by resorting to PID-like actions at the 6 DOFs independently. In view of these assumptions it is:

$$\boldsymbol{\tau}_v = \boldsymbol{\tau}_{PID} + \hat{\boldsymbol{\tau}}_m(\boldsymbol{q}, \dot{\boldsymbol{q}}, \ddot{\boldsymbol{q}}_d) \,.$$

Figure 7.2 reports some experimental results of the end-effector error commanded to a periodic motion. In the top-left plot the vehicle is not controlled at all; the influence of the arm on the vehicle can be observed. In the top-right and the bottom-left plots the sole decoupling force and vehicle feedback are considered. Finally, in the bottom-right plot the benefit of considering the proposed control strategy can be fully appreciated.

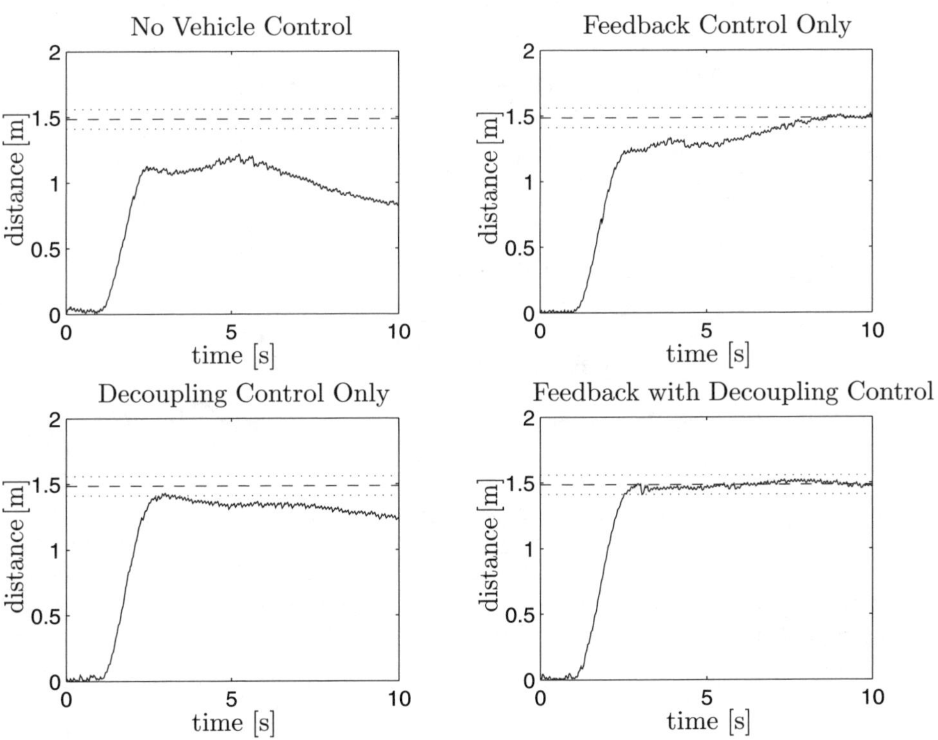


**Fig. 7.3.** Feedforward Decoupling control. End-effector step response. Top-left: without vehicle control; top-right: with the sole decoupling action; bottom-left: with the sole vehicle feedback control; bottom-right: the proposed approach (courtesy of T. McLain, Brigham Young University)

In figure 7.3 the end-effector step response is given and its settling time can be observed. It can be remarked that, without the proposed decoupling strategy, the end effector does not even reaches the given set point.

The overall improvement was of a factor 6 with respect to the vehicle without control and a factor 2.5 with respect to a separate control action. The applied thrust only showed an increase of about 5%.

## 7.3 Feedback Linearization

Reference [285] presents a model based control law. In detail, the symbolic dynamic model is derived using the Kane's equations [286]; this is further used in order to apply a full dynamic compensation. Simulations of a 6-DOF vehicle carrying two 3-link manipulators are provided. A similar approach has been presented in [255, 256].

From the mathematical point of view, the dynamics of an UVMS can be completely cancelled by resorting to the following control action:

$$\boxed{\boldsymbol{\tau} = \boldsymbol{M}(\boldsymbol{q})\dot{\boldsymbol{\zeta}}_a + \boldsymbol{C}(\boldsymbol{q},\boldsymbol{\zeta})\boldsymbol{\zeta} + \boldsymbol{D}(\boldsymbol{q},\boldsymbol{\zeta})\boldsymbol{\zeta} + \boldsymbol{g}(\boldsymbol{q},\boldsymbol{R}_B^I)} \tag{7.3}$$

where the $6+n$ dimensional vector

$$\dot{\boldsymbol{\zeta}}_a = \begin{bmatrix} \dot{\boldsymbol{\nu}}_a \\ \ddot{\boldsymbol{q}}_a \end{bmatrix}$$

is composed by a $6 \times 1$ vector defined as

$$\begin{aligned} \dot{\boldsymbol{\nu}}_a &= \boldsymbol{J}_e \ddot{\boldsymbol{\eta}}_e + \dot{\boldsymbol{J}}_e \dot{\boldsymbol{\eta}} \\ \ddot{\boldsymbol{\eta}}_e &= \ddot{\boldsymbol{\eta}}_d + \boldsymbol{K}_{pv}\tilde{\boldsymbol{\eta}} + \boldsymbol{K}_{vv}\dot{\tilde{\boldsymbol{\eta}}} + \boldsymbol{K}_{iv}\int_0^t \tilde{\boldsymbol{\eta}} \end{aligned}$$

and a $n \times 1$ vector defined as

$$\ddot{\boldsymbol{q}}_a = \ddot{\boldsymbol{q}}_d + \boldsymbol{K}_{pq}\tilde{\boldsymbol{q}} + \boldsymbol{K}_{vq}\dot{\tilde{\boldsymbol{q}}} + \boldsymbol{K}_{iq}\int_0^t \tilde{\boldsymbol{q}}\,.$$

The stability analysis is straightforward; assuming perfect dynamic compensation, in fact, gives two different linear models for the earth-fixed vehicle variables and the joint positions. With a proper choice of the matrix gains, moreover, the designer can shape the response of a second-order dynamic system.

## 7.4 Nonlinear Control for UVMSs with Composite Dynamics

In [67, 68, 69, 220] the singular perturbation theory has been considered due to the composite nature of UVMSs. The different bandwidth characteristics of the vehicle/manipulator dynamics are used as a basis for the control design. This has been developed having in mind a Vortex vehicle together with a PA10, 7-DOF, manipulator for which Tables 7.1 and 7.2 reports some time response of the subsystems.

**Table 7.1.** Vortex/PA10's sensor time response

| measured variable | time response |
|---|---|
| surge and sway $x$, $y$ | 400 ms |
| depth $z$ | 1000 ms |
| yaw $\psi$ | 1000 ms |
| pitch and roll $\phi$, $\theta$ | 100 ms |
| joint position $\boldsymbol{q}$ | 1 ms |

**Table 7.2.** Vortex/PA10's actuators time response

| type | time response |
|---|---|
| vehicle thruster | 160 ms |
| manipulator's electrical CD motor | 0.5 ms |

Let consider a state vector composed by the earth-fixed-frame-based coordinates of the vehicle and the joint position. In the following the dependencies will be dropped out to increase readability. It can be demonstrated that its $(6+n) \times (6+n)$ inertia matrix has the form [255]:

$$\begin{bmatrix} \boldsymbol{M}_v^{\star} + \boldsymbol{M}_{qq} & \boldsymbol{M}_{vq} \\ \boldsymbol{M}_{vq}^{\mathrm{T}} & \boldsymbol{M}_q \end{bmatrix}$$

where $\boldsymbol{M}_v^{\star} \in \mathbb{R}^{6\times 6}$ is the earth-fixed inertia matrix of the sole vehicle (see eq. (2.53)), $\boldsymbol{M}_q \in \mathbb{R}^{n\times n}$ is the inertia matrix of the sole manipulator, $\boldsymbol{M}_{qq} \in \mathbb{R}^{6\times 6}$ is the contribution of the manipulator on the inertia matrix *seen* from the vehicle, $\boldsymbol{M}_{vq} \in \mathbb{R}^{6\times n}$ is the coupling term between vehicle and manipulator; all the matrices include the added mass. Its inverse can be parameterized in:

$$\boldsymbol{M}^{-1} = \begin{bmatrix} \boldsymbol{M}_{11}^{-1} & -\boldsymbol{M}_{12} \\ -\boldsymbol{M}_{12}^{\mathrm{T}} & \boldsymbol{M}_{22}^{-1} \end{bmatrix} \tag{7.4}$$

where the block diagonal matrices $\boldsymbol{M}_{11}$ and $\boldsymbol{M}_{22}$ are of dimension $6 \times 6$ and $n \times n$, respectively, and $\boldsymbol{M}_{12} \in \mathbb{R}^{6\times n}$.

The equations of motion can be written as:

$$\ddot{\boldsymbol{\eta}} = \boldsymbol{M}_{11}^{-1}\left(\boldsymbol{\tau}_v^\star - \boldsymbol{n}^\star\right) - \boldsymbol{M}_{12}\left(\boldsymbol{\tau}_q - \boldsymbol{n}_q\right) \tag{7.5}$$

$$\ddot{\boldsymbol{q}} = \boldsymbol{M}_{22}^{-1}\left(\boldsymbol{\tau}_q - \boldsymbol{n}_q\right) - \boldsymbol{M}_{12}^{\mathrm{T}}\left(\boldsymbol{\tau}_v^\star - \boldsymbol{n}^\star\right) \tag{7.6}$$

where the nonlinear terms of the equations of motion have been collected in a $(6+n)$ dimensional vector:

$$\boldsymbol{n} = \begin{bmatrix} \boldsymbol{n}^\star \\ \boldsymbol{n}_q \end{bmatrix}$$

with $\boldsymbol{n}^\star \in \mathrm{I\!R}^6$ and $\boldsymbol{n}_q \in \mathrm{I\!R}^n$.

In [67] the following controller is assumed for the vehicle:

$$\boxed{\boldsymbol{\tau}_v^\star = \left(\hat{\boldsymbol{M}}_v^\star + \hat{\boldsymbol{M}}_{qq}\right)\left(\ddot{\boldsymbol{\eta}}_d + k_{vv}\dot{\tilde{\boldsymbol{\eta}}} + k_{pv}\tilde{\boldsymbol{\eta}}\right) + \Delta\boldsymbol{\tau}_v^\star\,,} \tag{7.7}$$

and for the manipulator

$$\boxed{\boldsymbol{\tau}_q = \hat{\boldsymbol{M}}_q\left(\ddot{\boldsymbol{q}}_d + k_{vq}\dot{\tilde{\boldsymbol{q}}} + k_{pq}\tilde{\boldsymbol{q}}\right)} \tag{7.8}$$

where, as usual, the symbol hat: $\hat{\cdot}$ denotes an estimate, positive definite in this case, of the corresponding matrix and the tilde: $\tilde{\cdot}$ represents the error defined as the desired minus the current variable. The scalar gains are chosen so that:

$$\begin{aligned} k_{vv} &= 2\xi\omega_{0v}\,, \\ k_{pv} &= \omega_{0v}^2\,, \\ k_{vq} &= 2\xi\omega_{0q}\,, \\ k_{pq} &= \omega_{0q}^2\,, \end{aligned}$$

i.e., they are defined by the damping ratio and the natural frequency of the linearized model. The bandwidth ratio

$$\varepsilon = \frac{\omega_{0v}}{\omega_{0q}} \ll 1$$

is given by the closed-loop bandwidths the the vehicle and the manipulator and it is small due to the dynamics of the two subsystems. The additional control action $\Delta\boldsymbol{\tau}_v^\star$ can be chosen in different ways, in [69] a *partial singular perturbed model-based compensation* and a *robust non-linear control* have been proposed.

A singular perturbation analysis can be carried out when $\varepsilon$ is small. It can be demonstrated the the manipulator is not affected by the slow vehicle dynamics and that, after the fast transient, the approximated model has an error $O(\varepsilon)$. On the other side, the vehicle dynamics is strongly affected by the manipulator's motion. For this reason the additional control action $\Delta\boldsymbol{\tau}_v^\star$ is required. In [67], a robust control is developed for the vehicle in order to counteract the coupling effects.

Simulations are carried out considering a dry weight of 150 kg and a length of 1 m for the vehicle and a dry weight of 40 kg and a length of 1 m for the manipulator. Moreover, the natural frequencies have been chosen as $\omega_{0v} = 1.5\,\text{rad/s}$ and $\omega_{0q} = 15\,\text{rad/s}$ leading to a value of $\varepsilon = 0.1$.

## 7.5 Non-regressor-Based Adaptive Control

In 1999 reference [185, 252] extend to UVMSs the non-regressor-based adaptive control developed by J. Yuh [316] and experimentally validated in [83, 216, 320, 323] with respect to AUVs. In [322], this controller is integrated with a disturbance observer to improve its tracking performance and simulated on a 1-DOF vehicle carrying a 2-DOF manipulator.

The main idea is to consider the vehicle subsystem separate from the manipulator subsystem and to develop two controllers with different bandwidth, independent one from the other. It is worth noticing that the controller has been developed together with a kinematic control approach (see Chapter 6).

The vehicle generalized force $\boldsymbol{\tau}_v^\star$, thus, can be computed by considering the controller

$$\boldsymbol{\tau}_v^\star = \boldsymbol{K}_{1,v}\ddot{\boldsymbol{\eta}}_d + \boldsymbol{K}_{2,v}\dot{\boldsymbol{\eta}} + \boldsymbol{K}_{3,v} + \boldsymbol{K}_{4,v}\dot{\tilde{\boldsymbol{\eta}}} + \boldsymbol{K}_{5,v}\tilde{\boldsymbol{\eta}} = \sum_{i=1}^{5} \boldsymbol{K}_{i,v}\boldsymbol{\phi}_{i,v}\,,$$

already defined and discussed in Section 3.3.

The manipulator is controlled with the following

$$\boldsymbol{\tau}_q = \boldsymbol{K}_{1,q}\ddot{\boldsymbol{q}}_d + \boldsymbol{K}_{2,q}\dot{\boldsymbol{q}} + \boldsymbol{K}_{3,q} + \boldsymbol{K}_{4,q}\dot{\tilde{\boldsymbol{q}}} + \boldsymbol{K}_{5,q}\tilde{\boldsymbol{q}} = \sum_{i=1}^{5} \boldsymbol{K}_{i,q}\boldsymbol{\phi}_{i,q}\,,$$

where $\tilde{\boldsymbol{q}} = \boldsymbol{q}_d - \boldsymbol{q}$, the gains $\boldsymbol{K}_{i,q} \in \mathrm{I\!R}^{n\times n}$ are computed as

$$\boldsymbol{K}_{i,q} = \frac{\hat{\gamma}_{i,q}\boldsymbol{s}_q\boldsymbol{\phi}_{i,q}^{\mathrm{T}}}{\|\boldsymbol{s}_q\|\,\|\boldsymbol{\phi}_{i,q}\|} \qquad i = 1,\ldots,5\,,$$

where

$$\boldsymbol{s}_q = \dot{\tilde{\boldsymbol{q}}} + \sigma\tilde{\boldsymbol{q}} \qquad \text{with } \sigma > 0\,,$$

and the factors $\hat{\gamma}_{i,q}$'s are updated by

$$\dot{\hat{\gamma}}_{i,q} = f_{i,q}\,\|\boldsymbol{s}_q\|\,\|\boldsymbol{\phi}_{i,q}\| \qquad \text{with } f_{i,q} > 0 \qquad i = 1,\ldots,5\,.$$

The stability analysis can be found in [185, 252]. In [252], this controller is used together with the kinematic control detailed in Section 6.4 in a simulation study involving a 6-DOF vehicle together with a 3-DOF planar manipulator subject to joint limits. The controller has been developed to be used with the UVMS SAUVIM under development and the Autonomous Systems Laboratory, University of Hawaii (Figure 7.4).

**Fig. 7.4.** SAUVIM under development and the Autonomous Systems Laboratory, University of Hawaii (courtesy of J. Yuh). The passive joint used for position measurement can be observed

## 7.6 Sliding Mode Control

Robust techniques such as Sliding Mode Control have been successfully applied in control of a wide class of mechanical systems. In this Section the application of a Sliding Mode based approach to motion control of UVMSs is discussed.

The basic idea of this approach is the definition of a sliding surface

$$\boldsymbol{s}(\boldsymbol{x},t) = \dot{\tilde{\boldsymbol{x}}} + \boldsymbol{\Lambda}\tilde{\boldsymbol{x}} = \boldsymbol{0} \tag{7.9}$$

where a second order mechanical system has been assumed, $\boldsymbol{x}$ is the state vector, $t$ is the time, $\tilde{\boldsymbol{x}} = \boldsymbol{x}_d - \boldsymbol{x}$ and $\boldsymbol{\Lambda}$ is a positive definite matrix. When the sliding condition is satisfied, the system is forced to *slide* toward the value $\tilde{\boldsymbol{x}} = \boldsymbol{0}$ with an exponential dynamic (for the scalar case, it is $\dot{\tilde{x}} = -\lambda\tilde{x}$). The control input, thus, has the objective to force the state laying in the sliding surface. With a proper choice of the sliding surface, $n$-order systems can be controlled considering a $1^{st}$-order problem in $\boldsymbol{s}$.

The only information required to design a stable sliding mode controller is a bound on the dynamic parameters. While this is an interesting property of the controller, one must pay the price of an high control activity. Typically, sliding mode controllers are based on a switching term that causes chattering in the control inputs.

While the first concepts on the sliding surface appeared in the Soviet literature in the end of the fifties, the first robotic applications of sliding mode control are given in [293, 313]. An introduction on Sliding Mode Control theory con be found in [268].

**Control law.** The vehicle attitude control problem has been addressed among the others in the paper [121] which extends the work in [108] and [265] to obtain a singularity-free tracking control of an underwater vehicle based on the use of the unit quaternion. Inspired by the work in [121], a control law is presented for the regulation problem of an UVMS. To overcome the occurrence of kinematic singularities, the control law is expressed in body-fixed and joint-space coordinates so as to avoid inversion of the system Jacobian. Further, to avoid representation singularities of the orientation, attitude control of the vehicle is achieved through a quaternion based error. The resulting control law is very simple and requires limited computational effort.

Let us recall the dynamic equations in matrix form (2.71):

$$\boldsymbol{M}(\boldsymbol{q})\dot{\boldsymbol{\zeta}} + \boldsymbol{C}(\boldsymbol{q},\boldsymbol{\zeta})\boldsymbol{\zeta} + \boldsymbol{D}(\boldsymbol{q},\boldsymbol{\zeta})\boldsymbol{\zeta} + \boldsymbol{g}(\boldsymbol{q},\boldsymbol{R}_B^I) = \boldsymbol{B}\boldsymbol{u}, \tag{2.71}$$

the control law is

$$\boxed{\boldsymbol{u} = \boldsymbol{B}^{\dagger}\left[\boldsymbol{K}_D\boldsymbol{s} + \hat{\boldsymbol{g}}(\boldsymbol{q},\boldsymbol{R}_B^I) + \boldsymbol{K}_S\,\mathrm{sign}(\boldsymbol{s})\right],} \tag{7.10}$$

where $\boldsymbol{B}^{\dagger}$ is the pseudoinverse of matrix $\boldsymbol{B}$, $\boldsymbol{K}_D$ is a positive definite matrix of gains, $\hat{\boldsymbol{g}}(\boldsymbol{q},\boldsymbol{R}_B^I)$ is the estimate of gravitational and buoyant forces, $\boldsymbol{K}_S$

is a positive definite matrix, and $\text{sign}(\boldsymbol{x})$ is the vector function whose $i$-th component is

$$\text{sign}(\boldsymbol{x})_i = \begin{cases} 1 & \text{if } x_i \geq 0 \\ -1 & \text{if } x_i < 0. \end{cases}$$

In (7.10), $\boldsymbol{s}$ is the $\big((6+n) \times 1\big)$ sliding manifold defined as follows

$$\boldsymbol{s} = \boldsymbol{\Lambda} \begin{bmatrix} \boldsymbol{R}_I^B \tilde{\boldsymbol{\eta}}_1 \\ \tilde{\boldsymbol{\varepsilon}} \\ \tilde{\boldsymbol{q}} \end{bmatrix} - \begin{bmatrix} \boldsymbol{\nu}_1 \\ \boldsymbol{\nu}_2 \\ \dot{\boldsymbol{q}} \end{bmatrix} = \boldsymbol{y} - \boldsymbol{\zeta}\,, \tag{7.11}$$

with $\boldsymbol{\Lambda} > \boldsymbol{O}$, $\tilde{\boldsymbol{\eta}}_1 = [x_d - x \quad y_d - y \quad z_d - z]^{\mathrm{T}}$, $\tilde{\boldsymbol{q}} = \boldsymbol{q}_d - \boldsymbol{q}$ where the subscript $d$ denotes desired values for the relevant variables.

### 7.6.1 Stability Analysis

In this Section it will be demonstrated that the discussed control law is asymptotically stable in a Lyapunov sense. Let us consider the function

$$V = \frac{1}{2}\boldsymbol{s}^{\mathrm{T}}\boldsymbol{M}(\boldsymbol{q})\boldsymbol{s}\,, \tag{7.12}$$

that is positive definite being $\boldsymbol{M}(\boldsymbol{q}) > \boldsymbol{O}$.

Differentiating $V$ with respect to time yields

$$\dot{V} = \frac{1}{2}\boldsymbol{s}^{\mathrm{T}}\dot{\boldsymbol{M}}\boldsymbol{s} + \boldsymbol{s}^{\mathrm{T}}\boldsymbol{M}\dot{\boldsymbol{s}}$$

that, taking into account the model (2.71), (7.11) and the skew-symmetry of $\dot{\boldsymbol{M}} - 2\boldsymbol{C}$, can be rewritten as

$$\dot{V} = -\boldsymbol{s}^{\mathrm{T}}\boldsymbol{D}\boldsymbol{s} + \boldsymbol{s}^{\mathrm{T}}[\boldsymbol{M}\dot{\boldsymbol{y}} - \boldsymbol{B}\boldsymbol{u} + \boldsymbol{C}\boldsymbol{y} + \boldsymbol{D}\boldsymbol{y} + \boldsymbol{g}]\,. \tag{7.13}$$

Plugging (7.10) into (7.13) gives

$$\dot{V} = -\boldsymbol{s}^{\mathrm{T}}(\boldsymbol{D} + \boldsymbol{K}_D)\boldsymbol{s} + \boldsymbol{s}^{\mathrm{T}}[\boldsymbol{M}\dot{\boldsymbol{y}} + (\boldsymbol{C} + \boldsymbol{D})\boldsymbol{y} + \tilde{\boldsymbol{g}} - \boldsymbol{K}_S\,\text{sign}(\boldsymbol{s})]$$

that, in view of positive definiteness of $\boldsymbol{K}_D$ and $\boldsymbol{D}$, can be upper bounded as follows

$$\begin{aligned} \dot{V} \leq & -\lambda_{\min}(\boldsymbol{K}_D + \boldsymbol{D})\,\|\boldsymbol{s}\|^2 - \lambda_{\min}(\boldsymbol{K}_S)\,\|\boldsymbol{s}\| + \\ & + \|\boldsymbol{M}\dot{\boldsymbol{y}} + (\boldsymbol{C} + \boldsymbol{D})\boldsymbol{y} + \tilde{\boldsymbol{g}}\|\,\|\boldsymbol{s}\|\,, \end{aligned}$$

where $\lambda_{\min}$ denotes the smallest eigenvalue of the corresponding matrix.

By choosing $\boldsymbol{K}_S$ such that

$$\lambda_{\min}(\boldsymbol{K}_S) \geq \|\boldsymbol{M}\dot{\boldsymbol{y}} + (\boldsymbol{C} + \boldsymbol{D})\boldsymbol{y} + \tilde{\boldsymbol{g}}\|\,, \tag{7.14}$$

the time derivative of $V$ is negative definite and thus $\boldsymbol{s}$ tends to zero asymptotically.

If an estimate of the dynamic parameters in (2.71) is available, it might be convenient to consider the control law

$$\boldsymbol{u} = \boldsymbol{B}^{\dagger}[\boldsymbol{K}_D \boldsymbol{s} + \hat{\boldsymbol{g}} + \hat{\boldsymbol{M}}\dot{\boldsymbol{y}} + (\hat{\boldsymbol{C}} + \hat{\boldsymbol{D}})\boldsymbol{y} + \boldsymbol{K}_S \operatorname{sign}(\boldsymbol{s})] \tag{7.15}$$

in lieu of (7.10). Starting from the function in (7.12) and plugging (7.15) in (7.13) gives

$$\dot{V} = -\boldsymbol{s}^{\mathrm{T}}(\boldsymbol{D} + \boldsymbol{K}_D)\boldsymbol{s} + \boldsymbol{s}^{\mathrm{T}}[\widetilde{\boldsymbol{M}}\dot{\boldsymbol{y}} + (\widetilde{\boldsymbol{C}} + \widetilde{\boldsymbol{D}})\boldsymbol{y} + \tilde{\boldsymbol{g}} - \boldsymbol{K}_S \operatorname{sign}(\boldsymbol{s})]$$

that, in view of positive definiteness of $\boldsymbol{K}_D$ and $\boldsymbol{D}$, leads to negative definiteness of $\dot{V}$ if

$$\lambda_{\min}(\boldsymbol{K}_S) \geq \left\| \widetilde{\boldsymbol{M}}\dot{\boldsymbol{y}} + (\widetilde{\boldsymbol{C}} + \widetilde{\boldsymbol{D}})\boldsymbol{y} + \tilde{\boldsymbol{g}} \right\| . \tag{7.16}$$

It is worth noting that condition (7.16) is weaker than condition (7.14) in that the matrix $\boldsymbol{K}_S$ must overcome the sole model parameters mismatching.

**Stability of the sliding manifold.** It was demonstrated that the discussed control law guarantees convergence of $\boldsymbol{s}$ to the sliding manifold $\boldsymbol{s} = \boldsymbol{0}$. In the following it will be demonstrated that, once the sliding manifold has been reached, the error vectors $\tilde{\boldsymbol{\eta}}_1$, $\tilde{\boldsymbol{\varepsilon}}$, $\tilde{\boldsymbol{q}}$ converge asymptotically to the origin, i.e., that regulation of the system variables to their desired values is achieved.

By taking $\boldsymbol{\Lambda} = \operatorname{blockdiag}\{\boldsymbol{\Lambda}_p, \boldsymbol{\Lambda}_o, \boldsymbol{\Lambda}_q\}$ where $\boldsymbol{\Lambda}_p \in \mathbb{R}^{3\times 3}$, $\boldsymbol{\Lambda}_o \in \mathbb{R}^{3\times 3}$, $\boldsymbol{\Lambda}_q \in \mathbb{R}^{n\times n}$, from (7.11) it is possible to notice that the stability analysis can be *decoupled* in 3 parts as follows.

**Vehicle position error dynamics.** The vehicle position error dynamics on the sliding manifold is described by the equation

$$-\boldsymbol{\nu}_1 + \boldsymbol{\Lambda}_p \boldsymbol{R}_I^B \tilde{\boldsymbol{\eta}}_1 = \boldsymbol{0} .$$

Notice that the rotation matrix $\boldsymbol{R}_I^B$ is a function of the vehicle orientation.

By considering

$$V = \frac{1}{2}\tilde{\boldsymbol{\eta}}_1^{\mathrm{T}}\tilde{\boldsymbol{\eta}}_1$$

as Lyapunov function candidate and observing that $\dot{\boldsymbol{\eta}}_1 = \boldsymbol{R}_B^I \boldsymbol{\nu}_1$, it is easily obtained

$$\dot{V} = -\tilde{\boldsymbol{\eta}}_1^{\mathrm{T}} \boldsymbol{R}_B^I \boldsymbol{\Lambda}_p \boldsymbol{R}_I^B \tilde{\boldsymbol{\eta}}_1 ,$$

that is negative definite for $\boldsymbol{\Lambda}_p > \boldsymbol{O}$. Hence, $\tilde{\boldsymbol{\eta}}_1$ converges asymptotically to the origin.

**Vehicle orientation error dynamics.** The vehicle orientation error dynamics on the sliding manifold is described by the equation

$$-\boldsymbol{\nu}_2 + \boldsymbol{\Lambda}_o \tilde{\boldsymbol{\varepsilon}} = \boldsymbol{0} \quad \Rightarrow \quad \boldsymbol{\nu}_2 = \boldsymbol{\Lambda}_o \tilde{\boldsymbol{\varepsilon}} . \tag{7.17}$$

Further, by taking into account the quaternion propagation and (7.17), it can be recognized that

$$\dot{\tilde{\eta}} = \frac{1}{2}\tilde{\boldsymbol{\varepsilon}}^{\mathrm{T}}\boldsymbol{\nu}_2 = \frac{1}{2}\tilde{\boldsymbol{\varepsilon}}^{\mathrm{T}}\boldsymbol{\Lambda}_o\tilde{\boldsymbol{\varepsilon}} . \tag{7.18}$$

Let consider the Lyapunov function candidate

$$V = \tilde{\varepsilon}^{\mathrm{T}} \tilde{\varepsilon} . \tag{7.19}$$

The time derivative of $V$ is:

$$\dot{V} = 2\tilde{\varepsilon}^{\mathrm{T}} \dot{\tilde{\varepsilon}} = -\tilde{\varepsilon}^{\mathrm{T}} \tilde{\eta} \boldsymbol{\nu}_2 - \tilde{\varepsilon}^{\mathrm{T}} \boldsymbol{S}(\tilde{\varepsilon}) \boldsymbol{\nu}_2 . \tag{7.20}$$

Plugging (7.17) into (7.20) and taking $\boldsymbol{\Lambda}_o = \lambda_o \boldsymbol{I}_3$ with $\lambda_o > 0$, gives

$$\dot{V} = -\tilde{\eta}\lambda_o \tilde{\varepsilon}^{\mathrm{T}} \tilde{\varepsilon} - \lambda_o \tilde{\varepsilon}^{\mathrm{T}} \boldsymbol{S}(\tilde{\varepsilon})\tilde{\varepsilon} = -\tilde{\eta}\lambda_o \tilde{\varepsilon}^{\mathrm{T}} \tilde{\varepsilon} .$$

which is negative semidefinite with $\tilde{\eta} \geq 0$. It must be noted that, in view of (7.18), $\tilde{\eta}$ is a not-decreasing function of time and thus it stays positive when starting from a positive initial value.

The set $R$ of all points $\tilde{\varepsilon}$ where $\dot{V} = 0$ is given by

$$R = \{\tilde{\varepsilon} = \mathbf{0}, \quad \tilde{\varepsilon} : \tilde{\eta} = 0\} ;$$

from (2.7), however, it can be recognized that

$$\tilde{\eta} = 0 \quad \Rightarrow \quad \|\tilde{\varepsilon}\| = 1$$

and thus $\dot{\tilde{\eta}} > 0$ in view of (7.18). Therefore, the largest invariant set in $R$ is

$$M = \{\tilde{\varepsilon} = \mathbf{0}\}$$

and the invariant set theorem ensures asymptotic convergence to the origin.

**Manipulator joint error dynamics.** The manipulator joint error dynamics on the sliding manifold is described by the equation

$$-\dot{\boldsymbol{q}} + \boldsymbol{\Lambda}_q \tilde{\boldsymbol{q}} = \mathbf{0}$$

whose convergence to $\tilde{\boldsymbol{q}} = \mathbf{0}$ is evident taking $\boldsymbol{\Lambda}_q > \boldsymbol{O}$.

### 7.6.2 Simulations

Dynamic simulations have been performed in order to show the effectiveness of the discussed control law. The UVMS simulator was developed in the MATLAB© /SIMULINK© environment.

For this simulations, the vehicle data are taken from [145]; they refer to the experimental Autonomous Underwater Vehicle NPS Phoenix. A two-link manipulator with rotational joints has been considered which is mounted under the vehicle body with the joint axes parallel to the fore-aft direction; since the vehicle inertia along that axis is minimum, this choice increases dynamic coupling between the vehicle and the manipulator. The length of each link is 1 m, the center of gravity is coincident with the center of buoyancy and it is supposed to be in the geometrical center of the link; each link is not neutrally buoyant. Dry and viscous joint frictions are also taken into account.

As for the control law, implementation of (7.10) was considered; however, it is well known that the sign function would lead to chattering in the system.

Practical implementation of (7.10), therefore, requires replacement of the sign function, e.g., with the sat function

$$\boldsymbol{u} = \boldsymbol{B}^{\dagger}[\boldsymbol{K}_D\boldsymbol{s} + \hat{\boldsymbol{g}}(\boldsymbol{q}, \boldsymbol{R}_I^B) + \boldsymbol{K}_S\mathrm{sat}(\boldsymbol{s}, \varepsilon)]\,, \tag{7.21}$$

where the $\mathrm{sat}(\boldsymbol{x}, \varepsilon)$ is the vector function whose $i$-th component is

$$\mathrm{sat}(\boldsymbol{x}, \varepsilon)_i = \begin{cases} 1 & \text{if } x_i > \varepsilon \\ -1 & \text{if } x_i < -\varepsilon \\ \frac{x_i}{\varepsilon} & \text{otherwise.} \end{cases}$$

Convergence to the equilibrium of the UVMS under this different control law can be easily demonstrated starting from (7.12) following the guidelines in [268]. In detail, it is obtained that $\dot{V} < 0$ in the region characterized by $\|\boldsymbol{s}\| \geq \varepsilon$, while the sign of $\dot{V}$ is undetermined in the boundary layer characterized by $\|\boldsymbol{s}\| < \varepsilon$. This approach is well established in sliding mode control and does not represent a practical drawback since $\varepsilon$ can be taken sufficiently small.

In the simulation $\boldsymbol{B}$ is supposed to be the identity matrix, meaning that direct control of forces and moments acting on the vehicle and joint torques is available. The control law parameters are

$$\begin{aligned} \boldsymbol{\Lambda}_o &= \boldsymbol{\Lambda}_p = \mathrm{diag}\{0.5, 0.5, 0.5\}\,, \\ \boldsymbol{\Lambda}_q &= \mathrm{diag}\{3, 2\}\,, \end{aligned}$$

and

$$\begin{aligned} \boldsymbol{K}_D &= \mathrm{blockdiag}\{10^4\boldsymbol{I}_6, 3000, 500\}\,, \\ \boldsymbol{K}_S &= 1000\boldsymbol{I}_8\,, \\ \varepsilon &= 0.1\,. \end{aligned}$$

A station keeping task for the vehicle in the initial location was considered

$$\boldsymbol{\eta}_i = [0 \quad 0 \quad 0 \quad 0 \quad 0 \quad 0]^{\mathrm{T}} \quad \mathrm{m,rad}$$

with the manipulator in the initial configuration $\boldsymbol{q}_i = \left[-\frac{\pi}{4} \quad \frac{\pi}{2}\right]^{\mathrm{T}}$ rad. The vehicle must be then kept still, i.e., $\boldsymbol{\eta}_d = \boldsymbol{\eta}_i$, while moving the manipulator arm to the desired final configuration $\boldsymbol{q}_f = [0 \quad 0]^{\mathrm{T}}$ rad according to a 5th order polynomial.

It should be noted that the vehicle orientation set point is assigned in terms of Euler angles; these must be converted into the corresponding rotation matrix so as to extract the quaternion expressing the orientation error from the rotation matrix computed as in Subsection 2.2.3. Remarkably, this procedure is free of singularities.

The obtained simulation results are reported in Figures 7.5–7.7 in terms of the time histories of the vehicle position, the vehicle control forces, the vehicle attitude expressed by Euler angles, the vehicle moments, the manipulator joint errors, and the manipulator joint torques, respectively.

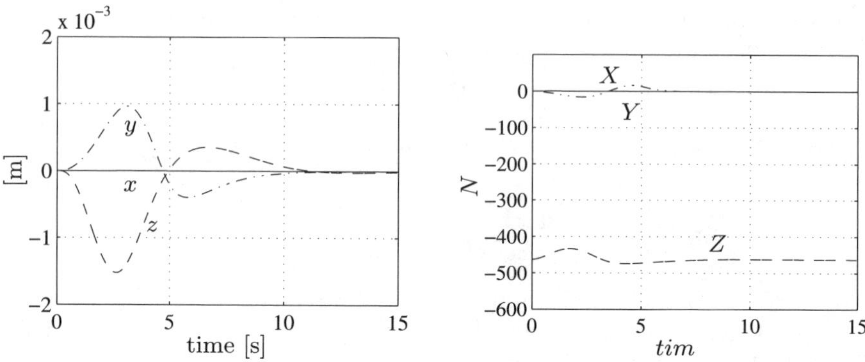


**Fig. 7.5.** Sliding mode control. Left: vehicle positions. Right: vehicle control forces

Figure 7.5 shows that, as expected, the vehicle position is affected by the manipulator motion; however, the displacements are small and the target position is recovered after a transient. It can be recognized that at steady state the force along $z$ is non null; this happens because the manipulator is not neutrally buoyant.

Figure 7.6 shows that the dynamic coupling is mostly experienced along the roll direction because of the chosen UVMS structure. This effect was intentional in order to test the control robustness. It can be recognized that vehicle control moments are zero at steady state; this happens because the center of gravity and the center of buoyancy of vehicle body and manipulator links are all aligned with the $z$-axis of the earth-fixed frame at the final system configuration.

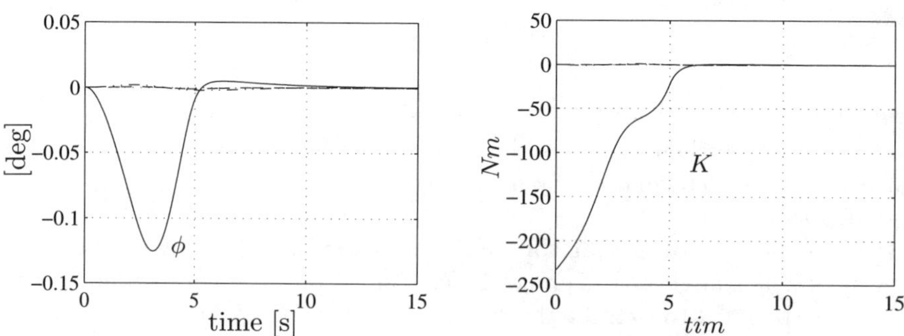


**Fig. 7.6.** Sliding mode control. Left: vehicle attitude in terms of Euler angles. Right: vehicle control moments

Figure 7.7 shows the time histories of manipulator joint errors and torques. It is worth noting that the initial value of the joint torques is non null because of gravity and buoyancy compensation, while they are null at steady state in view of the particular final system configuration. It can be recognized that control generalized forces are smooth while the task is successfully executed.

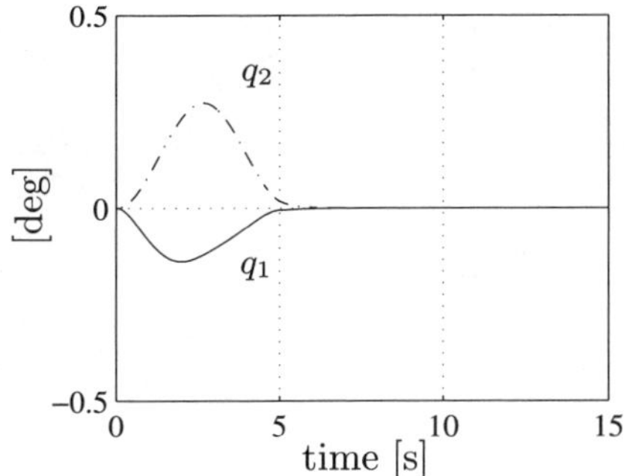


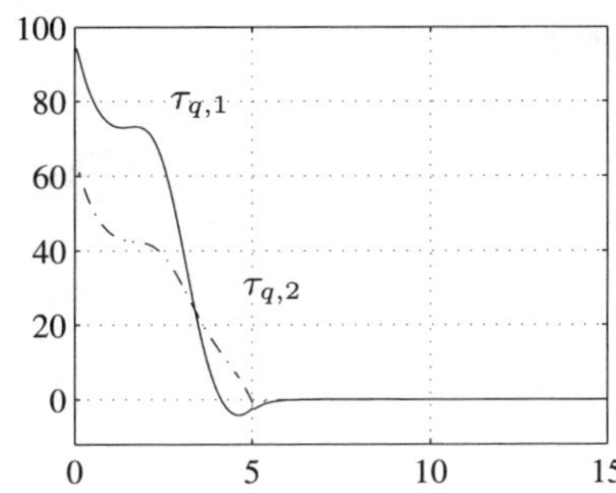


**Fig. 7.7.** Sliding mode control. Left: manipulator joint errors. Right: manipulator joint torques. The steady state vehicle moment and manipulator torques are null due to the restoring force characteristic of this specific UVMS

## 7.7 Adaptive Control

Adaptive Control is a wide topic in control theory. The basic idea of adaptive control is to modify on-line some control gains to *adapt* the controller to the plant's parameters that are supposed to be unknown or slowly varying. However, the term *adaptive control* can assume slightly different meanings. In this work, a controller is considered adaptive if it includes explicitly on-line system parameters estimation.

Often, the mathematical model of the system to be controlled is known but the dynamic parameters are not known or may depend from a load. In this case, the adaptive control is mainly based on a PD action plus a dynamic compensation the parameters of which are updated on-line. This dynamic compensation can be intended to cancel the system dynamics, thus achieving decoupling and linearization of the system, or preserve the passivity properties of the closed loop system [223].

While the first concepts of adaptive control appeared, without success, in the aircraft control in the early fifties, the first robotic applications appeared later [223, 267].

**Control law.** Based on the control law developed in the previous Section, a control law is presented for the tracking problem of UVMSs. As in the previous control law, to overcome the occurrence of kinematic singularities, the control law is expressed in body-fixed and joint-space coordinates so as to avoid inversion of the system Jacobian. Further, to avoid representation singularities of the orientation, attitude control of the vehicle is achieved through a quaternion based error. To achieve good tracking performance, the control law includes model-based compensation of the system dynamics. An adaptive estimate of the model parameters is provided, since they are uncertain and slowly varying.

Given the dynamic equations in matrix form (2.71)–(2.73):

$$\boldsymbol{M}(\boldsymbol{q})\dot{\boldsymbol{\zeta}} + \boldsymbol{C}(\boldsymbol{q},\boldsymbol{\zeta})\boldsymbol{\zeta} + \boldsymbol{D}(\boldsymbol{q},\boldsymbol{\zeta})\boldsymbol{\zeta} + \boldsymbol{g}(\boldsymbol{q},\boldsymbol{R}_B^I) = \boldsymbol{\Phi}(\boldsymbol{q},\boldsymbol{R}_I^B,\boldsymbol{\zeta},\dot{\boldsymbol{\zeta}})\boldsymbol{\theta} = \boldsymbol{B}\boldsymbol{u},$$

the control law is

$$\boldsymbol{u} = \boldsymbol{B}^{\dagger}[\boldsymbol{K}_D\boldsymbol{s}' + \boldsymbol{\Phi}(\boldsymbol{q},\boldsymbol{R}_I^B,\boldsymbol{\zeta},\boldsymbol{\zeta}_r,\dot{\boldsymbol{\zeta}}_r)\hat{\boldsymbol{\theta}}], \tag{7.22}$$

with the update law given by

$$\dot{\hat{\boldsymbol{\theta}}} = \boldsymbol{K}_\theta^{-1}\boldsymbol{\Phi}^{\mathrm{T}}(\boldsymbol{q},\boldsymbol{R}_I^B,\boldsymbol{\zeta},\boldsymbol{\zeta}_r,\dot{\boldsymbol{\zeta}}_r)\boldsymbol{s}, \tag{7.23}$$

where $\boldsymbol{B}^{\dagger}$ is the pseudoinverse of matrix $\boldsymbol{B}$, $\boldsymbol{K}_\theta > \boldsymbol{O}$ and $\boldsymbol{\Phi}$ is the system regressor defined in (2.73). The vectors $\boldsymbol{s}' \in R^{(6+n)\times 1}$ and $\boldsymbol{s} \in R^{(6+n)\times 1}$ are defined as follows

$$\boldsymbol{s}' = \begin{bmatrix} \tilde{\boldsymbol{\nu}}_1 \\ \tilde{\boldsymbol{\nu}}_2 \\ \dot{\tilde{\boldsymbol{q}}} \end{bmatrix} + (\boldsymbol{\Lambda} + \boldsymbol{K}_D^{-1}\boldsymbol{K}_P) \begin{bmatrix} {}^B\boldsymbol{R}_I\tilde{\boldsymbol{\eta}}_1 \\ \tilde{\boldsymbol{\varepsilon}} \\ \tilde{\boldsymbol{q}} \end{bmatrix} = \tilde{\boldsymbol{\zeta}} + (\boldsymbol{\Lambda} + \boldsymbol{K}_D^{-1}\boldsymbol{K}_P)\,\tilde{\boldsymbol{y}}, \tag{7.24}$$

$$\boldsymbol{s} = \tilde{\boldsymbol{\zeta}} + \boldsymbol{\Lambda}\tilde{\boldsymbol{y}}, \tag{7.25}$$

with $\tilde{\boldsymbol{\eta}}_1 = [x_d - x \quad y_d - y \quad z_d - z]^{\mathrm{T}}$, $\tilde{\boldsymbol{q}} = \boldsymbol{q}_d - \boldsymbol{q}$, $\tilde{\boldsymbol{\nu}}_1 = \boldsymbol{\nu}_{1,d} - \boldsymbol{\nu}_1$, $\dot{\tilde{\boldsymbol{q}}} = \dot{\boldsymbol{q}}_d - \dot{\boldsymbol{q}}$, where the subscript $d$ denotes desired values for the relevant variables.

$\boldsymbol{\Lambda}$ is defined as $\boldsymbol{\Lambda} = \mathrm{blockdiag}\{\lambda_p\boldsymbol{I}_3, \lambda_o\boldsymbol{I}_3, \boldsymbol{\Lambda}_q\}$ with $\boldsymbol{\Lambda}_q \in R^{n\times n}$, $\boldsymbol{\Lambda} > \boldsymbol{O}$. $\boldsymbol{K}_P$ is defined as $\boldsymbol{K}_P = \mathrm{blockdiag}\{k_p\boldsymbol{I}_3, k_o\boldsymbol{I}_3, \boldsymbol{K}_q\}$, with $\boldsymbol{K}_q \in R^{n\times n}$, $\boldsymbol{K}_P > \boldsymbol{O}$. $\boldsymbol{K}_q$ and $\boldsymbol{\Lambda}_q$ must be defined so as $\boldsymbol{K}_q\boldsymbol{\Lambda}_q > \boldsymbol{O}$. Finally, it is $\boldsymbol{\zeta}_r = \boldsymbol{\zeta}_d + \boldsymbol{\Lambda}\tilde{\boldsymbol{y}}$ and $\boldsymbol{K}_D > \boldsymbol{O}$.

### 7.7.1 Stability Analysis

In this Section it will be shown that the control law (7.22)–(7.23) is stable in a Lyapunov-Like sense. Let define the following partition for the variable $\boldsymbol{s}$ that will be useful later:

$$\boldsymbol{s} = \begin{bmatrix} \boldsymbol{s}_p \\ \boldsymbol{s}_o \\ \boldsymbol{s}_q \end{bmatrix} \tag{7.26}$$

with $\boldsymbol{s}_p \in R^3$, $\boldsymbol{s}_o \in R^3$, $\boldsymbol{s}_q \in R^n$ respectively.

Let us consider the scalar function

$$V = \frac{1}{2}\boldsymbol{s}^{\mathrm{T}}\boldsymbol{M}(\boldsymbol{q})\boldsymbol{s} + \frac{1}{2}\tilde{\boldsymbol{\theta}}^{\mathrm{T}}\boldsymbol{K}_{\theta}\tilde{\boldsymbol{\theta}} + \\ + \frac{1}{2}\begin{bmatrix}\tilde{\boldsymbol{\eta}}_1\\ \tilde{\boldsymbol{z}}\\ \tilde{\boldsymbol{q}}\end{bmatrix}^{\mathrm{T}}\begin{bmatrix}k_p\boldsymbol{I}_3 & \boldsymbol{O}_{3\times 4} & \boldsymbol{O}_{3\times n}\\ \boldsymbol{O}_{4\times 3} & 2k_o\boldsymbol{I}_4 & \boldsymbol{O}_{4\times n}\\ \boldsymbol{O}_{n\times 3} & \boldsymbol{O}_{n\times 4} & \boldsymbol{K}_q\end{bmatrix}\begin{bmatrix}\tilde{\boldsymbol{\eta}}_1\\ \tilde{\boldsymbol{z}}\\ \tilde{\boldsymbol{q}}\end{bmatrix} \tag{7.27}$$

where $\tilde{\boldsymbol{z}} = [1 \quad \boldsymbol{0}^{\mathrm{T}}]^{\mathrm{T}} - \boldsymbol{z} = [1-\tilde{\eta} \quad -\tilde{\boldsymbol{\varepsilon}}^{\mathrm{T}}]^{\mathrm{T}}$. $V \geq 0$ in view of positive definiteness of $\boldsymbol{M}(\boldsymbol{q})$, $\boldsymbol{K}_{\theta}$, $k_p$, $k_o$ and $\boldsymbol{K}_q$.

Differentiating $V$ with respect to time yields

$$\dot{V} = \frac{1}{2}\boldsymbol{s}^{\mathrm{T}}\dot{\boldsymbol{M}}\boldsymbol{s} + \boldsymbol{s}^{\mathrm{T}}\boldsymbol{M}\dot{\boldsymbol{s}} + \tilde{\boldsymbol{\theta}}^{\mathrm{T}}\boldsymbol{K}_{\theta}\dot{\tilde{\boldsymbol{\theta}}} + \\ + k_p\tilde{\boldsymbol{\eta}}_1^{\mathrm{T}}\boldsymbol{R}_B^I\tilde{\boldsymbol{\nu}}_1 - 2k_o\tilde{\boldsymbol{z}}^{\mathrm{T}}\boldsymbol{J}_{k,oq}(\boldsymbol{z})\tilde{\boldsymbol{\nu}}_2 + \tilde{\boldsymbol{q}}^{\mathrm{T}}\boldsymbol{K}_q\dot{\tilde{\boldsymbol{q}}}\,. \tag{7.28}$$

Observing that, in view of (7.25) and (7.26) it is

$$\tilde{\boldsymbol{\nu}}_1 = \boldsymbol{s}_p - \lambda_p\boldsymbol{R}_I^B\tilde{\boldsymbol{\eta}}_1, \tag{7.29}$$

$$\tilde{\boldsymbol{\nu}}_2 = \boldsymbol{s}_o - \lambda_o\tilde{\boldsymbol{\varepsilon}}, \tag{7.30}$$

$$\dot{\tilde{\boldsymbol{q}}} = \boldsymbol{s}_q - \boldsymbol{\Lambda}_q\tilde{\boldsymbol{q}}, \tag{7.31}$$

and taking into account (2.71), (2.16), (7.29)–(7.31), and the skew-symmetry of $\dot{\boldsymbol{M}} - 2\boldsymbol{C}$, (7.28) can be rewritten as

$$\dot{V} = -\boldsymbol{s}^{\mathrm{T}}\boldsymbol{D}\boldsymbol{s} - \tilde{\boldsymbol{\theta}}^{\mathrm{T}}\boldsymbol{K}_{\theta}\dot{\hat{\boldsymbol{\theta}}} + \boldsymbol{s}^{\mathrm{T}}[\boldsymbol{M}\dot{\boldsymbol{\zeta}}_r - \boldsymbol{B}\boldsymbol{u} + \boldsymbol{C}(\boldsymbol{\zeta})\boldsymbol{\zeta}_r + \boldsymbol{D}(\boldsymbol{\zeta})\boldsymbol{\zeta}_r + \boldsymbol{g}] + \\ + k_p\tilde{\boldsymbol{\eta}}_1^{\mathrm{T}}\boldsymbol{R}_B^I\boldsymbol{s}_p - k_p\lambda_p\tilde{\boldsymbol{\eta}}_1^{\mathrm{T}}\tilde{\boldsymbol{\eta}}_1 + k_o\tilde{\boldsymbol{\varepsilon}}^{\mathrm{T}}\boldsymbol{s}_o - \lambda_o k_o\tilde{\boldsymbol{\varepsilon}}^{\mathrm{T}}\tilde{\boldsymbol{\varepsilon}} + \\ \tilde{\boldsymbol{q}}^{\mathrm{T}}\boldsymbol{K}_q\boldsymbol{s}_q - \tilde{\boldsymbol{q}}^{\mathrm{T}}\boldsymbol{K}_q\boldsymbol{\Lambda}_q\tilde{\boldsymbol{q}} \tag{7.32}$$

where $\dot{\tilde{\theta}} = -\dot{\hat{\theta}}$ was assumed, i.e., the dynamic parameters are constant or slowly varying.

Exploiting (2.73), (7.32) can be rewritten in compact form:

$$\dot{V} = -\begin{bmatrix}\tilde{\boldsymbol{\eta}}_1\\ \tilde{\boldsymbol{\varepsilon}}\\ \tilde{\boldsymbol{q}}\end{bmatrix}^{\mathrm{T}}\begin{bmatrix}k_p\lambda_p\boldsymbol{I}_3 & \boldsymbol{O}_{3\times 3} & \boldsymbol{O}_{3\times n}\\ \boldsymbol{O}_{3\times 3} & k_o\lambda_o\boldsymbol{I}_3 & \boldsymbol{O}_{3\times n}\\ \boldsymbol{O}_{n\times 3} & \boldsymbol{O}_{n\times 3} & \boldsymbol{K}_q\boldsymbol{\Lambda}_q\end{bmatrix}\begin{bmatrix}\tilde{\boldsymbol{\eta}}_1\\ \tilde{\boldsymbol{\varepsilon}}\\ \tilde{\boldsymbol{q}}\end{bmatrix} + \\ + \boldsymbol{s}^{\mathrm{T}}\left[\boldsymbol{\Phi}(\boldsymbol{q},\boldsymbol{R}_I^B,\boldsymbol{\zeta},\boldsymbol{\zeta}_r,\dot{\boldsymbol{\zeta}}_r)\boldsymbol{\theta} - \boldsymbol{B}\boldsymbol{u} + \boldsymbol{K}_P\tilde{\boldsymbol{y}}\right] - \boldsymbol{s}^{\mathrm{T}}\boldsymbol{D}\boldsymbol{s} - \tilde{\boldsymbol{\theta}}^{\mathrm{T}}\boldsymbol{K}_{\theta}\dot{\hat{\boldsymbol{\theta}}}\,. \tag{7.33}$$

Plugging the control law (7.22)–(7.23) into (7.33), one finally obtains:

$$\dot{V} = -\tilde{\boldsymbol{y}}^{\mathrm{T}}\boldsymbol{K}'\tilde{\boldsymbol{y}} - \boldsymbol{s}^{\mathrm{T}}\left(\boldsymbol{K}_D + \boldsymbol{D}\right)\boldsymbol{s}$$

that is negative semi-definite over the state space $\{\tilde{\boldsymbol{y}}, \boldsymbol{s}, \tilde{\boldsymbol{\theta}}\}$.

It is now possible to prove the system stability in a Lyapunov-Like sense using the Barbălat's Lemma. Since $V$ is lower bounded, $\dot{V}(\tilde{\boldsymbol{y}}, \boldsymbol{s}, \tilde{\boldsymbol{\theta}}) \leq 0$ and $\dot{V}(\tilde{\boldsymbol{y}}, \boldsymbol{s}, \tilde{\boldsymbol{\theta}})$ is uniformly continuous then $\dot{V}(\tilde{\boldsymbol{y}}, \boldsymbol{s}, \tilde{\boldsymbol{\theta}}) \to 0$ as $t \to \infty$. Thus $\tilde{\boldsymbol{y}}, \boldsymbol{s} \to \boldsymbol{0}$ as $t \to \infty$. However it is not possible to prove asymptotic stability of the state, since $\tilde{\boldsymbol{\theta}}$ is only guaranteed to be bounded.

### 7.7.2 Simulations

By considering the same UVMS as the previous Section, numerical simulations have been performed to test the discussed control law.

The full simulated model includes a large number of dynamic parameters, thus the symbolic regressor $\boldsymbol{\Phi} \in R^{(6+n)\times n_\theta}$ has a complex expression. While the simulation is performed via the Newton-Euler based algorithm to overcome this complexity, the control law requires the expression of the symbolic regressor. Practical implementation of (7.22)–(7.23), then, might benefit from some simplifications. Considering that the vehicle is usually kept still during the manipulator motion it is first proposed to decouple the vehicle and manipulator dynamics in the regressor computation. The only coupling effect considered is the vehicle orientation in the manipulator restoring effects. Further, it is possible to simplify the hydrodynamic effects taking a linear and a quadratic velocity dependent term for each degree of freedom.

In this reduced form the vehicle regressor is the same as if it was without manipulator. On the other hand the manipulator regressor is the ground fixed regressor, except for the computation of the restoring forces in which the vehicle orientation cannot be omitted; this means that a reduced set of dynamic parameters has been obtained. In this simulation the vehicle parameter vector is $\hat{\boldsymbol{\theta}}_v \in R^{n_{\theta,v}}$ with $n_{\theta,v} = 23$, the manipulator parameter vector is $\hat{\boldsymbol{\theta}}_m \in R^{n_{\theta,m}}$ with $n_{\theta,m} = 13$; it can be recognized that $n_{\theta,v} + n_{\theta,m} \ll n_\theta$.

The regressor implemented in the control law (7.22)–(7.23) has then the form:

$$\boldsymbol{\Phi}(\boldsymbol{q}, \boldsymbol{R}_I^B, \boldsymbol{\zeta}, \boldsymbol{\zeta}_r, \dot{\boldsymbol{\zeta}}_r) = \begin{bmatrix} \boldsymbol{\Phi}_v(\boldsymbol{R}_I^B, \boldsymbol{\nu}, \boldsymbol{\zeta}_{r,v}, \dot{\boldsymbol{\zeta}}_{r,v}) & \boldsymbol{O}_{6\times 13} \\ \boldsymbol{O}_{2\times 23} & \boldsymbol{\Phi}_m(\boldsymbol{q}, \boldsymbol{R}_I^B, \dot{\boldsymbol{q}}, \boldsymbol{\zeta}_{r,m}, \dot{\boldsymbol{\zeta}}_{r,m}) \end{bmatrix}$$

where $\boldsymbol{\Phi}_v \in R^{6\times 23}$ is the vehicle regressor and $\boldsymbol{\Phi}_m \in R^{2\times 13}$ is the manipulator regressor. The corresponding parameter vector is $\hat{\boldsymbol{\theta}} = [\,\hat{\boldsymbol{\theta}}_v^{\mathrm{T}} \;\; \hat{\boldsymbol{\theta}}_m^{\mathrm{T}}\,]^{\mathrm{T}}$. The vectors $\boldsymbol{\zeta}_{r,v}$ and $\boldsymbol{\zeta}_{r,m}$ are the vehicle and manipulator components of $\boldsymbol{\zeta}$.

In the simulation the initial value of $\hat{\boldsymbol{\theta}}$ is affected by an error greater then the 50% of the true value. The control law parameters are

$$\begin{aligned} \boldsymbol{\Lambda} &= \text{blockdiag}\{0.5\boldsymbol{I}_3, 0.5\boldsymbol{I}_3, 3, 2\}\,, \\ \boldsymbol{K}_D &= 1000\boldsymbol{I}_8\,, \\ \boldsymbol{K}_P &= 100\boldsymbol{I}_8\,. \end{aligned}$$

$\dot{\tilde{\boldsymbol{y}}}$ is computed by a filtered numerical time derivative:

$$\dot{\tilde{\boldsymbol{y}}}_k = \alpha \dot{\tilde{\boldsymbol{y}}}_{k-1} + (1-\alpha)\frac{\tilde{\boldsymbol{y}}_k - \tilde{\boldsymbol{y}}_{k-2}}{2\Delta T}\,,$$

with $\Delta T$ being the simulation sampling time.

A station keeping task for the system in the initial configuration was considered

$$\begin{aligned} \boldsymbol{\eta}_i &= [0 \;\; 0 \;\; 0 \;\; 0 \;\; 0 \;\; 0]^{\mathrm{T}} && \text{m, rad}\,, \\ \boldsymbol{q}_i &= \begin{bmatrix} \dfrac{\pi}{4} & \dfrac{\pi}{6} \end{bmatrix}^{\mathrm{T}} && \text{rad}\,. \end{aligned}$$

The vehicle must be then kept still, i.e., $\boldsymbol{\eta}_d = \boldsymbol{\eta}_i$, while moving the manipulator arm to the desired final configuration $\boldsymbol{q}_f = [0 \quad 0]^{\mathrm{T}}$ rad according to a 5th order polynomial. The trajectory is executed two twice without resetting the parameter update.

It should be noted that, if the vehicle orientation trajectory was assigned in terms of Euler angles, these should be converted into the corresponding rotation matrix so as to extract the quaternion expressing the orientation error from the rotation matrix computed as described in Subsection 2.2.3. Remarkably, this procedure is free of singularities.

The obtained simulation results are reported in Figures 7.8–7.10 in terms of the time histories of the vehicle position, the vehicle control forces, the vehicle attitude expressed by Euler angles, the vehicle moments, the manipulator joint errors, and the manipulator joint torques, respectively.

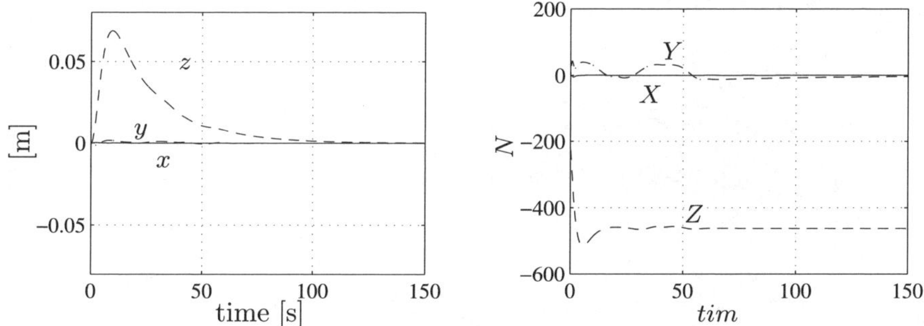


**Fig. 7.8.** Adaptive control. Left: vehicle positions. Right: vehicle control forces

Figure 7.8 shows that, as expected, the vehicle position is affected by the manipulator motion. The main displacement is observed along $z$; this is due to the intentional large initial error in the restoring force compensation. However, the displacements are small and the target position is recovered after a transient. It can be recognized that at steady state the force along $z$ is non null; this happens because the manipulator is not neutrally buoyant. The mismatching in the initial restoring force compensation is recovered by the update of the parameter estimation. The manipulator weight is not included in the vehicle regressor, nevertheless it is compensated as a gravitational vehicle parameter and a null steady state error is obtained.

Figure 7.9 shows that the dynamic coupling is mostly experienced along the roll direction because of the chosen UVMS structure. This effect was intentional in order to test the control robustness. It can be recognized that vehicle control moments are zero at steady state; this happens because the center of gravity and the center of buoyancy of vehicle body and manipulator links are all aligned with the $z$-axis of the earth-fixed frame at the final system configuration.

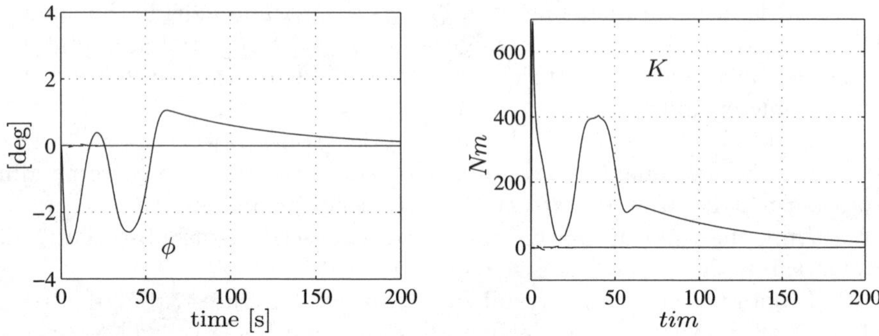


**Fig. 7.9.** Adaptive control. Left: vehicle attitude in terms of Euler angles. Right: vehicle control moments

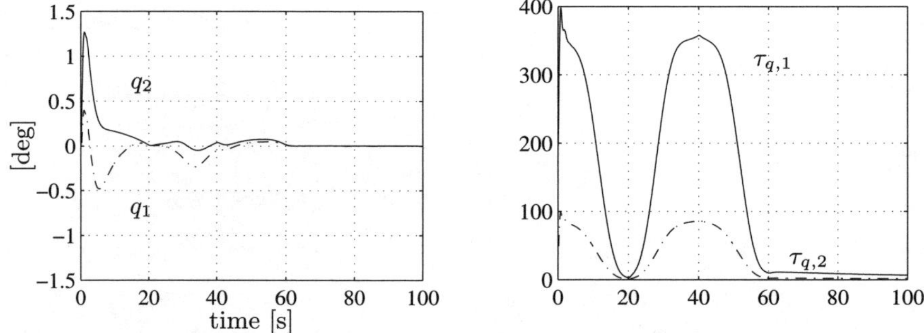


**Fig. 7.10.** Adaptive control. Left: joint position errors. Right: joint control torques

Figure 7.10 shows the time histories of manipulator joint errors and torques. It is worth noting that the initial value of the joint torques is non null because of gravity and buoyancy compensation, while they are null at steady state in view of the particular final system configuration. The large initial joint error is due to mismatching in the restoring torques compensation; the integral action provided by the parameters update gives a null steady state error.

Figure 7.11 finally shows a performance comparison between (7.22)–(7.23) with and without adaptation. Remarkably at the very beginning of the trajectory both control laws perform the same error; afterward the adaptive controller provides a significant error reduction.

## 7.8 Output Feedback Control

Underwater vehicles are typically equipped with acoustic sensors or video systems for position measurements, while the vehicle attitude can be obtained

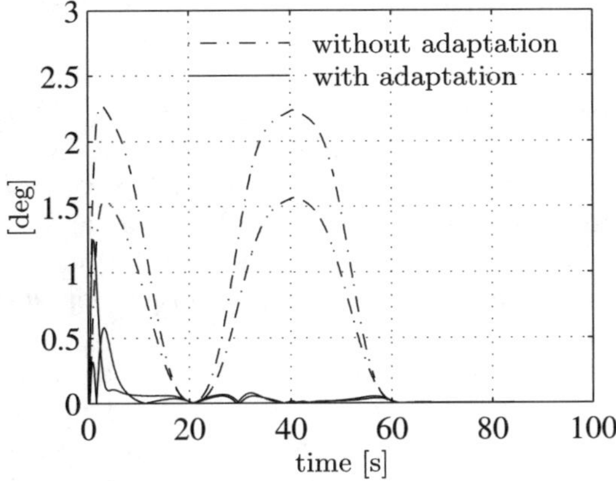


**Fig. 7.11.** Adaptive control. Comparison between adaptive and PD + dynamic compensation in terms of joint position error absolute values

from gyroscopic sensors and/or compasses. Velocity measurements are usually obtained from sensors based on the Doppler effect.

In the case of underwater vehicles operating close to off-shore structures, position and orientation measurements are fairly accurate, while velocity measurements are poor, especially during slow maneuvers. Hence, it is worth devising algorithms for position and attitude control that do not require direct velocity feedback.

A nonlinear observer for vehicle velocity and acceleration has been proposed in [125, 129], although a combined controller-observer design procedure is not developed. On the other hand, a passivity-based control law is proposed in [193], where the velocities are reconstructed via a lead filter; however, this control scheme achieves only regulation of position and orientation variables for an underwater vehicle-manipulator system.

In this Section, the problem of output feedback tracking control of UVMSs is addressed. The output of the controlled system is represented by the position and the attitude of the vehicle, together with the position of the manipulator's joints. Remarkably, the unit quaternion is used to express the orientation of a vehicle-fixed frame so as to avoid representation singularities when expressing the vehicle attitude.

The new control law here discussed is inspired by the work in [45] in that a model-based control law is designed together with a nonlinear observer for velocity estimation; the two structures are tuned to each other in order to achieve exponential convergence to zero of both motion tracking and estimation errors. It must be remarked that differently from the work in [45], where a simple time-derivative relates position and velocity variables at the joints, in the control problem considered in this Chapter, a nonlinear mapping exists between orientation variables (unit quaternion) and angular velocity of the vehicle; this makes the extension of the previous approach to our case not straightforward.

As a matter of fact, the use of numerical differentiation of noisy position/orientation measurements may lead to chattering of the control inputs, and thus to high energy consumption and reduced lifetime of the actuators. Moreover, low-pass filtering of the numerically reconstructed velocities may significantly degrade the system's dynamic behavior and, eventually, affect the closed-loop stability. In other words, such a filter has to be designed together with the controller so as to preserve closed-loop stability and good tracking performance.

This is the basic idea which inspired the approach described in the following: namely, a nonlinear filter (observer) on the position and attitude measures is designed together with a model-based controller so as to achieve exponential stability and ensure tracking of the desired position and attitude trajectories.

A Lyapunov stability analysis is developed to establish sufficient conditions on the control and observer parameters ensuring exponential convergence of tracking and estimation errors.

In view of the limited computational power available in real-time digital control hardware, simplified control laws are suggested aimed at suitably trading-off tracking performance against reduced computational load. Also, the problem of evaluating some dynamic compensation terms, to be properly estimated, is addressed.

A simulation case study is carried out to demonstrate practical application of the discussed control scheme to the experimental vehicle NPS AUV Phoenix [145]. The obtained performance is compared to that achieved with a control scheme in which velocity is reconstructed via numerical differentiation of position measurements.

**Controller-observer scheme.** The desired position for the vehicle is assigned in terms of the vector $\boldsymbol{\eta}_{1,d}(t)$, while the commanded attitude trajectory can be assigned in terms of the rotation matrix $\boldsymbol{R}^I_{B,d}(t)$ expressing the orientation of the desired vehicle frame $\Sigma_d$ with respect to $\Sigma_i$. Equivalently, the desired orientation can be expressed in terms of the unit quaternion $\mathcal{Q}_d(t)$ corresponding to $\boldsymbol{R}^I_{B,d}(t)$. Finally, the desired joint motion is assigned in terms of the vector of joint variables $\boldsymbol{q}_d(t)$.

The desired velocity vectors are denoted by $\dot{\boldsymbol{\eta}}_{1,d}(t)$, $\boldsymbol{\nu}^I_{2,d}(t)$, and $\dot{\boldsymbol{q}}_d(t)$, while the desired accelerations are assigned in terms of the vectors $\ddot{\boldsymbol{\eta}}_{1,d}(t)$, $\dot{\boldsymbol{\nu}}^I_{2,d}(t)$, and $\ddot{\boldsymbol{q}}_d(t)$.

Notice that all the desired quantities are naturally assigned with respect to the earth-fixed frame $\Sigma_i$; the corresponding position and velocity in the vehicle-fixed frame $\Sigma_b$ are computed as

$$\boldsymbol{\eta}^B_{1,d} = \boldsymbol{R}^B_I \boldsymbol{\eta}_{1,d}, \quad \boldsymbol{\zeta}_d = \begin{bmatrix} \boldsymbol{R}^B_I \dot{\boldsymbol{\eta}}_{1,d} \\ \boldsymbol{R}^B_I \boldsymbol{\nu}^I_{2,d} \\ \dot{\boldsymbol{q}}_d \end{bmatrix} = \begin{bmatrix} \boldsymbol{\nu}_{1,d} \\ \boldsymbol{\nu}_{2,d} \\ \dot{\boldsymbol{q}}_d \end{bmatrix}.$$

It is worth pointing out that the computation of the desired acceleration $\dot{\zeta}_d$ requires knowledge of the actual angular velocity $\nu_2$; in fact, in view of $\dot{R}_I^B = -S(\nu_2)R_I^B$, it is

$$\dot{\zeta}_d = \begin{bmatrix} R_I^B \ddot{\eta}_{1,d} - S(\nu_2)\nu_{1,d} \\ R_I^B \dot{\nu}_{2,d}^I - S(\nu_2)\nu_{2,d} \\ \ddot{q}_d \end{bmatrix} .$$

Hence, it is convenient to use in the control law the modified acceleration vector defined as

$$a_d = \begin{bmatrix} R_I^B \ddot{\eta}_{1,d} - S(\nu_{2,d})\nu_{1,d} \\ R_I^B \dot{\nu}_{2,d}^I - S(\nu_{2,d})\nu_{2,d} \\ \ddot{q}_d \end{bmatrix} ,$$

which can be evaluated without using the actual velocity; the two vectors are related by the equality

$$\dot{\zeta}_d = a_d + S_{PO}(\widetilde{\nu}_{2,d})\zeta_d ,$$

where $S_{PO}(\cdot) = \text{blockdiag}\{S(\cdot), S(\cdot), O_{n\times n}\}$, and $\widetilde{\nu}_{2,d} = \nu_{2,d} - \nu_2$.

Hereafter it is assumed that $\|\zeta_d(t)\| \le \zeta_{dM}$ for all $t \ge 0$.

A tracking control law is naturally based on the *tracking error*

$$e_d = \begin{bmatrix} \widetilde{\eta}_{1,d}^B \\ \widetilde{\varepsilon}_d \\ \widetilde{q}_d \end{bmatrix} , \tag{7.34}$$

where $\widetilde{\eta}_{1,d}^B = \eta_{1,d}^B - \eta_1^B$, $\widetilde{q}_d = q_d - q$ and $\widetilde{\varepsilon}_d$ is the vector part of the unit quaternion $\widetilde{\mathcal{Q}}_d = \mathcal{Q}^{-1} * \mathcal{Q}_d$.

It must be noticed that a derivative control action based on (7.34) would require velocity measurements in the control loop. In the absence of velocity measurements, a suitable estimate $\zeta_e$ of the velocity vector has to be considered. Let also $\eta_{1,e}^B$ and $\mathcal{Q}_e$ denote the estimated position and attitude of the vehicle, respectively; the estimated joint variables are denoted by $q_e$. Hence, the following error vector has to be considered

$$e_{de} = \begin{bmatrix} \widetilde{\eta}_{1,de}^B \\ \widetilde{\varepsilon}_{de}^e \\ \widetilde{q}_{de} \end{bmatrix} , \tag{7.35}$$

where $\widetilde{\eta}_{1,de}^B = \eta_{1,d}^B - \eta_{1,e}^B$, $\widetilde{q}_{de} = q_d - q_e$, and $\widetilde{\varepsilon}_{de}^e$ is the vector part of the unit quaternion $\widetilde{\mathcal{Q}}_{de} = \mathcal{Q}_e^{-1} * \mathcal{Q}_d$.

In order to avoid direct velocity feedback, the corresponding velocity error can be defined as

$$\widetilde{\zeta}_{de} = \begin{bmatrix} R_I^B \dot{\widetilde{\eta}}_{1,de} - S(\nu_{2,d})\widetilde{\eta}_{1,de}^B \\ \dot{\widetilde{\varepsilon}}_{de}^e \\ \dot{\widetilde{q}}_{de} \end{bmatrix} ,$$

which is related to the time derivative of $\boldsymbol{e}_{de}$ as follows

$$\dot{\boldsymbol{e}}_{de} = \widetilde{\boldsymbol{\zeta}}_{de} + \boldsymbol{S}_P(\widetilde{\boldsymbol{\nu}}_{2,d})\boldsymbol{e}_{de} \,,$$

where $\boldsymbol{S}_P(\cdot) = \text{blockdiag}\{\boldsymbol{S}(\cdot), \boldsymbol{O}_3, \boldsymbol{O}_{n\times n}\}$.

In order to design an observer providing velocity estimates, the *estimation error* has to be considered

$$\boldsymbol{e}_e = \begin{bmatrix} \widetilde{\boldsymbol{\eta}}_{1,e}^B \\ \widetilde{\boldsymbol{\varepsilon}}_e \\ \widetilde{\boldsymbol{q}}_e \end{bmatrix} ,$$

where $\widetilde{\boldsymbol{\eta}}_{1,e}^B = \boldsymbol{\eta}_{1,e}^B - \boldsymbol{\eta}_1^B$, $\widetilde{\boldsymbol{q}}_e = \boldsymbol{q}_e - \boldsymbol{q}$, and $\widetilde{\boldsymbol{\varepsilon}}_e$ is the vector part of the unit quaternion $\widetilde{\mathcal{Q}}_e = \mathcal{Q}^{-1} * \mathcal{Q}_e$.

Finally, consider the vectors

$$\boldsymbol{\zeta}_r = \boldsymbol{\zeta}_d + \boldsymbol{\Lambda}_d\boldsymbol{e}_{de} \tag{7.36}$$

$$\boldsymbol{\zeta}_o = \boldsymbol{\zeta}_e + \boldsymbol{\Lambda}_e\boldsymbol{e}_e . \tag{7.37}$$

where $\boldsymbol{\Lambda}_d = \text{blockdiag}\{\boldsymbol{\Lambda}_{dP}, \lambda_{dO}\boldsymbol{I}_3, \boldsymbol{\Lambda}_{dQ}\}$, $\boldsymbol{\Lambda}_e = \text{blockdiag}\{\boldsymbol{\Lambda}_{eP}, \lambda_{eO}\boldsymbol{I}_3, \boldsymbol{\Lambda}_{eQ}\}$ are diagonal and positive definite matrices. It is worth remarking that $\boldsymbol{\zeta}_r$ and $\boldsymbol{\zeta}_o$ can be evaluated without using the actual velocity $\boldsymbol{\zeta}$.

Let us recall the dynamic equations in matrix form (2.71):

$$\boldsymbol{M}(\boldsymbol{q})\dot{\boldsymbol{\zeta}} + \boldsymbol{C}(\boldsymbol{q},\boldsymbol{\zeta})\boldsymbol{\zeta} + \boldsymbol{D}(\boldsymbol{q},\boldsymbol{\zeta})\boldsymbol{\zeta} + \boldsymbol{g}(\boldsymbol{q},\boldsymbol{R}_B^I) = \boldsymbol{B}\boldsymbol{u}, \tag{2.71}$$

the control law is

$$\begin{aligned}\boldsymbol{u} = \boldsymbol{B}^\dagger \big[&\boldsymbol{M}(\boldsymbol{q})\boldsymbol{a}_r + \boldsymbol{C}(\boldsymbol{q},\boldsymbol{\zeta}_o)\boldsymbol{\zeta}_r + \boldsymbol{K}_v(\boldsymbol{\zeta}_r - \boldsymbol{\zeta}_o) + \\ &+\boldsymbol{K}_p\boldsymbol{e}_d + \boldsymbol{g}(\boldsymbol{q},\boldsymbol{R}_B^I) + \tfrac{1}{2}\boldsymbol{D}(\boldsymbol{q},\boldsymbol{\zeta}_r)(\boldsymbol{\zeta}_r + \boldsymbol{\zeta}_o)\big] ,\end{aligned} \tag{7.38}$$

where $\boldsymbol{K}_p = \text{blockdiag}\{k_{pP}\boldsymbol{I}_3, k_{pO}\boldsymbol{I}_3, \boldsymbol{K}_{pQ}\}$ is a diagonal positive definite matrix and $\boldsymbol{K}_v$ is a symmetric positive definite matrix. The reference acceleration vector $\boldsymbol{a}_r$ is defined as

$$\boldsymbol{a}_r = \boldsymbol{a}_d + \boldsymbol{\Lambda}_d\widetilde{\boldsymbol{\zeta}}_{de} \,, \tag{7.39}$$

and thus the control law (7.38) does not require feedback of the vehicle and/or manipulator velocities.

The estimated velocity vector $\boldsymbol{\zeta}_e$ is obtained via the observer defined by the equations:

$$\left\{\begin{aligned} \dot{\boldsymbol{z}} &= \boldsymbol{M}(\boldsymbol{q})\boldsymbol{a}_r - \big(\boldsymbol{L}_p + \boldsymbol{L}_v\boldsymbol{A}(\widetilde{\mathcal{Q}}_e)\boldsymbol{\Lambda}_e\big)\boldsymbol{e}_e + \boldsymbol{K}_p\boldsymbol{e}_d + \\ &\quad\ \boldsymbol{C}(\boldsymbol{q},\boldsymbol{\zeta}_o)\boldsymbol{\zeta}_r + \boldsymbol{C}^{\mathrm{T}}(\boldsymbol{q},\boldsymbol{\zeta}_r)\boldsymbol{\zeta}_o \\ \boldsymbol{\zeta}_e &= \boldsymbol{M}^{-1}(\boldsymbol{q})\,(\boldsymbol{z} - \boldsymbol{L}_v\boldsymbol{e}_e) - \boldsymbol{\Lambda}_e\boldsymbol{e}_e, \end{aligned}\right. \tag{7.40}$$

where the matrix $\boldsymbol{L}_p = \text{blockdiag}\{l_{pP}\boldsymbol{I}_3, l_{pO}\boldsymbol{I}_3, \boldsymbol{L}_{pQ}\}$ is diagonal positive definite. The matrix $\boldsymbol{L}_v = \text{blockdiag}\{\boldsymbol{L}_{vP}, l_{vO}\boldsymbol{I}_3, \boldsymbol{L}_{vQ}\}$ is symmetric and positive definite, and

$$\boldsymbol{A}(\widetilde{\mathcal{Q}}_e) = \text{blockdiag}\left\{\boldsymbol{I}_3,\ \boldsymbol{E}(\widetilde{\mathcal{Q}}_e)/2,\ \boldsymbol{I}_n\right\}.$$

The estimated quantities $\boldsymbol{\eta}_{1,e}$ and $\boldsymbol{q}_e$ are computed by integrating the corresponding estimated velocities $\dot{\boldsymbol{\eta}}_{1,e} = \boldsymbol{R}_B^I \boldsymbol{\nu}_{1,e}$ and $\dot{\boldsymbol{q}}_e$, respectively, whereas the estimated orientation $\mathcal{Q}_e$ is computed from the estimated angular velocity $\boldsymbol{\nu}_{2,e}^I = \boldsymbol{R}_B^I \boldsymbol{\nu}_{2,e}$ via the quaternion propagation rule.

**Implementation issues.** Implementation of the above controller-observer scheme (7.38),(7.40), requires computation of the dynamic compensation terms. While this can be done quite effectively for the terms related to rigid body dynamics, the terms related to hydrodynamic effects are usually affected by some degree of approximation and/or uncertainty. Besides the use of adaptive control schemes aimed at on-line estimation of relevant model parameters, e.g. [22, 197, 314], it is important to have an estimate of the main hydrodynamic coefficients.

An estimate of the added mass coefficients can be obtained via strip theory [127]. A rough approximation of the hydrodynamic damping is obtained by considering only the linear skin friction and the drag generalized forces.

Another important point concerns the computational complexity associated with dynamic compensation against the limited computing power typically available on board. This might suggest the adoption of a control law computationally lighter than the one derived above. A reasonable compromise between tracking performance and computational burden is achieved if the compensation of Coriolis, centripetal and damping terms is omitted resulting in the controller

$$\boldsymbol{u} = \boldsymbol{B}^{\dagger}\big(\boldsymbol{M}(\boldsymbol{q})\boldsymbol{a}_r + \boldsymbol{K}_v(\boldsymbol{\zeta}_r - \boldsymbol{\zeta}_o) + \boldsymbol{K}_p\boldsymbol{e}_d + \boldsymbol{g}(\boldsymbol{q}, \boldsymbol{R}_B^I)\big)\,, \tag{7.41}$$

with the simplified observer

$$\begin{cases} \dot{\boldsymbol{z}} = \boldsymbol{M}(\boldsymbol{q})\boldsymbol{a}_r - \big(\boldsymbol{L}_p + \boldsymbol{L}_v\boldsymbol{A}(\widetilde{\mathcal{Q}}_e)\boldsymbol{\Lambda}_e\big)\boldsymbol{e}_e + \boldsymbol{K}_p\boldsymbol{e}_d \\ \boldsymbol{\zeta} = \boldsymbol{M}^{-1}(\boldsymbol{q})\,(\boldsymbol{z} - \boldsymbol{L}_v\boldsymbol{e}_e) - \boldsymbol{\Lambda}_e\boldsymbol{e}_e, \end{cases} \tag{7.42}$$

The computational load can be further reduced if a suitable constant diagonal inertia matrix $\widehat{\boldsymbol{M}}$ is used in lieu of the matrix $\boldsymbol{M}(\boldsymbol{q})$, i.e.,

$$\boldsymbol{u} = \boldsymbol{B}^{\dagger}\big(\widehat{\boldsymbol{M}}\boldsymbol{a}_r + \boldsymbol{K}_v(\boldsymbol{\zeta}_r - \boldsymbol{\zeta}_o) + \boldsymbol{K}_p\boldsymbol{e}_d + \boldsymbol{g}(\boldsymbol{q}, \boldsymbol{R}_B^I)\big)\,, \tag{7.43}$$

with the observer

$$\begin{cases} \dot{\boldsymbol{z}} = \widehat{\boldsymbol{M}}\boldsymbol{a}_r - \big(\boldsymbol{L}_p + \boldsymbol{L}_v\boldsymbol{A}(\widetilde{\mathcal{Q}}_e)\boldsymbol{\Lambda}_e\big)\boldsymbol{e}_e + \boldsymbol{K}_p\boldsymbol{e}_d \\ \widehat{\boldsymbol{\zeta}} = \widehat{\boldsymbol{M}}^{-1}(\boldsymbol{z} - \boldsymbol{L}_v\boldsymbol{e}_e) - \boldsymbol{\Lambda}_e\boldsymbol{e}_e\,, \end{cases} \tag{7.44}$$

Table 7.3 shows the computational load of each control law, in terms of required floating point operations, in the case of a 6-DOF vehicle equipped with a 3-DOF manipulator. Where required, inversion of the inertia matrix has been obtained via the Cholesky factorization since $\boldsymbol{M}(\boldsymbol{q})$ is symmetric and positive definite; of course, the inverse of the constant matrix $\widehat{\boldsymbol{M}}$ is computed once off-line. As shown by the results in Table 7.3, the computational load is reduced by about 80% when the control law (7.43),(7.44) is considered.

**Table 7.3.** Computational burden for different output feedback controllers

| | mult/div | add/sub |
|---|---|---|
| Control law (7.38),(7.40) | 1831 | 1220 |
| Control law (7.41),(7.42) | 1216 | 849 |
| Control law (7.43),(7.44) | 354 | 147 |

### 7.8.1 Stability Analysis

In order to derive the closed-loop dynamic equations, it is useful to define the variables

$$\boldsymbol{\sigma}_d = \boldsymbol{\zeta}_r - \boldsymbol{\zeta} = \widetilde{\boldsymbol{\zeta}}_d + \boldsymbol{\Lambda}_d \boldsymbol{e}_{de} \tag{7.45}$$

$$\boldsymbol{\sigma}_e = \boldsymbol{\zeta}_o - \boldsymbol{\zeta} = \widetilde{\boldsymbol{\zeta}}_e + \boldsymbol{\Lambda}_e \boldsymbol{e}_e \,, \tag{7.46}$$

where

$$\widetilde{\boldsymbol{\zeta}}_d = \boldsymbol{\zeta}_d - \boldsymbol{\zeta} \tag{7.47}$$

$$\widetilde{\boldsymbol{\zeta}}_e = \boldsymbol{\zeta}_e - \boldsymbol{\zeta} \,. \tag{7.48}$$

Combining (2.71) with the control law (7.38), (7.39), and using the equality

$$\boldsymbol{a}_r = \dot{\boldsymbol{\zeta}}_r + \boldsymbol{S}_{PO}(\widetilde{\boldsymbol{\nu}}_{2,d})\boldsymbol{\zeta}_d + \boldsymbol{\Lambda}_d \boldsymbol{S}_P(\widetilde{\boldsymbol{\nu}}_{2,d})\boldsymbol{e}_{de} \,,$$

the tracking error dynamics can be derived

$$\begin{aligned} \boldsymbol{M}(\boldsymbol{q})\dot{\boldsymbol{\sigma}}_d + \boldsymbol{C}(\boldsymbol{q},\boldsymbol{\zeta})\boldsymbol{\sigma}_d + \boldsymbol{K}_v\boldsymbol{\sigma}_d - \boldsymbol{K}_p\boldsymbol{e}_d = {} & \boldsymbol{K}_v\boldsymbol{\sigma}_e - \boldsymbol{C}(\boldsymbol{q},\boldsymbol{\sigma}_e)\boldsymbol{\zeta}_r + \\ & -\boldsymbol{M}(\boldsymbol{q})\boldsymbol{S}_{PO}(\widetilde{\boldsymbol{\nu}}_{2,d})\boldsymbol{\zeta}_d - \boldsymbol{M}(\boldsymbol{q})\boldsymbol{\Lambda}_d\boldsymbol{S}_P(\widetilde{\boldsymbol{\nu}}_{2,d})\boldsymbol{e}_{de} + \\ & \boldsymbol{D}(\boldsymbol{q},\boldsymbol{\zeta})\boldsymbol{\zeta} - \frac{1}{2}\boldsymbol{D}(\boldsymbol{q},\boldsymbol{\zeta}_r)(\boldsymbol{\zeta}_r + \boldsymbol{\zeta}_o) \,. \end{aligned} \tag{7.49}$$

The observer equation (7.40), together with (7.49), yields the estimation error dynamics

$$\boldsymbol{M}(\boldsymbol{q})\dot{\boldsymbol{\sigma}}_e + \left(\boldsymbol{L}_v\boldsymbol{A}(\widetilde{\mathcal{Q}}_e) - \boldsymbol{K}_v\right)\boldsymbol{\sigma}_e - \boldsymbol{L}_p\boldsymbol{e}_e = -\boldsymbol{K}_v\boldsymbol{\sigma}_d - \boldsymbol{C}(\boldsymbol{q},\boldsymbol{\zeta})\boldsymbol{\sigma}_e +$$

$$C^{\mathrm{T}}(\boldsymbol{q},\boldsymbol{\sigma}_d)\boldsymbol{\zeta}_o + \boldsymbol{D}(\boldsymbol{q},\boldsymbol{\zeta})\boldsymbol{\zeta} - \frac{1}{2}\boldsymbol{D}(\boldsymbol{q},\boldsymbol{\zeta}_r)(\boldsymbol{\zeta}_r + \boldsymbol{\zeta}_o)\,. \tag{7.50}$$

A state vector for the closed-loop system (7.49),(7.50) is then

$$\underline{\boldsymbol{x}} = \begin{bmatrix} \boldsymbol{\sigma}_d \\ \boldsymbol{e}_d \\ \boldsymbol{\sigma}_e \\ \boldsymbol{e}_e \end{bmatrix}.$$

Notice that perfect tracking of the desired motion together with exact estimate of the system velocities results in $\underline{\boldsymbol{x}} = \mathbf{0}$. Therefore, the control objective is fulfilled if the closed loop system (7.49), (7.50) is asymptotically stable at the origin of its state space. This is ensured by the following theorem:

**Theorem.** *There exists a choice of the controller gains* $\boldsymbol{K}_p$, $\boldsymbol{K}_v$, $\boldsymbol{\Lambda}_d$ *and of the observer parameters* $\boldsymbol{L}_p$, $\boldsymbol{L}_v$, $\boldsymbol{\Lambda}_e$ *such that the origin of the state space of system (7.49),(7.50) is locally exponentially stable.*

Consider the positive definite Lyapunov function candidate

$$\begin{aligned} V = {} & \frac{1}{2}\boldsymbol{\sigma}_d^{\mathrm{T}}\boldsymbol{M}(\boldsymbol{q})\boldsymbol{\sigma}_d + \frac{1}{2}\boldsymbol{\sigma}_e^{\mathrm{T}}\boldsymbol{M}(\boldsymbol{q})\boldsymbol{\sigma}_e + \\ & + \frac{1}{2}k_{pP}\widetilde{\boldsymbol{\eta}}_{1,d}^{B\mathrm{T}}\widetilde{\boldsymbol{\eta}}_{1,d}^{B} + k_{pO}\left((1-\widetilde{\eta}_d)^2 + \widetilde{\boldsymbol{\varepsilon}}_d^{\mathrm{T}}\widetilde{\boldsymbol{\varepsilon}}_d\right) + \frac{1}{2}\widetilde{\boldsymbol{q}}_d^{\mathrm{T}}\boldsymbol{K}_{pQ}\widetilde{\boldsymbol{q}}_d + \\ & + \frac{1}{2}l_{pP}\widetilde{\boldsymbol{\eta}}_{1,e}^{B\mathrm{T}}\widetilde{\boldsymbol{\eta}}_{1,e}^{B} + l_{pO}\left((1-\widetilde{\eta}_e)^2 + \widetilde{\boldsymbol{\varepsilon}}_e^{\mathrm{T}}\widetilde{\boldsymbol{\varepsilon}}_e\right) + \frac{1}{2}\widetilde{\boldsymbol{q}}_e^{\mathrm{T}}\boldsymbol{L}_{pQ}\widetilde{\boldsymbol{q}}_e\,. \end{aligned} \tag{7.51}$$

The time derivative of $V$ along the trajectories of the closed-loop system (7.49),(7.50) is given by

$$\begin{aligned} \dot{V} = {} & -\boldsymbol{\sigma}_d^{\mathrm{T}}\boldsymbol{K}_v\boldsymbol{\sigma}_d - \boldsymbol{e}_{de}^{\mathrm{T}}\boldsymbol{\Lambda}_d\boldsymbol{K}_p\boldsymbol{e}_d - \boldsymbol{e}_e^{\mathrm{T}}\boldsymbol{\Lambda}_e\boldsymbol{L}_p\boldsymbol{e}_e + \\ & -\boldsymbol{\sigma}_e^{\mathrm{T}}\left(\boldsymbol{L}_v\boldsymbol{A}(\widetilde{\mathcal{Q}}_e) - \boldsymbol{K}_v\right)\boldsymbol{\sigma}_e - \boldsymbol{\sigma}_d^{\mathrm{T}}\boldsymbol{C}(\boldsymbol{q},\boldsymbol{\sigma}_e)\boldsymbol{\zeta}_r + \\ & -\boldsymbol{\sigma}_e^{\mathrm{T}}\boldsymbol{C}(\boldsymbol{q},\boldsymbol{\zeta})\boldsymbol{\sigma}_e + \boldsymbol{\sigma}_e^{\mathrm{T}}\boldsymbol{C}^{\mathrm{T}}(\boldsymbol{q},\boldsymbol{\sigma}_d)\boldsymbol{\zeta}_o + \\ & +(\boldsymbol{\sigma}_d+\boldsymbol{\sigma}_e)^{\mathrm{T}}\boldsymbol{D}(\boldsymbol{q},\boldsymbol{\zeta})\boldsymbol{\zeta} + -\frac{1}{2}(\boldsymbol{\sigma}_d+\boldsymbol{\sigma}_e)^{\mathrm{T}}\boldsymbol{D}(\boldsymbol{q},\boldsymbol{\zeta}_r)(\boldsymbol{\zeta}_r+\boldsymbol{\zeta}_o) + \\ & -\boldsymbol{\sigma}_d^{\mathrm{T}}\boldsymbol{M}(\boldsymbol{q})\boldsymbol{S}_{PO}(\widetilde{\boldsymbol{\nu}}_{2,d})\boldsymbol{\zeta}_d - \boldsymbol{\sigma}_d^{\mathrm{T}}\boldsymbol{M}(\boldsymbol{q})\boldsymbol{\Lambda}_d\boldsymbol{S}_P(\widetilde{\boldsymbol{\nu}}_{2,d})\boldsymbol{e}_{de}\,. \end{aligned} \tag{7.52}$$

In the following it is assumed that $\widetilde{\eta}_d > 0$, $\widetilde{\eta}_e > 0$; in view of the angle/axis interpretation of the unit quaternion, the above assumption corresponds to considering orientation errors characterized by angular displacements in the range $]-\pi,\pi[$.

From the equality $\widetilde{\mathcal{Q}}_{de} = \widetilde{\mathcal{Q}}_e^{-1} * \widetilde{\mathcal{Q}}_d$, the following equality follows

$$\widetilde{\boldsymbol{\varepsilon}}_{de}^{eT}\widetilde{\boldsymbol{\varepsilon}}_d = \widetilde{\eta}_e\widetilde{\boldsymbol{\varepsilon}}_d^{\mathrm{T}}\widetilde{\boldsymbol{\varepsilon}}_d - \widetilde{\eta}_d\widetilde{\boldsymbol{\varepsilon}}_d^{\mathrm{T}}\widetilde{\boldsymbol{\varepsilon}}_e\,,$$

where $\widetilde{\eta}_d$ and $\widetilde{\eta}_e$ are the scalar parts of the quaternions $\widetilde{\mathcal{Q}}_d$ and $\widetilde{\mathcal{Q}}_e$, respectively. The above equation, in view of $\widetilde{\boldsymbol{\eta}}_{1,de}^{B} = \widetilde{\boldsymbol{\eta}}_{1,d}^{B} - \widetilde{\boldsymbol{\eta}}_{1,e}^{B}$ and $\widetilde{\boldsymbol{q}}_{de} = \widetilde{\boldsymbol{q}}_d - \widetilde{\boldsymbol{q}}_e$, implies that

$$\boldsymbol{e}_{de}^{\mathrm{T}}\boldsymbol{\Lambda}_d\boldsymbol{K}_p\boldsymbol{e}_d = k_{pP}\widetilde{\boldsymbol{\eta}}_{1,d}^{B\mathrm{T}}\boldsymbol{\Lambda}_{dP}\widetilde{\boldsymbol{\eta}}_{1,d}^{B} + \lambda_{dO}k_{pO}\widetilde{\eta}_e \left\|\widetilde{\boldsymbol{\varepsilon}}_d\right\|^2 + \widetilde{\boldsymbol{q}}_d^{\mathrm{T}}\boldsymbol{\Lambda}_{dQ}\boldsymbol{K}_{pQ}\widetilde{\boldsymbol{q}}_d + \\ -k_{pP}\widetilde{\boldsymbol{\eta}}_{1,d}^{B\mathrm{T}}\boldsymbol{\Lambda}_{dP}\widetilde{\boldsymbol{\eta}}_{1,e}^{B} - \lambda_{dO}k_{pO}\widetilde{\eta}_d\widetilde{\boldsymbol{\varepsilon}}_d^{\mathrm{T}}\widetilde{\boldsymbol{\varepsilon}}_e + \widetilde{\boldsymbol{q}}_d^{\mathrm{T}}\boldsymbol{\Lambda}_{dQ}\boldsymbol{K}_{pQ}\widetilde{\boldsymbol{q}}_e\,,$$

and thus

$$\boldsymbol{e}_{de}^{\mathrm{T}}\boldsymbol{\Lambda}_d\boldsymbol{K}_p\boldsymbol{e}_d \geq \lambda_{\min}(\boldsymbol{\Lambda}_d\boldsymbol{K}_p)\widetilde{\eta}_e\left\|\boldsymbol{e}_d\right\|^2 - \lambda_{\max}(\boldsymbol{\Lambda}_d\boldsymbol{K}_p)\left\|\boldsymbol{e}_d\right\|\left\|\boldsymbol{e}_e\right\|\,, \quad (7.53)$$

where $\lambda_{\min}(\boldsymbol{\Lambda}_d\boldsymbol{K}_p)$ ($\lambda_{\max}(\boldsymbol{\Lambda}_d\boldsymbol{K}_p)$) is the minimum (maximum) eigenvalue of the matrix $\boldsymbol{\Lambda}_d\boldsymbol{K}_p$. Moreover, in view of the block diagonal structure of the matrix $\boldsymbol{L}_v$ and of the skew-symmetry of the matrix $\boldsymbol{S}(\cdot)$, the following inequality holds

$$\boldsymbol{\sigma}_e^{\mathrm{T}}\boldsymbol{L}_v\boldsymbol{A}(\widetilde{\mathcal{Q}}_e)\boldsymbol{\sigma}_e \geq \frac{1}{2}\lambda_{\min}(\boldsymbol{\Lambda}_v)\widetilde{\eta}_e\left\|\boldsymbol{\sigma}_e\right\|^2\,, \quad (7.54)$$

where $\lambda_{\min}(\boldsymbol{\Lambda}_v)$ is the minimum eigenvalue of the matrix $\boldsymbol{\Lambda}_v$.

Moreover, the following two terms in (7.52) can be rewritten as:

$$(\boldsymbol{\sigma}_d+\boldsymbol{\sigma}_e)^{\mathrm{T}}\boldsymbol{D}(\boldsymbol{q},\boldsymbol{\zeta})\boldsymbol{\zeta} - \frac{1}{2}(\boldsymbol{\sigma}_d+\boldsymbol{\sigma}_e)^{\mathrm{T}}\boldsymbol{D}(\boldsymbol{q},\boldsymbol{\zeta}_r)(\boldsymbol{\zeta}_r+\boldsymbol{\zeta}_o) = \\ -\frac{1}{2}(\boldsymbol{\sigma}_d+\boldsymbol{\sigma}_e)^{\mathrm{T}}\boldsymbol{D}(\boldsymbol{q},\boldsymbol{\zeta})(\boldsymbol{\sigma}_d+\boldsymbol{\sigma}_e) + \\ -\frac{1}{2}(\boldsymbol{\sigma}_d+\boldsymbol{\sigma}_e)^{\mathrm{T}}(\boldsymbol{D}(\boldsymbol{q},\boldsymbol{\zeta}_r)-\boldsymbol{D}(\boldsymbol{q},\boldsymbol{\zeta}))(\boldsymbol{\zeta}_r+\boldsymbol{\zeta}_o)\,. \quad (7.55)$$

In view of the properties of the model (2.71) and equations (7.45), (7.46), (7.53), (7.54), by taking into account that $\boldsymbol{\zeta} = \boldsymbol{\zeta}_d - \widetilde{\boldsymbol{\zeta}}_d$ with $\|\boldsymbol{\zeta}_d\| \leq \zeta_{dM}$ and $\|\boldsymbol{e}_{de}\| \leq \|\boldsymbol{e}_d\| + \|\boldsymbol{e}_e\|$, the function $\dot{V}$ can be upper bounded as follows

$$\dot{V} = -\lambda_{\min}(\boldsymbol{K}_v)\left\|\boldsymbol{\sigma}_d\right\|^2 - \lambda_{\min}(\boldsymbol{\Lambda}_d\boldsymbol{K}_p)\widetilde{\eta}_e\left\|\boldsymbol{e}_d\right\|^2 - \frac{\lambda_{\min}(\boldsymbol{\Lambda}_v)}{2}\widetilde{\eta}_e\left\|\boldsymbol{\sigma}_e\right\|^2 + \\ +\lambda_{\max}(\boldsymbol{K}_v)\left\|\boldsymbol{\sigma}_d\right\|^2 - \lambda_{\min}(\boldsymbol{L}_p)\left\|\boldsymbol{e}_e\right\|^2 + \lambda_{\max}(\boldsymbol{\Lambda}_d\boldsymbol{K}_p)\left\|\boldsymbol{e}_d\right\|\left\|\boldsymbol{e}_e\right\| + \\ +C_M\left\|\boldsymbol{\sigma}_d\right\|\left\|\boldsymbol{\sigma}_e\right\|\left(2\left\|\widetilde{\boldsymbol{\zeta}}_d\right\| + 2\zeta_{dM} + \left\|\boldsymbol{\sigma}_d\right\| + \left\|\boldsymbol{\sigma}_e\right\|\right) + \\ +C_M\left\|\boldsymbol{\sigma}_e\right\|^2\left(\left\|\widetilde{\boldsymbol{\zeta}}_d\right\| + \zeta_{dM}\right) + \\ +\frac{D_M}{2}(\left\|\boldsymbol{\sigma}_d\right\|^2 + \left\|\boldsymbol{\sigma}_d\right\|\left\|\boldsymbol{\sigma}_e\right\|)(2\left\|\widetilde{\boldsymbol{\zeta}}_d\right\| + 2\zeta_{dM} + \left\|\boldsymbol{\sigma}_d\right\| + \left\|\boldsymbol{\sigma}_e\right\|) + \\ +\lambda_{\max}(\boldsymbol{M})\left\|\boldsymbol{\sigma}_d\right\|\left\|\widetilde{\boldsymbol{\zeta}}_d\right\|(\zeta_{dM} + \lambda_{\max}(\boldsymbol{\Lambda}_d)(\left\|\boldsymbol{e}_d\right\| + \left\|\boldsymbol{e}_e\right\|)) \quad (7.56)$$

where $\lambda_{\min}(\boldsymbol{K}_v)$ ($\lambda_{\max}(\boldsymbol{K}_v)$) denotes the minimum (maximum) eigenvalue of the matrix $\boldsymbol{K}_v$, $\lambda_{\min}(\boldsymbol{L}_p)$ denotes the minimum eigenvalue of $\boldsymbol{L}_p$ and $\lambda_{\max}(\boldsymbol{\Lambda}_d)$ denotes the maximum eigenvalue of the matrix $\boldsymbol{\Lambda}_d$.

Consider the state space domain defined as follows

$$B_\rho = \{\underline{\boldsymbol{x}} : \ \|\underline{\boldsymbol{x}}\| < \rho,\, \rho < 1\}\,,$$

with $\widetilde{\eta}_d > 0$, $\widetilde{\eta}_e > 0$. It can be recognized that in the domain $B_\rho$ the following inequalities hold

$$0 < \sqrt{1-\rho^2} < \widetilde{\eta}_e < 1, \tag{7.57}$$

$$\left\|\widetilde{\boldsymbol{\zeta}}_d\right\| = \|\boldsymbol{\sigma}_d - \boldsymbol{\Lambda}_d \boldsymbol{e}_{de}\| \leq (1 + 2\lambda_{\max}(\boldsymbol{\Lambda}_d))\rho. \tag{7.58}$$

By completing the squares in (7.56) and using (7.57),(7.58), it can be shown that there exists a scalar $\kappa > 0$ such that

$$\dot{V} \leq -\kappa \|\underline{\boldsymbol{x}}\|^2 \tag{7.59}$$

in the domain $B_\rho$, provided that the controller and observer parameters satisfy the inequalities

$$\begin{aligned}
\lambda_{\min}(\boldsymbol{K}_v) &> \alpha_1\left(C_M + \frac{3D_M}{2}\right) + \alpha_2\lambda_{\max}(\boldsymbol{M})(1 + \lambda_{\max}(\boldsymbol{\Lambda}_d)) \\
\lambda_{\min}(\boldsymbol{\Lambda}_d\boldsymbol{K}_p) &> \frac{\alpha_2\lambda_{\max}(\boldsymbol{M})\lambda_{\max}(\boldsymbol{\Lambda}_d)}{\sqrt{1-\rho^2}}, \\
\lambda_{\min}(\boldsymbol{L}_p) &> \max\left\{\alpha_2\lambda_{\max}(\boldsymbol{M})\lambda_{\max}(\boldsymbol{\Lambda}_d), \frac{\lambda_{\max}(\boldsymbol{\Lambda}_d\boldsymbol{K}_p)^2}{\lambda_{\min}(\boldsymbol{\Lambda}_d\boldsymbol{K}_p)\sqrt{1-\rho^2}}\right\}, \\
\lambda_{\min}(\boldsymbol{\Lambda}_v) &> \frac{2}{\sqrt{1-\rho^2}}\left(\lambda_{\max}(\boldsymbol{K}_v) + (2\alpha_1 + \rho)C_M + \frac{\alpha_1 D_M}{2}\right)
\end{aligned}$$

where $\alpha_1 = 2(1 + \lambda_{\max}(\boldsymbol{\Lambda}_d))\rho + \zeta_{dM}$ and $\alpha_2 = \zeta_{dM} + 2\lambda_{\max}(\boldsymbol{\Lambda}_d)\rho$.

Therefore, given a domain $B_\rho$ characterized by any $\rho < 1$, there always exists a set of observer and controller gains such that $\dot{V} \leq 0$ in $B_\rho$. Moreover, for $\widetilde{\eta}_d \geq 0$, $\widetilde{\eta}_e \geq 0$ the following inequality holds

$$0 \leq (1 - \widetilde{\eta}_d)^2 \leq (1 - \widetilde{\eta}_d)(1 + \widetilde{\eta}_d) = \|\widetilde{\boldsymbol{\varepsilon}}_d\|^2 ,$$

and a similar inequality can be written in terms of $\widetilde{\eta}_e$ and $\widetilde{\boldsymbol{\varepsilon}}_e$. Hence, function $V$ can be bounded as

$$c_m \|\underline{\boldsymbol{x}}\|^2 \leq V(\underline{\boldsymbol{x}}) \leq c_M \|\underline{\boldsymbol{x}}\|^2 , \tag{7.60}$$

with

$$\begin{aligned}
c_m &= \frac{1}{2}\min\{\lambda_{\min}(\boldsymbol{M}), \lambda_{\min}(\boldsymbol{K}_p), \lambda_{\min}(\boldsymbol{L}_p)\} \\
c_M &= \frac{1}{2}\max\{\lambda_{\max}(\boldsymbol{M}), 4\lambda_{\max}(\boldsymbol{K}_p), 4\lambda_{\max}(\boldsymbol{L}_p)\} ,
\end{aligned}$$

where $\lambda_{\min}(\boldsymbol{K}_p)$ ($\lambda_{\max}(\boldsymbol{K}_p)$) is the minimum (maximum) eigenvalue of the matrix $\boldsymbol{K}_p$, $\lambda_{\max}(\boldsymbol{L}_p)$ is the maximum eigenvalue of the matrix $\boldsymbol{L}_p$.

Since $V(t)$ is a decreasing function along the system trajectories, the inequality (7.60) guarantees that, for a given $0 < \rho < 1$, all the trajectories $\underline{\boldsymbol{x}}(t)$ starting in the domain

$$\Omega_\rho = \left\{\underline{\boldsymbol{x}} : \|\underline{\boldsymbol{x}}\| < \rho\sqrt{\frac{c_m}{c_M}}\right\} ,$$

remain in the domain $B_\rho$ for all $t > 0$ provided that $\widetilde{\eta}_d(t) > 0$, $\widetilde{\eta}_e(t) > 0$ for all $t > 0$. The latter condition is fulfilled when $\widetilde{\eta}_d(0)$ and $\widetilde{\eta}_e(0)$ are positive; in fact, $\|\widetilde{\boldsymbol{\varepsilon}}_d\| < \rho < 1$ and $\|\widetilde{\boldsymbol{\varepsilon}}_e\| < \rho < 1$ for all $t > 0$ implies that $\widetilde{\eta}_d(t)$ and $\widetilde{\eta}_e(t)$ cannot change their sign.

Moreover, from (7.59) and (7.60), the convergence in the domain $B_\rho$ is exponential [170], which implies exponential convergence of $\boldsymbol{e}_d$, $\boldsymbol{e}_e$, $\widetilde{\boldsymbol{\zeta}}_d$ and $\widetilde{\boldsymbol{\zeta}}_e$.

The condition $\rho < 1$ is due to the unit norm constraint on the quaternion components, and gives a rather conservative estimate of the domain of attraction. However, it must be pointed out that this limitation arises when spheres are used to estimate the domain of attraction; better estimates can be obtained by using domain of different shapes, e.g. ellipsoids.

### 7.8.2 Simulations

As for the previous sections, numerical simulations have been performed resorting to the simulator described in [23]. To test this control law, however, a different manipulator has been considered. A three-link manipulator with elbow kinematic structure has been simulated that is mounted under the vehicle body. Since the vehicle is neutrally buoyant, but the arm is not neutrally buoyant, the whole system results to be not neutrally buoyant. The 3 links are cylindrical, thus hydrodynamic effects can be computed by simplified relations as in [255]. Matrix $\boldsymbol{B}$ is supposed to be constant and full-rank (for simplicity it has been set to identity), meaning that direct control of forces and moments acting on the vehicle and joint torques is available.

A task involving motion of both the vehicle and the manipulator has been considered. At the initial time, the initial vehicle configuration is $\boldsymbol{\eta}_i = [0 \;\; 0 \;\; 0.1 \;\; 15 \;\; 0 \;\; -15]^{\mathrm{T}}$ m,deg and the initial manipulator configuration is $\boldsymbol{q}_i = [20 \;\; -30 \;\; 40]^{\mathrm{T}}$ deg. The vehicle must move to the final location $\boldsymbol{\eta}_f = [0 \;\; 0 \;\; 0 \;\; 0 \;\; 0 \;\; 0]^{\mathrm{T}}$ m, deg in 20 s according to a 5th order polynomial time law. The manipulator must move to $\boldsymbol{q}_f = [0 \;\; 0 \;\; 0]^{\mathrm{T}}$ deg in 3 s according to a 5th order polynomial time law. Notice that the assigned trajectories correspond to a fast desired motion for the manipulator while the vehicle is kept almost in hovering. Figure 7.12 shows the desired trajectories. It must be noticed that the vehicle orientation set point is assigned in terms of Euler angles, as usual in navigation planning; these are converted into the corresponding rotation matrix so as to extract the quaternion expressing the orientation error. Remarkably, this procedure is free of singularities [261].

**First case study.** The performance of the control law (7.41),(7.42) has been compared to that obtained with a control scheme of similar structure in which the velocity feedback is implemented through numerical differentiation of position measurements. The following control law has then been considered

$$\boldsymbol{u} = \boldsymbol{B}^{\dagger}\left(\boldsymbol{M}(\boldsymbol{q})\dot{\boldsymbol{\zeta}}_r + \boldsymbol{C}(\boldsymbol{q},\boldsymbol{\zeta})\boldsymbol{\zeta}_r + \boldsymbol{K}_v(\boldsymbol{\zeta}_r - \boldsymbol{\zeta}) + \boldsymbol{K}_p\boldsymbol{e}_d + \boldsymbol{g}(\boldsymbol{q},\boldsymbol{R}_B^I)\right), \quad (7.61)$$

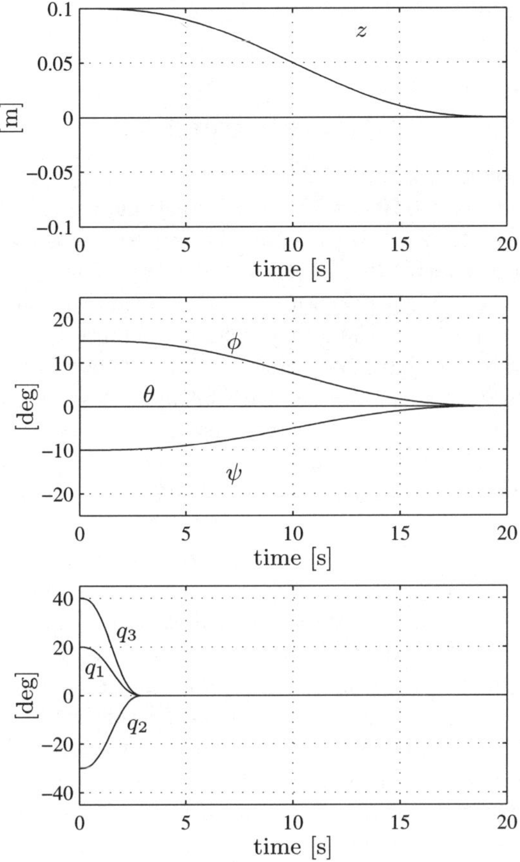


**Fig. 7.12.** Output feedback control. Desired trajectories used in the case studies. Top: vehicle position; Middle: vehicle orientation (RPY angles); Bottom: joint positions

where $\boldsymbol{\zeta}_r = \boldsymbol{\zeta}_d + \boldsymbol{\Lambda}_d \boldsymbol{e}_d$. The vectors $\dot{\boldsymbol{\zeta}}$ and $\dot{\boldsymbol{\zeta}}_r$ are computed via first-order difference. The above control law is analogous to the operational space control law proposed in [266] and extended in [108] in the framework of quaternion-based attitude control. To obtain a control law of computational complexity similar to that of (7.41), the algorithm (7.61) has been modified into the simpler form

$$\boldsymbol{u} = \boldsymbol{B}^{\dagger}\left(\boldsymbol{M}(\boldsymbol{q})\dot{\boldsymbol{\zeta}}_r + \boldsymbol{K}_v(\boldsymbol{\zeta}_r - \boldsymbol{\zeta}) + \boldsymbol{K}_p\boldsymbol{e}_d + \boldsymbol{g}(\boldsymbol{q}, \boldsymbol{R}_B^I)\right) . \tag{7.62}$$

The parameters in the control laws are set to

$$\boldsymbol{\Lambda}_d = \mathrm{blockdiag}\{0.005\boldsymbol{I}_3, 0.01\boldsymbol{I}_3, 0.01\boldsymbol{I}_3\} ,$$
$$\boldsymbol{\Lambda}_e = \mathrm{blockdiag}\{5\boldsymbol{I}_3, 10\boldsymbol{I}_3, 10, 10, 5\} ,$$

$$\begin{aligned}
\boldsymbol{L}_v &= \text{blockdiag}\{\boldsymbol{I}_3, 20\boldsymbol{I}_3, 160, 160, 900\}\,,\\
\boldsymbol{L}_p &= \text{blockdiag}\{5000\boldsymbol{I}_3, 10^5\boldsymbol{I}_3, 10^3, 10^3, 2\cdot 10^3\}\,,\\
\boldsymbol{K}_p &= \text{blockdiag}\{400\boldsymbol{I}_3, 500\boldsymbol{I}_3, 1500\boldsymbol{I}_3\}\,,\\
\boldsymbol{K}_v &= \text{blockdiag}\{4000\boldsymbol{I}_3, 4000\boldsymbol{I}_3, 400\boldsymbol{I}_3\}\,.
\end{aligned}$$

A digital implementation of the control laws has been considered. The sensor update rate is 100 Hz for the joint positions and 20 Hz for the vehicle position and orientation, while the control inputs to the actuators are updated at 100 Hz, i.e., the control law is computed every 10 ms. The relatively high update rate for the vehicle position and orientation measurements has been chosen so as to achieve a satisfactory tracking accuracy.

Quantization effects have been introduced in the simulation by assuming a 16-bit A/D converter on the sensors outputs. Also, Gaussian zero-mean noise has been added to the signals coming from the sensors.

Figure 7.13 shows the time history of the norm of the tracking and estimation errors obtained with the control laws (7.41),(7.42) and (7.62), respectively.

Figures 7.14 to 7.16 show the corresponding control forces, moments and torques. It can be recognized that good tracking is achieved in both cases, although the performance in terms of tracking errors is slightly better for the control law (7.62), where numerical derivatives are used. On the other hand, the presence of measurement noise and quantization effects results in chattering of the control commands to the actuators; this is much lower when the control law (7.41),(7.42) is adopted, as compared to control law (7.62). An indicator of the energy consumption due to the chattering at steady state is the variance of the control commands reported in Table 7.4 for each component; this data clearly show the advantage of using the controller-observer scheme. Of course, the improvement becomes clear when the noise and quantization effects are larger than a certain threshold. The derivation of such a threshold would require a stochastic analysis of a nonlinear system, which is beyond the scope of the present work. Moreover, it can be easily recognized that such a bound is strongly dependent on the characteristics of the actuators.

**Second case study.** In this case study the same task as above is executed by adopting the control law (7.43),(7.44) and its counterpart using numerical differentiation of the measured position/orientation, i.e.,

$$\boldsymbol{u} = \boldsymbol{B}^{\dagger}\left(\widehat{\boldsymbol{M}}\dot{\boldsymbol{\zeta}}_r + \boldsymbol{K}_v(\boldsymbol{\zeta}_r - \boldsymbol{\zeta}) + \boldsymbol{K}_p\boldsymbol{e}_d + \boldsymbol{g}(\boldsymbol{q}, \boldsymbol{R}_B^I)\right)\,, \tag{7.63}$$

where the same parameters as in the previous case study have been used. Also, the same measurement update rates, quantization resolution and sensory noise have been considered in the simulation.

The results are shown in Figure 7.17 in terms of tracking and estimation errors. It can be recognized that the errors are comparable to those obtained with the control scheme (7.41),(7.42) in spite of the extremely simplified con-

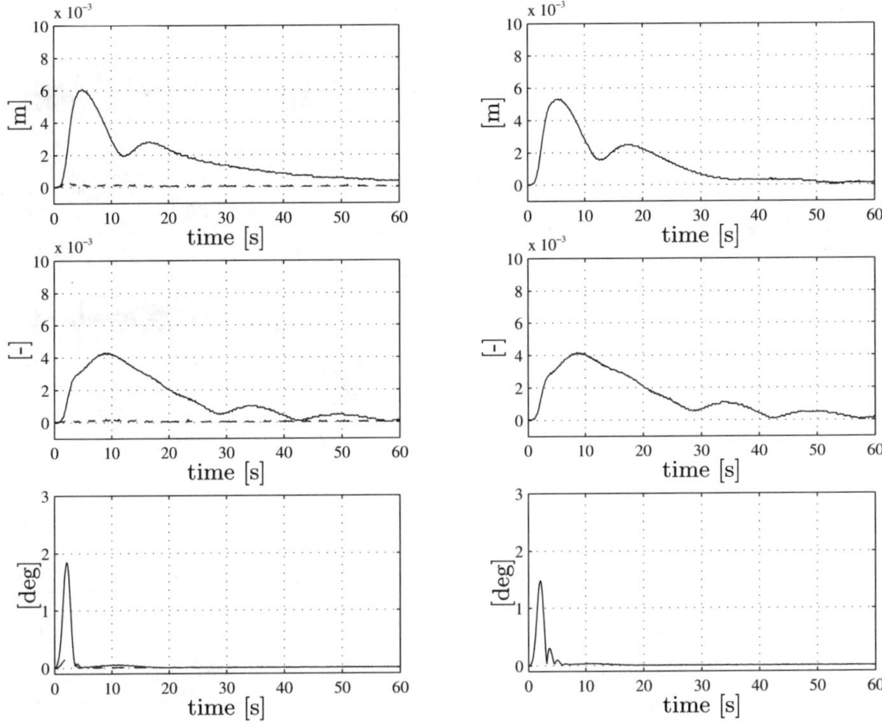


**Fig. 7.13.** Output feedback control; first case study. Comparison between the control laws (7.41),(7.42) and (7.62): norm of the tracking (solid) and estimation (dashed) errors. Left: control law (7.41),(7.42). Right: control law (7.62). Top: vehicle position error; Middle: vehicle orientation error (vector part of the quaternion); Bottom: joint position errors

**Table 7.4.** Variance of control commands

| | Control law (7.41),(7.42) | Control law (7.62) |
|---|---|---|
| $X$ [$\text{N}^2$] | 0.0149 | 14.6326 |
| $Y$ [$\text{N}^2$] | 0.0077 | 11.4734 |
| $Z$ [$\text{N}^2$] | 0.0187 | 3.0139 |
| $K$ [$\text{N}^2\text{m}^2$] | 0.2142 | 10.6119 |
| $M$ [$\text{N}^2\text{m}^2$] | 2.8156 | 23.2798 |
| $N$ [$\text{N}^2\text{m}^2$] | 1.6192 | 26.0350 |
| $\tau_{q,1}$ [$\text{N}^2\text{m}^2$] | 0.3432 | 4.0455 |
| $\tau_{q,2}$ [$\text{N}^2\text{m}^2$] | 0.0725 | 3.2652 |
| $\tau_{q,3}$ [$\text{N}^2\text{m}^2$] | 0.3393 | 1.8259 |

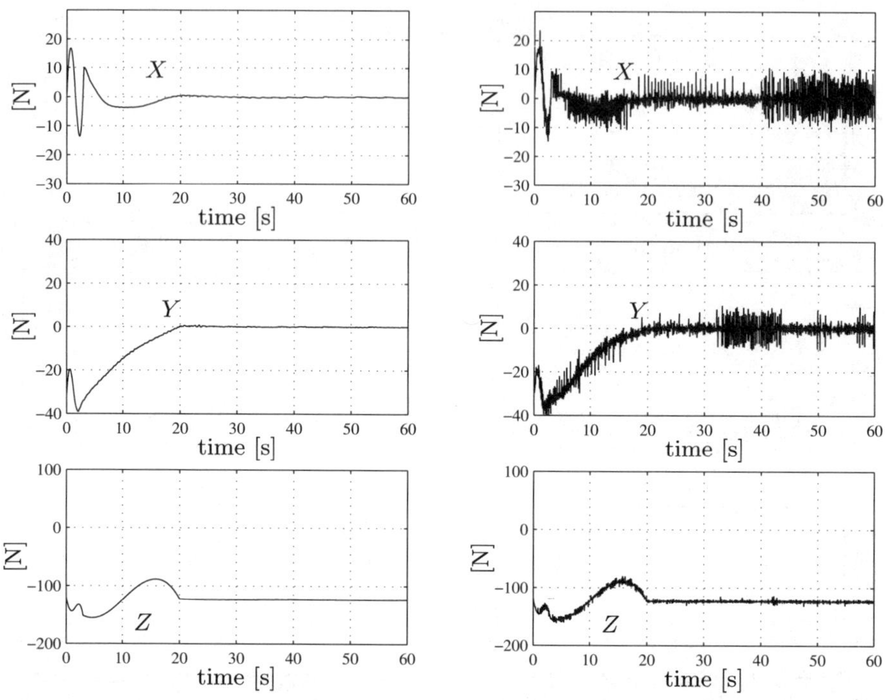


**Fig. 7.14.** Output feedback control; first case study. Comparison between the control laws (7.41),(7.42) and (7.62): vehicle control forces. Left: control law (7.41),(7.42). Right: control law (7.62)

trol structure; also, the control inputs remain free of chattering phenomena and are not reported for brevity.

The tracking performance obtained with the simplified control law confirms that the controller-observer approach is intrinsically robust with respect to uncertain knowledge of the system's dynamics, thanks to the exponential stability property. Hence, perfect compensation of inertia, Coriolis and centripetal terms, as well as of hydrodynamic damping terms, is not required.

**Third case study.** The control laws (7.43),(7.44) and (7.63) have been tested in more severe operating conditions. Namely, the update rate for the vehicle position/orientation measurements has been lowered to 5 Hz and the A/D word length has been set to 12 bit for all the sensor output signals. Gaussian zero-mean noise is still added to the measures. The parameters in the control laws are the same as in the previous case studies.

Figure 7.18 shows a small degradation of the tracking performance for both the control schemes. In fact, the tracking errors remain of the same order of magnitude as in the previous case studies, because the computing rate of the control law is unchanged (100 Hz). Namely, the update rate of

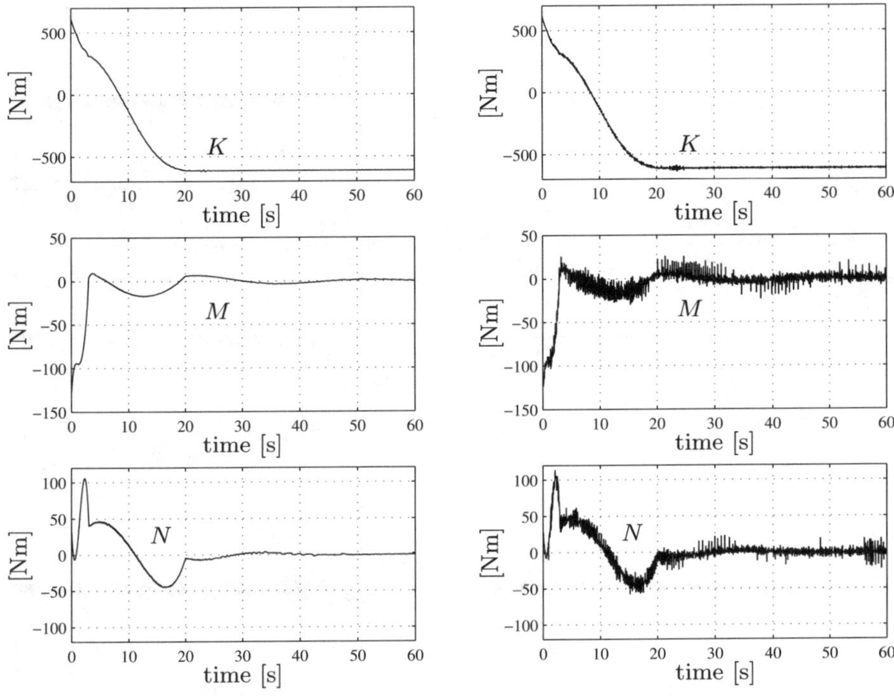


**Fig. 7.15.** Output feedback control; first case study. Comparison between the control laws (7.41),(7.42) and (7.62): vehicle control moments. Left: control law (7.41),(7.42); Right: control law (7.62)

the measurements relative to the subsystem with faster dynamics (i.e., the manipulator) remains the same (100 Hz), while the update rate of the measurements relative to the vehicle (5 Hz) is still adequate to its slower dynamics.

Figures 7.19 to 7.21 show the corresponding control forces, moments and torques. It can be recognized that unacceptable chattering on the control inputs is experienced when numerical derivatives are used, which is almost completely cancelled when the controller-observer scheme is adopted.

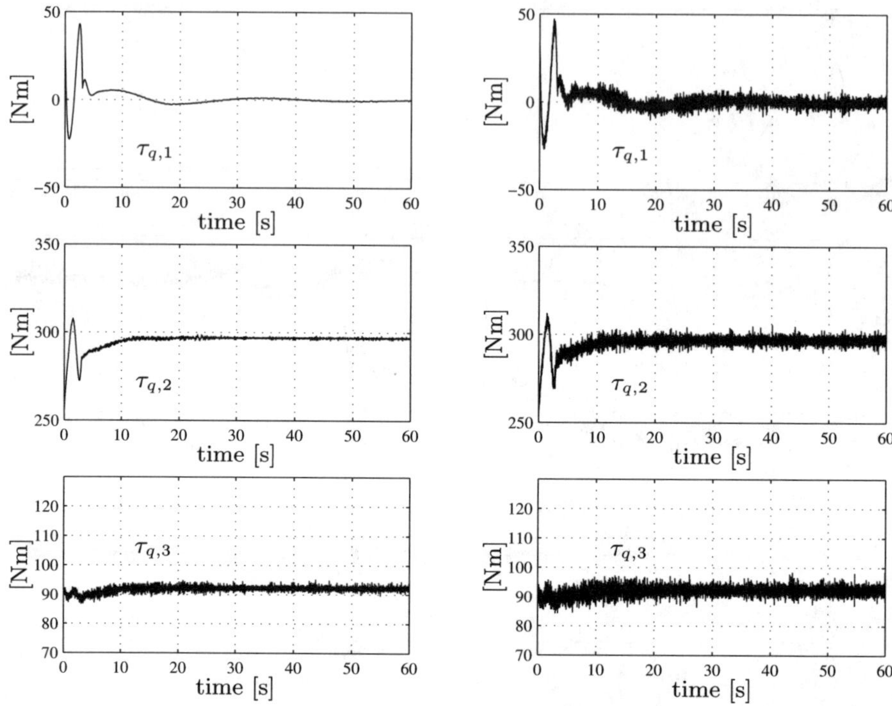


**Fig. 7.16.** Output feedback control; first case study. Comparison between the control laws (7.41),(7.42) and (7.62): joint control torques. Left: control law (7.41),(7.42); Right: control law (7.62)

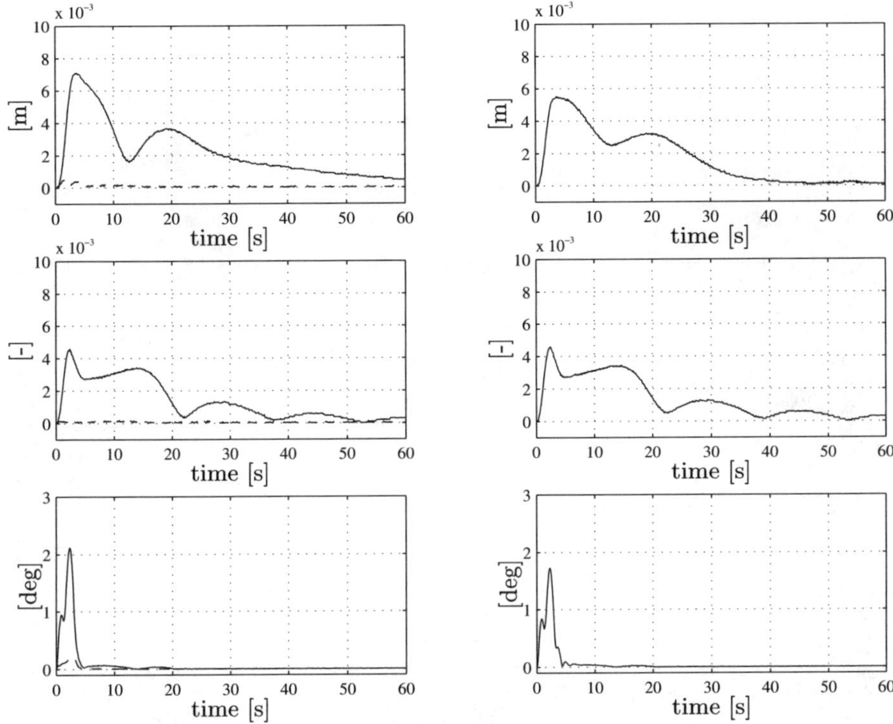


**Fig. 7.17.** Output feedback control; second case study. Comparison between the control laws (7.43),(7.44) and (7.63): norm of the tracking (solid) and estimation (dashed) errors. Top: vehicle position error; Middle: vehicle orientation error (vector part of the quaternion); Bottom: joint position errors

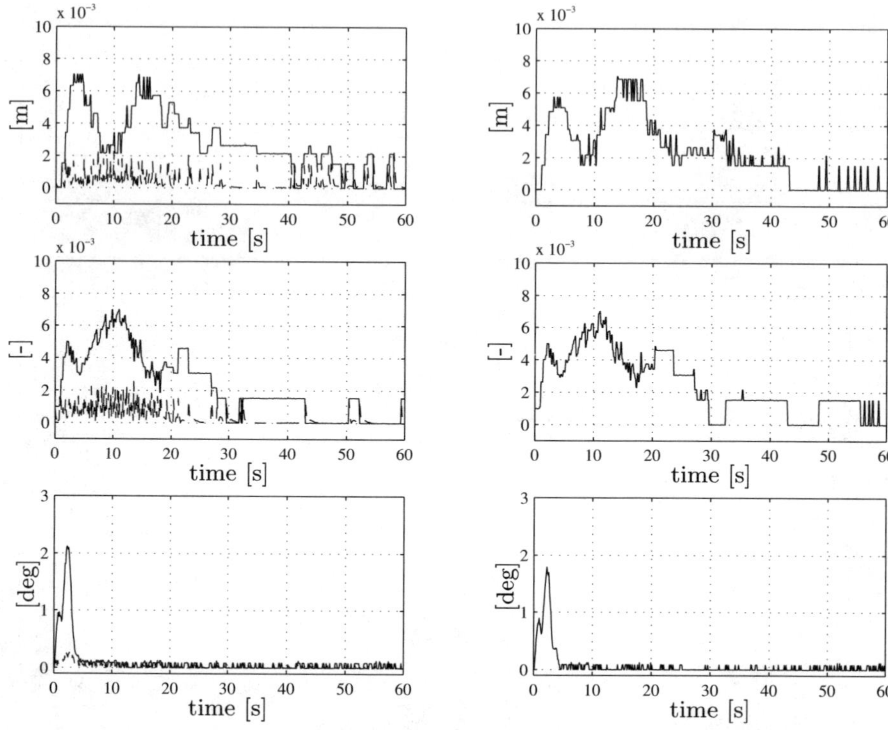


**Fig. 7.18.** Output feedback control; third case study. Comparison between the control laws (7.43),(7.44) and (7.63): norm of the tracking (solid) and estimation (dashed) errors. Left: control law (7.43),(7.44). Right: control law (7.63). Top: vehicle position error; Middle: vehicle orientation error (vector part of the quaternion); Bottom: joint position errors

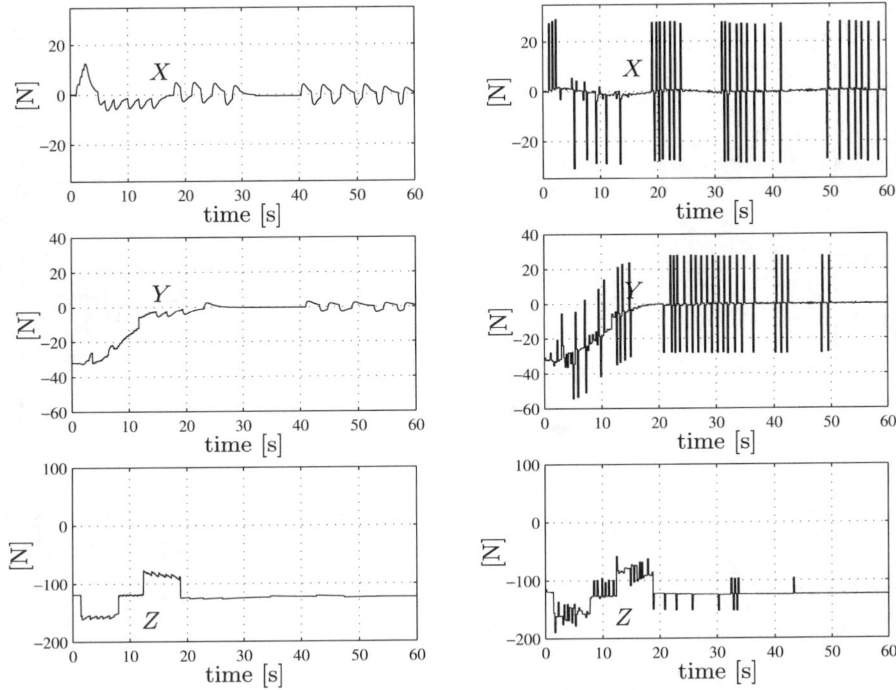


**Fig. 7.19.** Output feedback control; third case study. Comparison between the control laws (7.43),(7.44) and (7.63): vehicle control forces. Left: control law (7.43),(7.44). Right: control law (7.63)

## 7.9 Virtual Decomposition Based Control

*Divide et Impera.*
*Anonymous from the middle age.*

Usually, adaptive control approaches for UVMSs look at the system as a whole, giving rise to high-dimensional problems: differently from the case of earth-fixed manipulators, in the case of UVMSs it is not possible to achieve a reduction of the number of dynamic parameters to be adapted, since the base of the manipulator —i.e., the vehicle— has full mobility. As a matter of fact, the computational load of such control algorithms grows as much as the fourth-order power of the number of the system's degrees of freedom. For this reason, practical application of adaptive control to UVMS has been limited, even in simulation, to vehicles carrying arms with very few joints (i.e., two or three) and usually performing planar tasks.

In this Section an adaptive control scheme for the tracking problem of UVMS is discussed, which is based on the approach in [325]. Differently from previously proposed schemes, the serial-chain structure of the UVMS

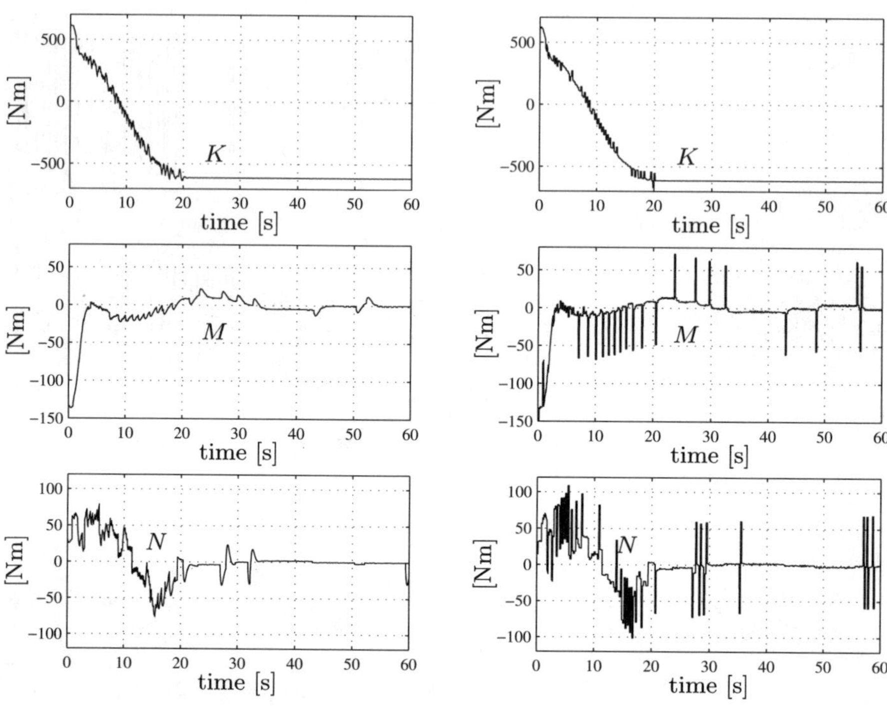


**Fig. 7.20.** Output feedback control; third case study. Comparison between the control laws (7.43),(7.44) and (7.63): vehicle control moments. Left: control law (7.43),(7.44). Right: control law (7.63)

is exploited to decompose the overall motion control problem in a set of elementary control problems regarding the motion of each rigid body in the system, namely the manipulator's links and the vehicle. For each body, a control action is designed to assign the desired motion, to adaptively compensate for the body dynamics, and to counteract force/moment exchanged with its neighborhoods along the chain.

The resulting control scheme has a modular structure which greatly simplifies its application to systems with a large number of links; furthermore, it reduces the required computational burden by replacing one high-dimensional problem with many low-dimensional ones; finally it allows efficient implementation on distributed computing architecture; it can be embedded in a kinematic control scheme, which allows handling of kinematic redundancy, i.e., to achieve joint limits avoidance and dexterity optimization; finally, it reduces the size of the control software code and improves its flexibility, i.e., its structure is not modified by changing the system's mechanical structure.

Remarkably, the control law is expressed in terms of body-fixed coordinates so as to overcome the occurrence of kinematic singularities. Moreover, a non-minimal representation of the orientation —i.e., the unit quater-

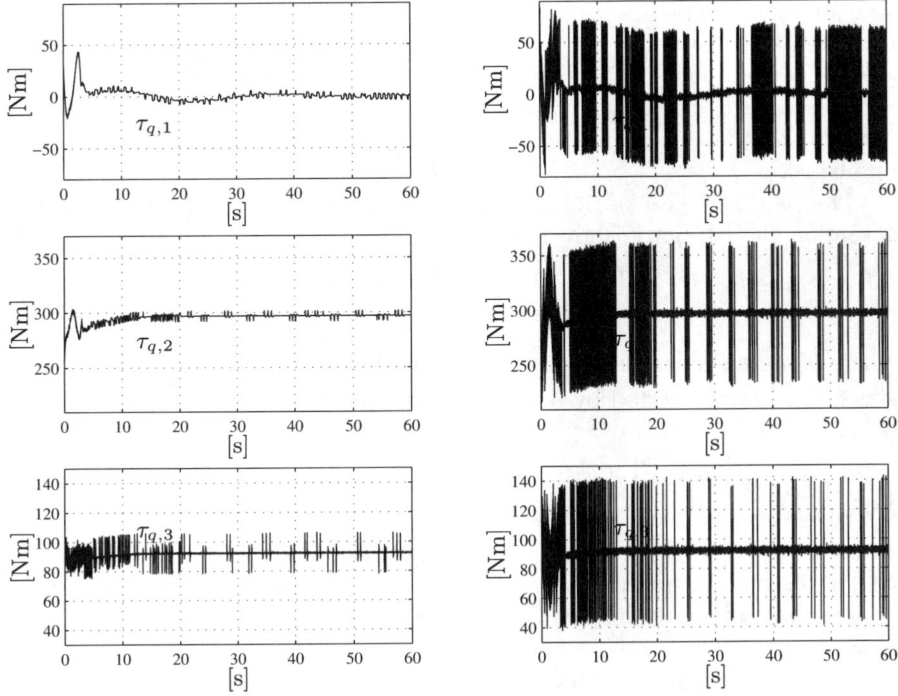


**Fig. 7.21.** Output feedback control; third case study. Comparison between the control laws (7.43),(7.44) and (7.63): joint control torques. Left: control law (7.43),(7.44). Right: control law (7.63)

nion [246]— is used in the control law; this allows overcoming the occurrence of representation singularities.

The discussed control scheme is tested in a numerical case study. A manipulation task is assigned in terms of a desired position and orientation trajectory for the end effector of a 6-DOF manipulator mounted on a 6-DOF vehicle. Then, the system's behavior under the discussed control law is verified in simulation.

**Control law.** The dynamics of an UVMS is rewritten in a way to remark the interaction between the different rigid bodies, i.e., between the links and between links and the vehicle. Consider an UVMS composed of a vehicle and of a $n$-DOF manipulator mounted on it.

The vehicle and the manipulator's links are assumed to be rigid bodies numbered from 0 (the vehicle) to $n$ (the last link, i.e., the end effector). Hence, the whole system can be regarded as an open kinematic chain with floating base.

A reference frame $\mathcal{T}_i$ is attached to each body according to the Denavit-Hartenberg formalism, while $\Sigma_i$ is the earth-fixed inertial reference frame.

Hereafter, a superscript will denote the frame to which a vector is referred to, the superscript will be dropped for quantities referred to the inertial frame.

Notice that some differences may arise in the symbology of the vehicle's variables due to the different approach followed in this Section. Coherently with the virtual decomposition approach, the vehicle is considered as link number 0.

The $(6 \times 1)$ vector of the total generalized force (i.e., force and moment) acting on the $i$th body is given by

$$\begin{aligned} \boldsymbol{h}_{t,i}^i &= \boldsymbol{h}_i^i - \boldsymbol{U}_{i+1}^i \boldsymbol{h}_{i+1}^{i+1}, \qquad i = 0, \ldots, n-1 \\ \boldsymbol{h}_{t,n}^n &= \boldsymbol{h}_n^n , \end{aligned} \tag{7.64}$$

where $\boldsymbol{h}_i^i$ is the generalized force exerted by body $i-1$ on body $i$, $\boldsymbol{h}_{i+1}^{i+1}$ is the generalized force exerted by body $i+1$ on body $i$. The matrix $\boldsymbol{U}_{i+1}^i \in \mathbb{R}^{6\times 6}$ is defined as

$$\boldsymbol{U}_{i+1}^i = \begin{bmatrix} \boldsymbol{R}_{i+1}^i & \boldsymbol{O}_{3\times 3} \\ \boldsymbol{S}(\boldsymbol{r}_{i,i+1}^i)\boldsymbol{R}_{i+1}^i & \boldsymbol{R}_{i+1}^i \end{bmatrix} .$$

where $\boldsymbol{R}_{i+1}^i \in \mathbb{R}^{3\times 3}$ is the rotation matrix from frame $\mathcal{T}_{i+1}$ to frame $\mathcal{T}_i$, $\boldsymbol{S}(\cdot)$ is the matrix operator performing the cross product between two $(3 \times 1)$ vectors, and $\boldsymbol{r}_{i,i+1}^i$ is the vector pointing from the origin of $\mathcal{T}_i$ to the origin of $\mathcal{T}_{i+1}$.

The equations of motion of each rigid body can be written in body-fixed reference frame in the form [314, 127]:

$$\boldsymbol{M}_i \dot{\boldsymbol{\nu}}_i^i + \boldsymbol{C}_i(\boldsymbol{\nu}_i^i)\boldsymbol{\nu}_i^i + \boldsymbol{D}_i(\boldsymbol{\nu}_i^i)\boldsymbol{\nu}_i^i + \boldsymbol{g}_i(\boldsymbol{R}_i) = \boldsymbol{h}_{t,i}^i , \tag{7.65}$$

where $\boldsymbol{\nu}_i^i \in \mathbb{R}^6$ is the vector of generalized velocity (i.e., linear and angular velocities defined in Section 2.8), $\boldsymbol{R}_i$ is the rotation matrix expressing the orientation of $\mathcal{T}_i$ with respect to the inertial reference frame, $\boldsymbol{M}_i \in \mathbb{R}^{6\times 6}$, $\boldsymbol{C}_i(\boldsymbol{\nu}_i^i)\boldsymbol{\nu}_i^i \in \mathbb{R}^6$, $\boldsymbol{D}_i(\boldsymbol{\nu}_i^i)\boldsymbol{\nu}_i^i \in \mathbb{R}^6$ and $\boldsymbol{g}_i(\boldsymbol{R}_i) \in \mathbb{R}^6$ are the quantities introduced in (2.51) referred to the generic rigid body. In Chap. 2, the details on the dynamics of a rigid body moving in a fluid are given.

According to the property of linearity in the parameters (7.65) can be rewritten as:

$$\boldsymbol{Y}(\boldsymbol{R}_i, \boldsymbol{\nu}_i^i, \dot{\boldsymbol{\nu}}_i^i)\boldsymbol{\theta}_i = \boldsymbol{h}_{t,i}^i$$

where $\boldsymbol{\theta}_i$ is the vector of dynamic parameters of the $i$th rigid body. Notice that, for the vehicle, i.e., for the body numbered as 0, the latter is exactly (2.54); only for this Section, however, the notation of the vehicle forces and regressor will be slightly different from the rest of the book.

The input torque $\tau_{q,i}$ at the $i$th joint of the manipulator can be obtained by projecting $\boldsymbol{h}_i$ on the corresponding joint axis via

$$\tau_{q,i} = \boldsymbol{z}_{i-1}^{i\,\mathrm{T}} \boldsymbol{h}_i^i , \tag{7.66}$$

where $\boldsymbol{z}_{i-1}^i = \boldsymbol{R}_i^{\mathrm{T}} \boldsymbol{z}_{i-1}$ is the $z$-axis of the frame $\mathcal{T}_{i-1}$ expressed in the frame $\mathcal{T}_i$.

The input force and moment acting on vehicle are instead given by the vector $\boldsymbol{h}_{t,0}^{0}$. Notice that this vector was introduced in Section 2.6 with the symbol $\boldsymbol{\tau}_v$, in this Section, however, it was preferred to modify the notation consistently with the serial chain formulation adopted here.

Let $\boldsymbol{p}_{d,o}(t)$, $\mathcal{Q}_{d,0}(t)$, $\boldsymbol{q}_d(t)$, $\boldsymbol{\nu}_{d,0}^{0}(t)$, $\dot{\boldsymbol{q}}_d(t)$, $\dot{\boldsymbol{\nu}}_{d,0}^{0}(t)$, $\ddot{\boldsymbol{q}}_d(t)$ represent the desired trajectory. Let define

$$\boldsymbol{\nu}_{r,0}^{0} = \boldsymbol{\nu}_{d,0}^{0} + \begin{bmatrix} \lambda_{p,0}\boldsymbol{I}_3 & \boldsymbol{O}_3 \\ \boldsymbol{O}_3 & \lambda_{o,0}\boldsymbol{I}_3 \end{bmatrix} \boldsymbol{e}_0, \tag{7.67}$$

$$\dot{q}_{r,i} = \dot{q}_{d,i} + \lambda_i \tilde{q}_i \qquad i = 1, \ldots, n \tag{7.68}$$

$$\boldsymbol{\nu}_{r,i+1}^{i+1} = \boldsymbol{U}_{i+1}^{i^{\mathrm{T}}} \boldsymbol{\nu}_{r,i}^{i} + \dot{q}_{r,i+1} \boldsymbol{z}_i^{i+1} \qquad i = 0, \ldots, n-1\,, \tag{7.69}$$

where the $(6 \times 1)$ vector

$$\boldsymbol{e}_0 = \begin{bmatrix} \boldsymbol{R}_0^{\mathrm{T}} \tilde{\boldsymbol{p}}_0 \\ \tilde{\boldsymbol{\varepsilon}}_0^{0} \end{bmatrix}$$

denotes position and orientation errors for the vehicle, $\lambda_{p,0}, \lambda_{o,0}, \lambda_i$ are positive design gains.

It is useful considering the following variables:

$$\begin{aligned} \boldsymbol{s}_i^i &= \boldsymbol{\nu}_{r,i}^{i} - \boldsymbol{\nu}_i^i & i &= 0, \ldots, n \\ s_{q,i} &= \dot{q}_{r,i} - \dot{q}_i & i &= 1, \ldots, n \\ \boldsymbol{s}_q &= \begin{bmatrix} s_{q,1} & \ldots & s_{q,n} \end{bmatrix}^{\mathrm{T}} \end{aligned}$$

The discussed control law is based on the computation of the *required* generalized force for each rigid body in the system. Then, the input torques for the manipulator and the input generalized force for the vehicle are computed from the required forces according to (7.64) and (7.66).

In the following it is assumed that only a nominal estimate $\hat{\boldsymbol{\theta}}_i$ of the vector of dynamic parameters is available for the $i$th rigid body. Hence, a suitable update law for the estimates has to be adopted so as to ensure asymptotic tracking of the desired trajectory.

For the generic rigid body (including the vehicle) the required force has the following structure

$$\boldsymbol{h}_{r,i}^{i} = \boldsymbol{h}_{c,i}^{i} - \boldsymbol{U}_{i+1}^{i} \boldsymbol{h}_{c,i+1}^{i+1}$$

that, including the designed required force, implies

$$\boldsymbol{h}_{c,i}^{i} = \boldsymbol{Y}\left(\boldsymbol{R}_i, \boldsymbol{\nu}_i^i, \boldsymbol{\nu}_{r,i}^{i}, \dot{\boldsymbol{\nu}}_{r,i}^{i}\right) \hat{\boldsymbol{\theta}}_i + \boldsymbol{K}_{v,i} \boldsymbol{s}_i^i + \boldsymbol{U}_{i+1}^{i} \boldsymbol{h}_{c,i+1}^{i+1} \tag{7.70}$$

with $\boldsymbol{K}_{v,i} > \boldsymbol{O}$. The parameters estimate $\hat{\boldsymbol{\theta}}_i$ is dynamically updated via

$$\dot{\hat{\boldsymbol{\theta}}}_i = \boldsymbol{K}_{\theta,i}^{-1} \boldsymbol{Y}^{\mathrm{T}} \left(\boldsymbol{R}_i, \boldsymbol{\nu}_i^i, \boldsymbol{\nu}_{r,i}^{i}, \dot{\boldsymbol{\nu}}_{r,i}^{i}\right) \boldsymbol{s}_i^i \tag{7.71}$$

with $\boldsymbol{K}_{\theta,i} > \boldsymbol{O}$.

The control torque at the $i$th manipulator's joint is given by

$$\tau_{q,i} = \boldsymbol{z}_{i-1}^{i}{}^{\mathrm{T}} \boldsymbol{h}_{c,i}^{i}\,. \tag{7.72}$$

Finally, the generalized force for the vehicle needed to achieve the corresponding required force is computed as

$$\boldsymbol{h}^0_{c,0} = \boldsymbol{h}^0_{r,0} + \boldsymbol{U}^0_1 \boldsymbol{h}^1_{c,1} \, .$$

### 7.9.1 Stability Analysis

In this Section, the stability analysis for the discussed control law is provided.

Let consider the following scalar function

$$V_i(\boldsymbol{s}^i_i, \tilde{\boldsymbol{\theta}}_i) = \frac{1}{2} \boldsymbol{s}^{i\,\mathrm{T}}_i \boldsymbol{M}_i \boldsymbol{s}^i_i + \frac{1}{2} \tilde{\boldsymbol{\theta}}^{\mathrm{T}}_i \boldsymbol{K}_{\theta,i} \tilde{\boldsymbol{\theta}}_i \, . \tag{7.73}$$

The scalar $V_i(\boldsymbol{s}^i_i, \tilde{\boldsymbol{\theta}}_i) > 0$ in view of positive definiteness of $\boldsymbol{M}_i$ and $\boldsymbol{K}_{\theta,i}$.

By differentiating $V_i$ with respect to time yields

$$\dot{V}_i = \boldsymbol{s}^{i\,\mathrm{T}}_i \boldsymbol{M}_i (\dot{\boldsymbol{\nu}}^i_{r,i} - \dot{\boldsymbol{\nu}}^i_i) - \tilde{\boldsymbol{\theta}}^{\mathrm{T}}_i \boldsymbol{K}_{\theta,i} \dot{\hat{\boldsymbol{\theta}}}_i \, ,$$

where the parameters was considered constant or slowly varying, i.e.,

$$\dot{\tilde{\boldsymbol{\theta}}}_i = -\dot{\hat{\boldsymbol{\theta}}}_i \, .$$

Taking into account the equations of motions (7.65), and considering the vector $\boldsymbol{n}_i = \boldsymbol{C}_i(\boldsymbol{\nu}^i_i)\boldsymbol{\nu}^i_{r,i} + \boldsymbol{D}_i(\boldsymbol{\nu}^i_i)\boldsymbol{\nu}^i_{r,i} + \boldsymbol{g}^i_i(\boldsymbol{R}_i)$ it is:

$$\dot{V}_i = -\boldsymbol{s}^{i\,\mathrm{T}}_i \boldsymbol{D}_i(\boldsymbol{\nu}^i_i)\boldsymbol{s}^i_i + \boldsymbol{s}^{i\,\mathrm{T}}_i \left( \boldsymbol{M}_i \dot{\boldsymbol{\nu}}^i_{r,i} + \boldsymbol{n}_i(\boldsymbol{\nu}^i_i, \boldsymbol{\nu}^i_{r,i}, \boldsymbol{R}_i) - \boldsymbol{h}^i_{t,i} \right) - \tilde{\boldsymbol{\theta}}^{\mathrm{T}}_i \boldsymbol{K}_\theta \dot{\hat{\boldsymbol{\theta}}}_i \, .$$

By adding and subtracting the term $\boldsymbol{s}^{i\,\mathrm{T}}_i \boldsymbol{h}^i_{c,i}$, where $\boldsymbol{h}^i_{c,i}$ is the control law as introduced in (7.70), and by exploiting the linearity in the parameters, the previous equation can be rewritten as

$$\begin{aligned} \dot{V}_i = & -\boldsymbol{s}^{i\,\mathrm{T}}_i \boldsymbol{D}_i \boldsymbol{s}^i_i + \boldsymbol{s}^{i\,\mathrm{T}}_i \left( \boldsymbol{Y}_i \boldsymbol{\theta}_i - \boldsymbol{h}^i_{t,i} - \boldsymbol{Y}_i \hat{\boldsymbol{\theta}}_i - \boldsymbol{K}_{v,i} \boldsymbol{s}^i_i - \boldsymbol{U}^i_{i+1} \boldsymbol{h}^{i+1}_{c,i+1} \right) + \\ & - \tilde{\boldsymbol{\theta}}^{\mathrm{T}}_i \boldsymbol{K}_{\theta,i} \dot{\hat{\boldsymbol{\theta}}}_i + \boldsymbol{s}^{i\,\mathrm{T}}_i \boldsymbol{h}^i_{c,i}, \end{aligned}$$

where $\boldsymbol{Y}_i = \boldsymbol{Y}(\boldsymbol{R}_i, \boldsymbol{\nu}^i_i, \boldsymbol{\nu}^i_{r,i}, \dot{\boldsymbol{\nu}}^i_{r,i})$ and $\boldsymbol{D}_i = \boldsymbol{D}_i(\boldsymbol{\nu}^i_i)$.

By rearranging the terms one obtains:

$$\dot{V}_i = -\boldsymbol{s}^{i\,\mathrm{T}}_i (\boldsymbol{K}_{v,i} + \boldsymbol{D}_i)\boldsymbol{s}^i_i + \boldsymbol{s}^{i\,\mathrm{T}}_i \left( \boldsymbol{Y}_i \tilde{\boldsymbol{\theta}}_i - \boldsymbol{h}^i_{t,i} + \boldsymbol{h}^i_{r,i} \right) - \tilde{\boldsymbol{\theta}}^{\mathrm{T}}_i \boldsymbol{K}_{\theta,i} \dot{\hat{\boldsymbol{\theta}}}_i \, ,$$

that, taking the update law for the dynamic parameters (7.71), finally gives

$$\dot{V}_i = -\boldsymbol{s}^{i\,\mathrm{T}}_i (\boldsymbol{K}_{v,i} + \boldsymbol{D}_i)\boldsymbol{s}^i_i + \boldsymbol{s}^{i\,\mathrm{T}}_i \tilde{\boldsymbol{h}}^i_{t,i} \, , \tag{7.74}$$

where $\tilde{\boldsymbol{h}}^i_{t,i} = \boldsymbol{h}^i_{r,i} - \boldsymbol{h}^i_{t,i}$. Equation (7.74) does not have any significant property with respect to its sign.

The Lyapunov candidate function for the UVMS is given by

$$V(\boldsymbol{s}^0_0, \ldots, \boldsymbol{s}^n_n, \tilde{\boldsymbol{\theta}}_0, \ldots, \tilde{\boldsymbol{\theta}}_n) = \sum_{i=0}^{n} V_i(\boldsymbol{s}^i_i, \tilde{\boldsymbol{\theta}}_i) \, ,$$

that is positive definite in view of positive definitiveness of $V_i$, $i = 0, \ldots, n$.

Its time derivative is simply given by the time derivatives of all the scalar functions:

$$\dot{V} = \sum_{i=0}^{n} \left[ -\boldsymbol{s}_i^{i\mathrm{T}} (\boldsymbol{K}_{v,i} + \boldsymbol{D}_i) \boldsymbol{s}_i^i + \boldsymbol{s}_i^{i\mathrm{T}} \tilde{\boldsymbol{h}}_{t,i}^i \right],$$

where the last term is null. In fact, let us consider the two equations:

$$\begin{aligned} \boldsymbol{h}_{t,i}^i &= \boldsymbol{h}_i^i - \boldsymbol{U}_{i+1}^i \boldsymbol{h}_{i+1}^{i+1}, \\ \boldsymbol{\nu}_{i+1}^{i+1} &= \boldsymbol{U}_{i+1}^{i\mathrm{T}} \boldsymbol{\nu}_i^i + \dot{q}_{i+1} \boldsymbol{z}_i^{i+1} \end{aligned}$$

it is possible to observe that the same relationships hold for $\tilde{\boldsymbol{h}}_{t,i}^i$ and $\boldsymbol{s}_i^i$:

$$\begin{aligned} \tilde{\boldsymbol{h}}_{t,i}^i &= \tilde{\boldsymbol{h}}_i^i - \boldsymbol{U}_{i+1}^i \tilde{\boldsymbol{h}}_{i+1}^{i+1}, \\ \boldsymbol{s}_{i+1}^{i+1} &= \boldsymbol{U}_{i+1}^{i\mathrm{T}} \boldsymbol{s}_i^i + s_{q,i+1} \boldsymbol{z}_i^{i+1}. \end{aligned}$$

where $\tilde{\boldsymbol{h}}_i^i = \boldsymbol{h}_{c,i}^i - \boldsymbol{h}_i^i$. Recalling that $\boldsymbol{\tau} = \boldsymbol{J}_w^{\mathrm{T}} \boldsymbol{h}_e$, the last term of the time derivative of the Lyapunov function candidate is then given by:

$$\begin{aligned} \sum_{i=0}^{n} \boldsymbol{s}_i^{i\mathrm{T}} \tilde{\boldsymbol{h}}_{t,i}^i &= \boldsymbol{s}_0^{0\mathrm{T}} \tilde{\boldsymbol{h}}_0^0 + \sum_{i=1}^{n} s_{q,i} \boldsymbol{z}_{i-1}^{i}{}^{\mathrm{T}} \tilde{\boldsymbol{h}}_i^i \\ &= \boldsymbol{s}_0^{0\mathrm{T}} \tilde{\boldsymbol{h}}_0^0 + \sum_{i=1}^{n} s_{q,i} \tilde{\tau}_{q,i} \\ &= \boldsymbol{s}_0^{0\mathrm{T}} \tilde{\boldsymbol{h}}_0^0 + \boldsymbol{s}_q^{\mathrm{T}} \tilde{\boldsymbol{\tau}}_q \\ &= \begin{bmatrix} \boldsymbol{s}_0^0 \\ \boldsymbol{s}_q \end{bmatrix}^{\mathrm{T}} \boldsymbol{J}_w^{\mathrm{T}} \tilde{\boldsymbol{h}}_e \\ &= 0 \end{aligned} \tag{7.75}$$

since, in absence of contact at the end effector, $\tilde{\boldsymbol{h}}_e = \boldsymbol{0}$. To understand the first equality let rewrite the first term for two consecutive links:

$$\begin{aligned} \boldsymbol{s}_i^{i\mathrm{T}} \tilde{\boldsymbol{h}}_{t,i}^i + \boldsymbol{s}_{i+1}^{i+1\mathrm{T}} \tilde{\boldsymbol{h}}_{t,i+1}^{i+1} &= (\boldsymbol{U}_i^{i-1\mathrm{T}} \boldsymbol{s}_{i-1}^{i-1} + s_{q,i} \boldsymbol{z}_{i-1}^i)^{\mathrm{T}} \tilde{\boldsymbol{h}}_i^i - \boldsymbol{s}_i^{i\mathrm{T}} \boldsymbol{U}_{i+1}^i \tilde{\boldsymbol{h}}_{i+1}^{i+1} + \\ &\quad (\boldsymbol{U}_{i+1}^{i}{}^{\mathrm{T}} \boldsymbol{s}_i^i + s_{q,i+1} \boldsymbol{z}_i^{i+1})^{\mathrm{T}} \tilde{\boldsymbol{h}}_{i+1}^{i+1} - \boldsymbol{s}_{i+1}^{i+1\mathrm{T}} \boldsymbol{U}_{i+2}^{i+1} \tilde{\boldsymbol{h}}_{i+2}^{i+2} \\ &= (\boldsymbol{U}_i^{i-1\mathrm{T}} \boldsymbol{s}_{i-1}^{i-1} + s_{q,i} \boldsymbol{z}_{i-1}^i)^{\mathrm{T}} \tilde{\boldsymbol{h}}_i^i + \\ &\quad (s_{q,i+1} \boldsymbol{z}_i^{i+1})^{\mathrm{T}} \tilde{\boldsymbol{h}}_{i+1}^{i+1} - \boldsymbol{s}_{i+1}^{i+1\mathrm{T}} \boldsymbol{U}_{i+2}^{i+1} \tilde{\boldsymbol{h}}_{i+2}^{i+2}, \end{aligned}$$

that, considering that the first and the last term are null for the first and the last rigid body respectively, gives the relation required in (7.75).

It is now possible to prove the system stability in a Lyapunov-Like sense using the Barbălat's Lemma. Since

- $V(\boldsymbol{s}_0^0, \ldots, \boldsymbol{s}_n^n, \tilde{\boldsymbol{\theta}}_0, \ldots, \tilde{\boldsymbol{\theta}}_n)$ is lower bounded;

- $\dot{V} \leq 0$;
- $\dot{V}$ is uniformly continuous.

then

- $\dot{V} \rightarrow 0$ as $t \rightarrow \infty$.

Thus $\boldsymbol{s}_0^0, \ldots, \boldsymbol{s}_n^n \rightarrow \mathbf{0}$ as $t \rightarrow \infty$. Due to the recursive definition of the vectors $\boldsymbol{s}_i^i$, the position errors converge to the null value as well. In fact, the vector $\boldsymbol{s}_0^0 \rightarrow \mathbf{0}$ implies $\boldsymbol{\nu}_0^0 \rightarrow \mathbf{0}$ and $\boldsymbol{e}_0 \rightarrow \mathbf{0}$ ($\lambda_{p,0} > 0$ and $\lambda_{o,0} > 0$). Moreover, since the rotation matrix has full rank, the vehicle position error $\tilde{\boldsymbol{\eta}}_1$ too decreases to the null value. Convergence to zero of $\boldsymbol{s}_i^i$ directly implies convergence to zero of $s_{q,1}, \ldots, s_{q,n}$; consequently, $\dot{\tilde{\boldsymbol{q}}} \rightarrow \mathbf{0}$ and $\tilde{\boldsymbol{q}} \rightarrow \mathbf{0}$. As usual in adaptive control technique, it is not possible to prove asymptotic stability of the whole state, since $\tilde{\boldsymbol{\theta}}_i$ is only guaranteed to be bounded.

**Remarks.**

- Achieving null error of $n+1$ rigid bodies with 6-DOF with only $6+n$ inputs is physically coherent since the control law is computed by taking into account the kinematic constraints of the system.
- In [325] the control law performs an implicit kinematic inversion. This approach requires to work with a 6-DOF robot for the stability analysis of a position/orientation control of the end effector. In case of a redundant robot, the Authors suggest the implementation of an augmented Jacobian approach in order to have a square Jacobian to work with. However, the possible occurrence of algorithmic singularities is not avoided (see also Section 6.2). On the other hand, if the kinematic control is kept separate from the dynamic loop, as in the discussed approach, it is possible to use inverse kinematic techniques robust to the occurrence of algorithmic singularities.

### 7.9.2 Simulations

Dynamic simulations have been performed to show the effectiveness of the discussed control law based on the simulation tool described in [23]. The vehicle data are taken from [145] and are referred to the experimental Autonomous Underwater Vehicle NPS Phoenix. For simulation purposes, a six-degree-of-freedom manipulator with large inertia has been considered which is mounted under the vehicle's body. The manipulator structure and the dynamic parameters are those of the Smart-3S manufactured by COMAU. Its weight is about 5% of the vehicle weight, its length is about 2 m stretched while the vehicle is 5 m long. The overall structure, thus, has 12 degrees of freedom. The relevant physical parameters of the system are reported in Table 7.5.

Notice one of the features of the controller: by changing the manipulator does not imply redesigning the control but it simply modifies the Denavit-Hartenberg table and the initial estimate of the dynamic parameters. As a

**Table 7.5.** Masses, vehicle length and Denavit-Hartenberg parameters [m,rad] of the manipulator mounted on the underwater vehicle

| | mass (kg) | length (m) | $a$ | $d$ | $\theta$ | $\alpha$ |
|---|---|---|---|---|---|---|
| vehicle | 5454 | 5.3 | - | - | - | - |
| link 1 | 80 | - | 0.150 | 0 | $q_1$ | $-\pi/2$ |
| link 2 | 80 | - | 0.610 | 0 | $q_2$ | 0 |
| link 3 | 30 | - | 0.110 | 0 | $q_3$ | $-\pi/2$ |
| link 4 | 50 | - | 0 | 0.610 | $q_4$ | $\pi/2$ |
| link 5 | 20 | - | 0 | $-0.113$ | $q_5$ | $-\pi/2$ |
| link 6 | 25 | - | 0 | 0.103 | $q_6$ | 0 |

matter of fact, the code would be exactly the same, just reading the parameters' values from different data files, with obvious advantages for software debugging and maintenance.

Notice, that, for all the simulations the following conditions have been considered:

- full degrees-of-freedom dynamic simulations;
- detailed mathematical model used in the simulation;
- inaccurate initial parameters estimates used by the controller (about 15% for each parameter);
- reduced order regressor used in the control law;
- digital implementation of the control law (at a sampling rate of 200 Hz);
- quantization of the sensor outputs and measurement noise.

The only significant simplification is the neglect of the thruster dynamics that causes limit cycle in the vehicle behavior [300]. It is worth remarking that the adaptive control law uses a reduced number of dynamic parameters for each body: body mass and mass of displaced fluid (1 parameter), first moment of gravity and buoyancy (3 parameters). Those are the parameters that, in absence of current, can affect a steady state error. Thus, each vector $\hat{\boldsymbol{\theta}}_i$ is composed of 4 elements. In Chapter 3, a detailed analysis of adaptive/integral actions in presence of the ocean current is discussed.

The desired end-effector trajectory is a straight line the projection of which on the inertial axis is a segment of 0.2 m. The line has to be executed four times, with duration of the single trial of 4 s with a 5th order polynomial time law. Then, 14 s of rest are imposed. The attitude of the end effector must be kept constant during the motion.

The initial configuration is (see Figure 7.22):

$$\begin{aligned} \boldsymbol{\eta} &= [0 \quad 0 \quad 0 \quad 0 \quad 0 \quad 0]^{\mathrm{T}} && \text{m, deg} \\ \boldsymbol{q} &= [0 \quad 180 \quad 0 \quad 0 \quad 90 \quad 180]^{\mathrm{T}} && \text{deg} \end{aligned}$$

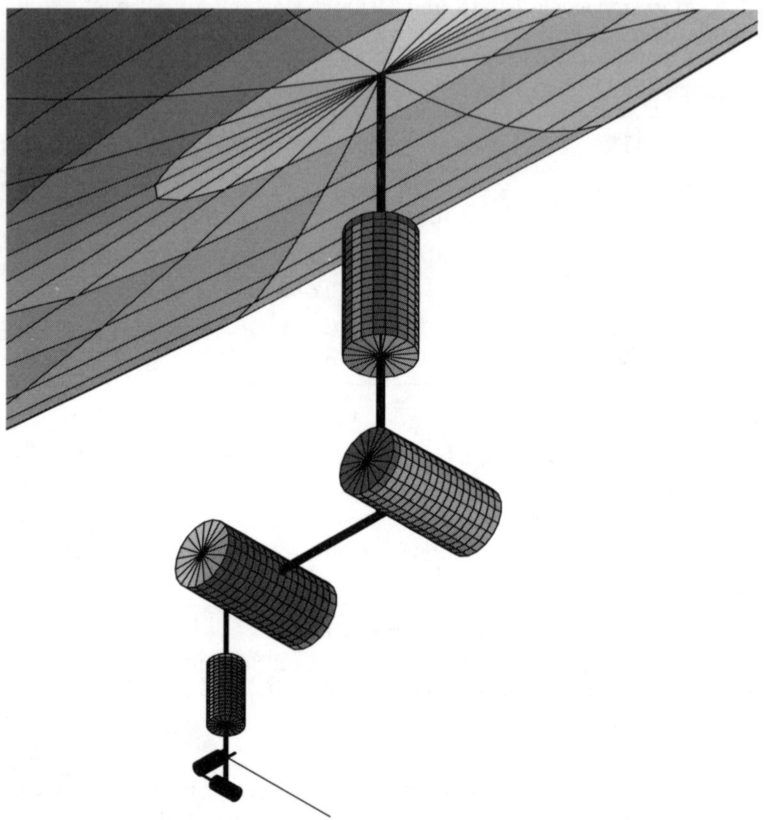

**Fig. 7.22.** Virtual decomposition control. Sketch of the initial configuration of the system and trace of the desired path

The redundancy of the system is used to keep still the vehicle. Notice, that this choice is aimed only at simplifying the analysis of the discussed controller. More complex secondary tasks can easily be given with the inverse kinematics algorithm [20, 24, 25].

The control law parameters are:

$$
\begin{aligned}
\lambda_{p,0} &= 0.4 \\
\lambda_{o,0} &= 0.6 \\
\lambda_i &= 0.9 && i = 1, \dots, n \\
\boldsymbol{K}_{v,0} &= \text{blockdiag}\left\{15000\boldsymbol{I}_3, 16000\boldsymbol{I}_3\right\} \\
\boldsymbol{K}_{v,i} &= \text{blockdiag}\left\{900\boldsymbol{I}_3, 1100\boldsymbol{I}_3\right\} && i = 1, \dots, n \\
\boldsymbol{K}_{\theta,0} &= 14\boldsymbol{I}_4 \\
\boldsymbol{K}_{\theta,i} &= 21\boldsymbol{I}_4 && i = 1, \dots, n
\end{aligned}
$$

Figures 7.23 and 7.24 report the time history of the end-effector position and orientation errors. It can be noted that, since the trajectory is repeated 4 times in the first 16 seconds, after a large initial error, the adaptive action can significantly reduce the tracking error. At steady state the errors tend to zero.

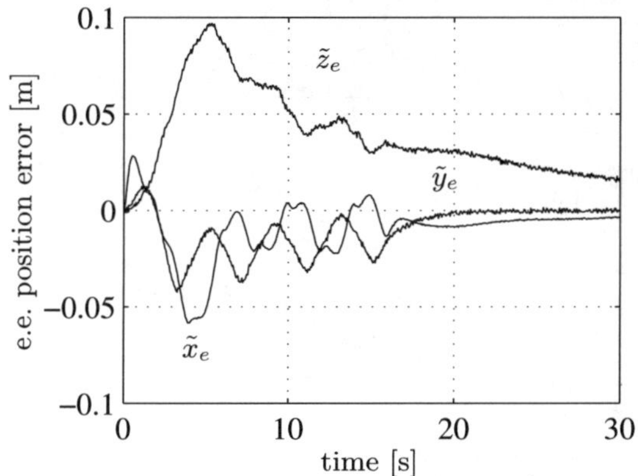


**Fig. 7.23.** Virtual decomposition control. Time history of the end-effector position error. The periodic desired trajectory makes it clear the advantage of the adaptive action

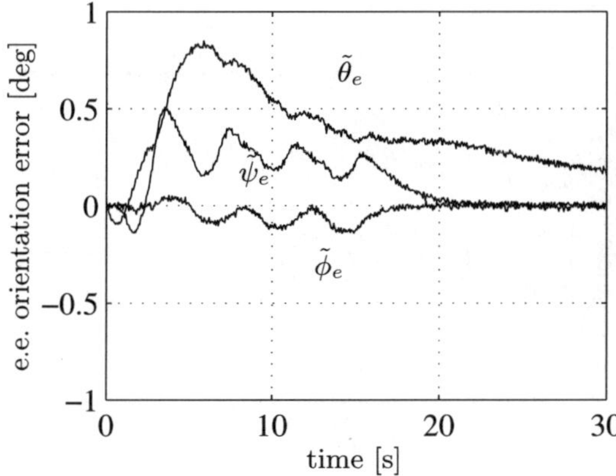


**Fig. 7.24.** Virtual decomposition control. Time history of the end-effector attitude error in terms of Euler angles

In Figure 7.25 the time history of the vehicle position is reported. It can be noted that the main tracking error is observed along $z$. This is due to the fact that the vehicle is not neutrally buoyant and at the beginning the estimation of the restoring forces has to wait for the adaptation of the control law. This can be observed also from Figure 7.26, where the time history of the vehicle linear forces is reported. Since, the manipulator is not neutrally buoyant, a large force is experienced along the $z$-axis at rest.

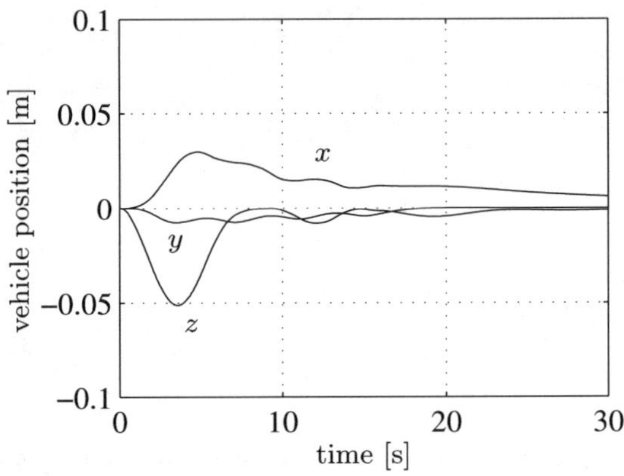


**Fig. 7.25.** Virtual decomposition control. Time history of the vehicle position

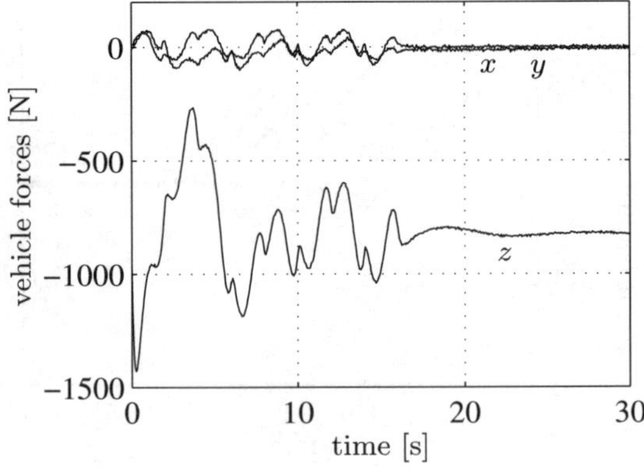


**Fig. 7.26.** Virtual decomposition control. Time history of the vehicle linear forces

In Figure 7.27 the time history of the vehicle attitude in terms of Euler angles is shown. It can be noted that the main tracking error is observed along the vehicle pitch angle $\theta$. The manipulator, in fact, interacts with the vehicle mainly in this direction. Also, Figure 7.28 shows the time history of the vehicle moments; notice the large initial overestimate due to the intentional wrong model compensation. Also, being the manipulator not neutrally buoyant, a large moment is experienced along the $y$-axis at rest.

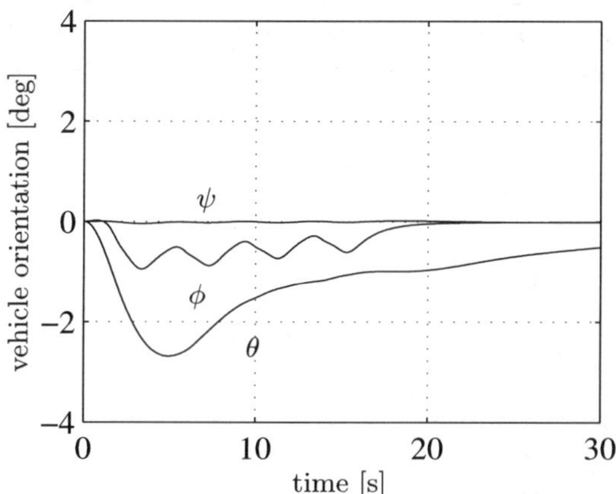


**Fig. 7.27.** Virtual decomposition control. Time history of the vehicle attitude in terms of Euler angles

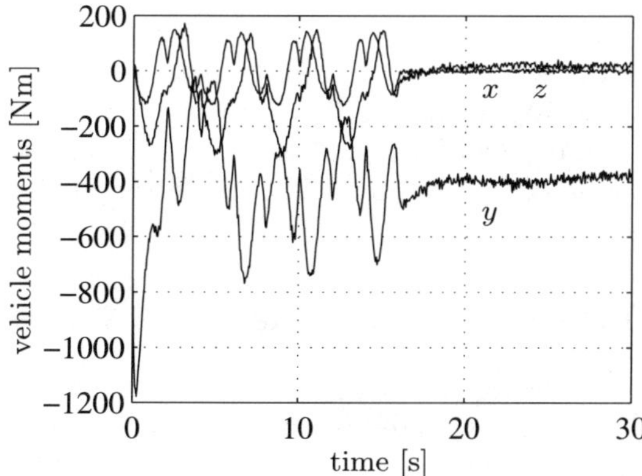


**Fig. 7.28.** Virtual decomposition control. Time history of the vehicle moments

Figure 7.29 shows the time history of the joint errors. Those are computed with respect to the desired joint positions as output from the inverse kinematics algorithm. The errors are quite large since the aim of the simulation was to show the benefit of the adaptive action in a virtual decomposition approach. The end-effector errors are the composition of all the tracking errors along the structure; the manipulator, characterized by smaller inertia and higher precision with respect to the vehicle, will be in charge of compensating for the effects of the vehicle tracking errors on the end-effector. For this specific simulation, thus, the end-effector tracking error is comparable with the vehicle position/orientation tracking error.

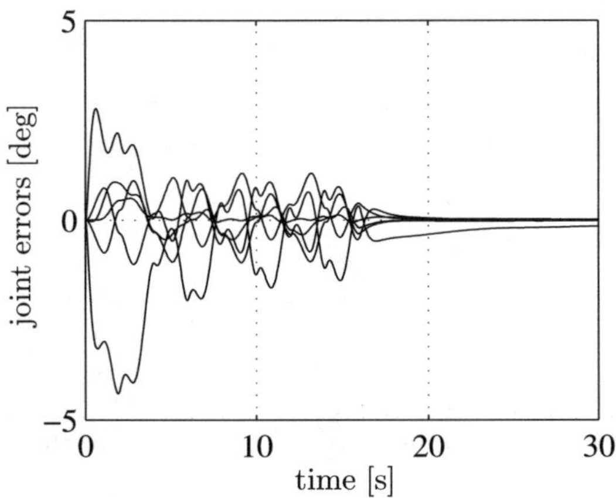


**Fig. 7.29.** Virtual decomposition control. Time history of the joint errors

In Figure 7.30 the time history of the joint torques is reported. Notice, that only the torque of the second joint is mainly affected by the manipulator restoring forces. At the very beginning the restoring compensation ($\approx 200\,\mathrm{Nm}$) presents a large difference with respect to the final value ($\approx 500\,\mathrm{Nm}$) corresponding to the same configuration due to the error in the parameter estimation.

### 7.9.3 Virtual Decomposition with the Proper Adapting Action

In the previous section, an adaptive control law in which the serial-chain structure of the UVMS is exploited is discussed. The overall motion control problem is decomposed in a set of elementary control problems regarding the motion of each rigid body in the system, namely the manipulator's links and the vehicle. For each body, a control action is designed to assign the desired motion, to adaptively compensate for the body dynamics, and to counteract force/moment exchanged with its neighborhoods along the chain.

On the other hand, in Chapter 3 it is shown, that, for a single rigid body, a suitable regressor-based adaptive control law (the mixed earth/vehicle-fixed-

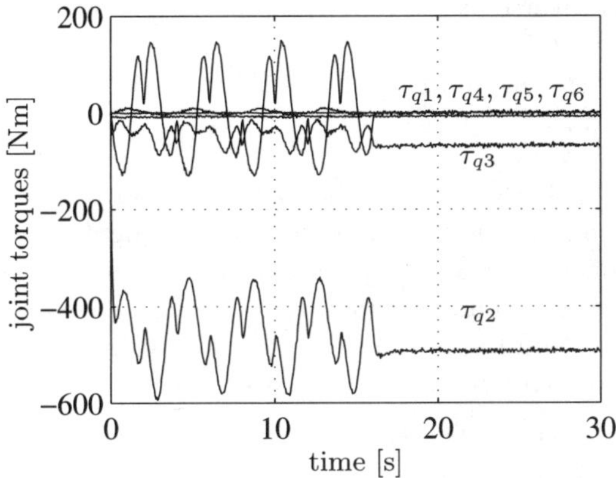


**Fig. 7.30.** Virtual decomposition control. Time history of the joint torques

frame-based, model-based controller presented in Section 3.6) gets improvement in the tracking error by considering the proper adaptation on the sole persistent terms, i.e., the current and the restoring forces. It is worth noticing that the rigid bodies of the manipulator are subject to a fast dynamics and thus the effects numerically shown in Chapter 3 for a single rigid body (the vehicle) are magnified for a fast movement of one of the arm.

It is advisable, thus, to merge these two approaches as done by G. Antonelli *et al.* in [15] in order to get benefit from both of them. The resulting control scheme has a modular structure which greatly simplifies its application to systems with a large number of links; furthermore, it reduces the required computational burden by replacing one high-dimensional problem with many low-dimensional ones, finally allows efficient implementation on distributed computing architectures.

Numerical simulations have been performed to show the effectiveness of the discussed control law with the same model used for the virtual decomposition approach; the simulation, thus, uses $\approx 400$ dynamic parameters. The overall number of parameters of the controller is $9 * (n + 1) = 63$. Moreover, the software to implement the controller is modular, the same function is used to compensate for all the rigid bodies, either the vehicle or links of the manipulator, with different parameters as inputs. This makes easier the debugging procedure.

The desired end-effector path is a straight line with length of 35 cm, to be executed 4 times according to a 5th order polynomial time law; the duration of each cycle is 4 s. The desired end-effector orientation is constant along the commanded path. The vehicle is commanded to keep its initial position and orientation during the task execution. Therefore, the inverse kinematics is needed to compute the sole joint vectors $\boldsymbol{q}_d(t)$, $\dot{\boldsymbol{q}}_d(t)$ and $\ddot{\boldsymbol{q}}_d(t)$ in real time, and thus only the inverse of the $(6 \times 6)$ manipulator Jacobian $\boldsymbol{J}_m$ is required in the inverse kinematics algorithm.

A constant ocean current affect the motion with the following components:

$$\boldsymbol{\nu}_c^I = [0.1 \quad 0.3 \quad 0 \quad 0 \quad 0 \quad 0]^{\mathrm{T}} \quad \mathrm{m/s}\,.$$

As reported in [14, 13], some of the parameters of the controllers are related to the knowledge of the mass and the first moment of gravity/buoyancy. Their initial estimate are set so as to give an estimation error larger then 20% of the true values. The other parameters are related to the presence of the current. Those, will be initialized to the null value.

The control law parameters have been set to:

$$\begin{aligned}
\boldsymbol{\Lambda}_0 &= \mathrm{blockdiag}\{0.4\boldsymbol{I}_3, 0.6\boldsymbol{I}_3\}\,,\\
\boldsymbol{\Lambda}_{i=1,6} &= \mathrm{blockdiag}\{0.9\boldsymbol{I}_3, 0.9\boldsymbol{I}_3\}\,,\\
\boldsymbol{K}_{v,0} &= \mathrm{blockdiag}\{9100\boldsymbol{I}_3, 9800\boldsymbol{I}_3\}\,,\\
\boldsymbol{K}_{v,i=1,6} &= \mathrm{blockdiag}\{600\boldsymbol{I}_3, 800\boldsymbol{I}_3\}\,,\\
\boldsymbol{K}_{\theta,0} &= 100\boldsymbol{I}_9\,,\\
\boldsymbol{K}_{\theta,i=1,6} &= 40\boldsymbol{I}_9\,,
\end{aligned}$$

where $\boldsymbol{I}_9$ is the $(9 \times 9)$ identity matrix.

The results are reported in Figures 7.31–7.32 in terms of end-effector position/orientation tracking errors; it can be recognized that, in spite of the demanding task commanded to the system, the errors are kept small in the transients and reach zero values at steady state.

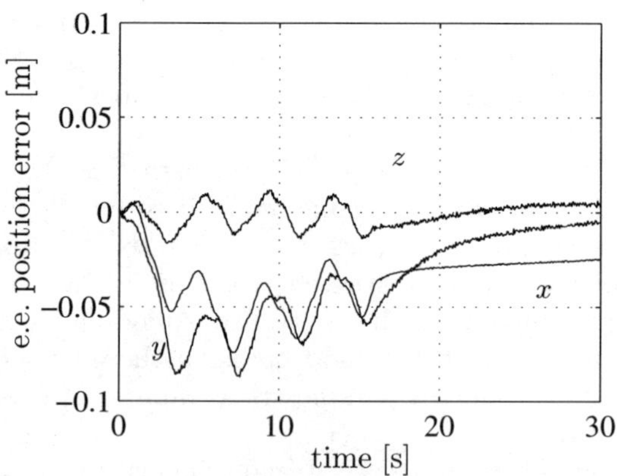


**Fig. 7.31.** Virtual decomposition control + proper adaptive action. End-effector position error

Moreover, the results in Figures 7.33–7.35 show that the control force and moments acting on the vehicle, as well as the control torques applied

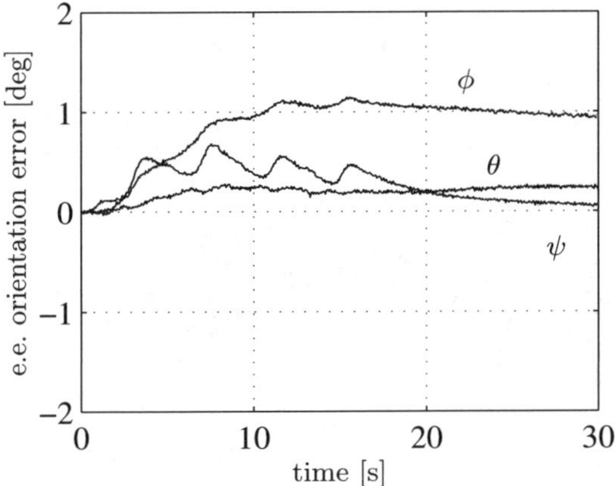


**Fig. 7.32.** Virtual decomposition control + proper adaptive action. End-effector orientation error

at the manipulator's joints, are kept limited along all the trajectory and are characterized by a smooth profile. It is worth noticing that, at the beginning of the task, the controller is not aware of the presence of the current; a compensation can be observed mainly along the $x$ and $y$ vehicle linear forces and moments.

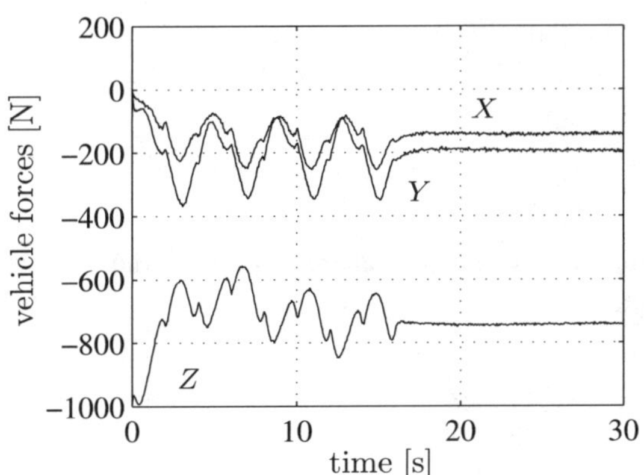


**Fig. 7.33.** Virtual decomposition control + proper adaptive action. Vehicle control forces

In Figures 7.36–7.38, the vehicle position and orientation and the joint tracking errors are reported. The tracking errors along the yaw direction can

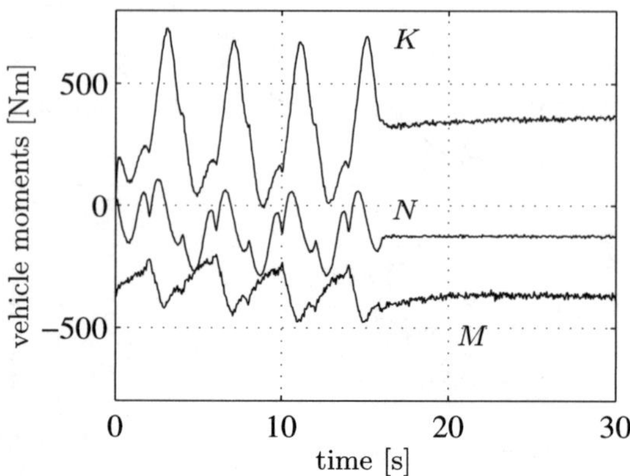


**Fig. 7.34.** Virtual decomposition control + proper adaptive action. Vehicle control moments

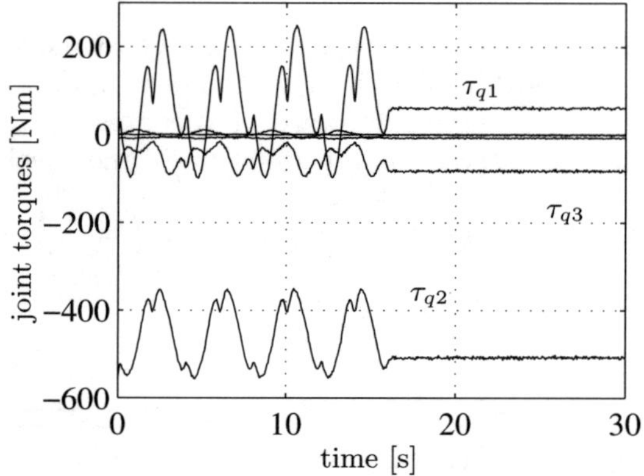


**Fig. 7.35.** Virtual decomposition control + proper adaptive action. Joint control torques

be motivated by the effect of the current, its value is decreasing to the null value according to the control gains.

## 7.10 Conclusions

In this Chapter an overview of possible control strategies for UVMSs has been presented. In view of these first, preliminary, results, it can be observed that the simple translation of control strategies developed for industrial robotics is not possible. The reason can be found both in the different technical characteristics of actuating/sensing system and in the different nature of the

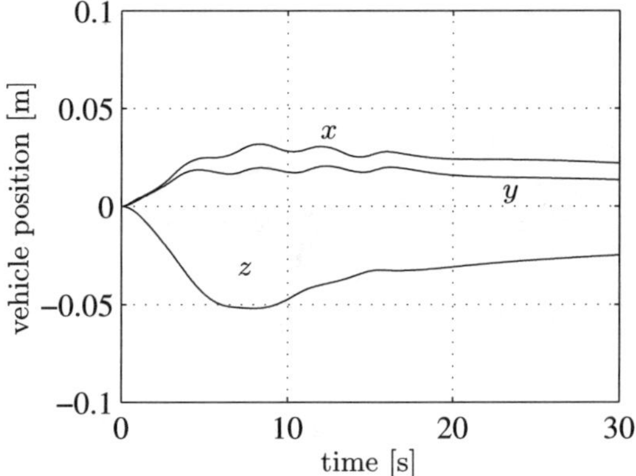


**Fig. 7.36.** Virtual decomposition control + proper adaptive action. Vehicle position

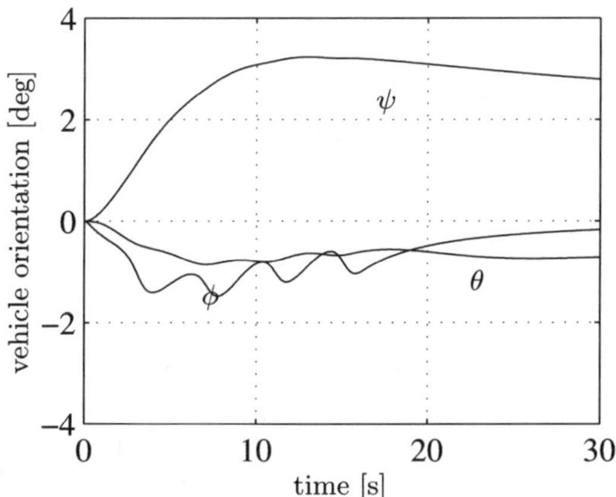


**Fig. 7.37.** Virtual decomposition control + proper adaptive action. Vehicle orientation

forces that act on a submerged body. As shown in Chapter 3, neglecting the physical nature of these forces can cause the controller to feed the system with a disturbance rather than a proper control action.

At the same time, an UVMS is a complex system, neglecting the computational aspect can lead the designer to develop a controller for which the tuning and wet-test phase might be unpractical. Starting from the simulation phase it might be appropriate to test simplified version of the controllers.

Among the control strategies analyzed the Virtual Decomposition approach, merged with the proper adaptation action, has several appealing characteristic: it is adaptive in the dynamic parameters; it is based on a Newton-Euler formulation that keep limited the computational burden; it is

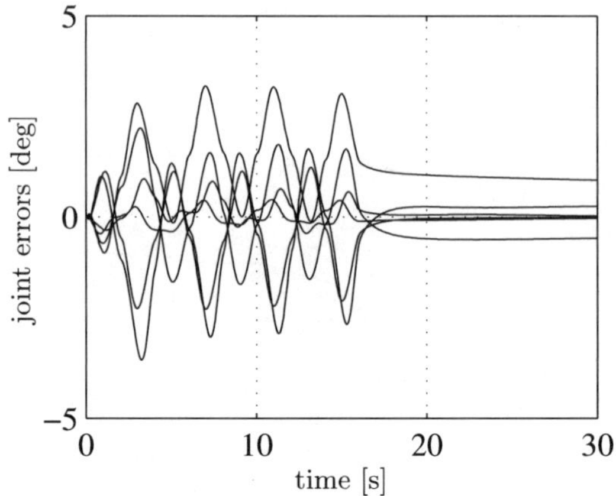


**Fig. 7.38.** Virtual decomposition control + proper adaptive action. Joint position errors

compatible with kinematic control strategies; it avoids representation singularities; it is modular, thus simplifying the software debugging and maintenance.

Motion control of UVMSs was addressed in literature in several papers. However, few experimental set-up exist which can perform autonomous vehicle/manipulator control.

The next step in understanding and evaluating the efficiency and reliability of the different approaches passes necessarily through the practical implementation of such algorithms.

In this sense, because of the symbolic analogy, numerical simulation analysis only can confirm the results already obtained in industrial robotics. It is obvious, on the other hand, that a rigorous numerical simulation of all the subsystems of the UVMSs (sensors, thrusters, interaction, ecc.), is crucial to tune the control law parameters before starting wet tests.

# 8. Interaction Control of UVMSs

## 8.1 Introduction to Interaction Control of Robots

In view of the development of an underwater vehicle able to perform a completely autonomous mission the capability of the vehicle to interact with the environment by the use of a manipulator is of greatest interest. To this aim, control of the force exchanged with the environment must be properly investigated.

Underwater Vehicle-Manipulator Systems are complex systems characterized by several strong constraints that must be taken into account when designing a force control scheme:

- Uncertainty in the model knowledge, mainly due to the poor description of the hydrodynamic effects;
- Complexity of the mathematical model;
- Structural redundancy of the system;
- Difficulty to control the vehicle in hovering, mainly due to the poor thrusters performance;
- Dynamic coupling between vehicle and manipulator;
- Low sensors' bandwidth.

Limiting our attention to elastically compliant, frictionless environments several control schemes have been proposed in the literature. An overview of interaction control schemes can be found, e.g., in [70, 303].

*Stiffness control* is obtained by adopting a suitable position control scheme when in contact with the environment [248]. In this case, it is not possible to give a reference force value; instead, a desired stiffness attitude of the tip of the manipulator is assigned. Force and position at steady state depend on the relative compliance between the environment and the manipulator.

*Impedance control* allows to achieve the behavior of a given mechanical impedance at the end effector rather than a simple compliance attitude [149]. In this case, while it is not possible to give a reference force value, the force measure at the end effector is required to achieve a decoupled behavior.

To allow the implementation of a control scheme that fulfills contact force regulation one should rather consider *direct force control* [302]. This can be effectively obtained by closing an *external force feedback* loop around a position/velocity feedback loop [101], since the output of the force controller becomes the reference input to the standard motion controller of the manipulator.

Nevertheless, many manipulation tasks require simultaneous control of both the end-effector position and the contact force. This in turn demands exact knowledge of the environment geometry: the force reference, in fact, must be consistent with the contact constraints [202].

A first strategy is the *hybrid force/position control* [239]: the force and position controllers are structurally decoupled according to the analysis of the geometric constraints to be satisfied during the task execution. These control schemes require a detailed knowledge of the environment geometry, and therefore are unsuitable for use in poorly structured environments and for handling the occurrence of unplanned impacts.

To overcome this problem, the *parallel force/position control* can be adopted [79]. In this case, position and force loops are closed in all task-space directions, while structural properties of the controller ensure that a properly assigned force reference value is reached at steady state. Since the two loops are not decoupled, a drawback of parallel control is the mutual disturbance of position and force variables during the transient.

Most force control schemes proposed in the literature do not take into account the possible presence of kinematic redundancy in the robotic system, which is always the case of UVMSs. Reference [171] presents a unified approach for motion and force control, extending the formulation to kinematically redundant manipulators. Reference [226] proposes two control schemes, the *extended hybrid control* and the *extended impedance control*, and some experimental results for a 3-link manipulator are provided. In [172] the Operational Space Formulation is experimentally applied to a coordinated task of two mobile manipulators. Reference [213] presents an hybrid force control, in which redundancy is used together with the dynamically consistent pseudo-inverse. Reference [219] proposes an *extended hybrid impedance control.* Finally, [211] develops a spatial impedance control for redundant manipulators, and reports experiments with a 7-DOF industrial manipulator. All the above schemes are based on dynamic compensation. However, while dynamic model parameters of land-based manipulators are usually well known, this is not the case of UVMSs for which the application of these schemes is not straightforward.

In [165] a very specific problem is approach, a floating vehicle, with a manipulator mounted on board, uses its thrusters and the resorting forces in order to apply the required force at the end effector.

## 8.2 Dexterous Cooperating Underwater 7-DOF Manipulators

AMADEUS (Advanced MAnipulator for DEep Underwater Sampling) was a project funded by the European Community within the MAST III framework program on Marine Technology development. One of the project's objectives includes the realization of a set-up composed by two 7-DOF Ansaldo manipulators to be used in cooperative mode [71]. The controller has been developed by G. Casalino *et al.* and it is based on a hierarchy of interacting functional subsystems.

At the lowest level there is the joint motion control, grouped in a functional scheme defined VLLC (Very Low Level Control), one for each manipulator, that receive as input the desired joint velocities and outputs the real joint positions. In case of high bandwidth loop this block behaves as an integrator.

On the top there is the LLC (Low Level Control) that receives as input the desired homogeneous transformation for the end-effector and output the desired joint velocities. Details on the handling of the kinematic singularities can be found in [71, 72] and reference therein.

At an higher level works the MLC (Medium Level Control) that operates completely in the operational space and outputs the desired homogeneous transformation matrix for the LLC. This block has been designed to interact with an operator via an human-computer interface and thus implements specific attention to the use of a expressly developed mouse. Moreover, the case of manipulators cooperation is properly taken into account.

The above approach has been developed using the RealTime Workshop tool of the MATLAB© software package. The controllers are based on C-modules running on distributed CPUs with VxWorks operative system properly synchronized at 5 ms.

Figure 8.1 reports a photo taken during an experiment run in a pool, the two manipulators have a fixed base and cooperate in the transportation of a rigid object. This experiment has been the unique of this kind ever reported in the literature.

In [282] the coordinated control of several UVMSs holding a rigid object is also taken into account by properly extending the UVMS dynamic control proposed in [281].

## 8.3 Impedance Control

Y. Cui *et al.* , in [94], propose an application for a classical impedance controller [149, 254].

It is required that the *desired* impedance at the end effector is described by the following

**Fig. 8.1.** Snapshot of the two seven-link Ansaldo manipulators during a wet test in a pool (courtesy of G. Casalino, Genoa Robotics And Automation Laboratory, Università di Genova and G. Veruggio, National Research Council-ISSIA, Italy)

$$\boldsymbol{f}_{e,d} = \boldsymbol{M}_d\ddot{\tilde{\boldsymbol{x}}} + \boldsymbol{D}_d\dot{\tilde{\boldsymbol{x}}} + \boldsymbol{K}_d\tilde{\boldsymbol{x}} \tag{8.1}$$

where $\boldsymbol{M}_d \in \mathbb{R}^{3\times3}$, $\boldsymbol{D}_d \in \mathbb{R}^{3\times3}$ and $\boldsymbol{K}_d \in \mathbb{R}^{3\times3}$ are positive definite matrices. In other words it is required to assign a desired impedance at the end effector expressed as a desired behavior of the linear e.e. force in the earth-fixed frame.

Let recall eq. (2.75)

$$\boldsymbol{M}(\boldsymbol{q})\dot{\boldsymbol{\zeta}} + \boldsymbol{C}(\boldsymbol{q},\boldsymbol{\zeta})\boldsymbol{\zeta} + \boldsymbol{D}(\boldsymbol{q},\boldsymbol{\zeta})\boldsymbol{\zeta} + \boldsymbol{g}(\boldsymbol{q},\boldsymbol{R}_B^I) = \boldsymbol{\tau} + \boldsymbol{J}_{pos}^{\mathrm{T}}(\boldsymbol{q},\boldsymbol{R}_B^I)\boldsymbol{f}_e \tag{2.75}$$

that, by defining

$$\boldsymbol{\tau}_n = \boldsymbol{C}(\boldsymbol{q},\boldsymbol{\zeta})\boldsymbol{\zeta} + \boldsymbol{D}(\boldsymbol{q},\boldsymbol{\zeta})\boldsymbol{\zeta} + \boldsymbol{g}(\boldsymbol{q},\boldsymbol{R}_B^I)$$

can be written as

$$\boldsymbol{M}(\boldsymbol{q})\dot{\boldsymbol{\zeta}} + \boldsymbol{\tau}_n = \boldsymbol{\tau} + \boldsymbol{J}_{pos}^{\mathrm{T}}(\boldsymbol{q},\boldsymbol{R}_B^I)\boldsymbol{f}_e\,. \tag{8.2}$$

The second-order differential relationship between the linear end-effector velocity in the earth-fixed frame and the system velocity is given by

$$\ddot{\boldsymbol{x}} = \boldsymbol{J}_{pos}(\boldsymbol{q}, \boldsymbol{R}_B^I)\dot{\boldsymbol{\zeta}} + \dot{\boldsymbol{J}}_{pos}(\boldsymbol{q}, \boldsymbol{R}_B^I)\boldsymbol{\zeta}$$

that, neglecting dependencies, can be inverted in the simplest way as

$$\dot{\boldsymbol{\zeta}} = \boldsymbol{J}_{pos}^{\dagger}\left(\ddot{\boldsymbol{x}} - \dot{\boldsymbol{J}}_{pos}\boldsymbol{\zeta}\right)$$

that, plugged into (8.2), and defining

$$\boldsymbol{\tau} = \boldsymbol{J}_{pos}^{\mathrm{T}}\boldsymbol{f} \tag{8.3}$$

where $\boldsymbol{f} \in \mathrm{I\!R}^3$ is to be defined, leads to

$$\overline{\boldsymbol{M}}\ddot{\boldsymbol{x}} + \boldsymbol{f}_n = \boldsymbol{f} - \boldsymbol{f}_e \tag{8.4}$$

where $\overline{\boldsymbol{M}} = \boldsymbol{J}_{pos}^{\mathrm{T}}\boldsymbol{M}\boldsymbol{J}_{pos}^{\dagger}$, and $\boldsymbol{f}_n = \boldsymbol{J}_{pos}^{\mathrm{T}}\boldsymbol{\tau}_n - \overline{\boldsymbol{M}}\dot{\boldsymbol{J}}_{pos}\boldsymbol{J}_{pos}^{\dagger}\dot{\boldsymbol{x}}$. The vector $\boldsymbol{f} \in \mathrm{I\!R}^3$ can be selected as

$$\boldsymbol{f} = \hat{\overline{\boldsymbol{M}}}\boldsymbol{x}_c + \boldsymbol{f}_n + \boldsymbol{f}_e\,. \tag{8.5}$$

where

$$\boldsymbol{x}_c = \ddot{\boldsymbol{x}}_d + \boldsymbol{M}_d^{-1}\left(\boldsymbol{D}_d\dot{\tilde{\boldsymbol{x}}} + \boldsymbol{K}_d\tilde{\boldsymbol{x}} - \boldsymbol{f}_e\right) \tag{8.6}$$

Details can be found in [94] as well as [254]; Figure 8.2 represents a block scheme of the impedance approach. In [95, 96] a unified version with the hybrid approach is also proposed.

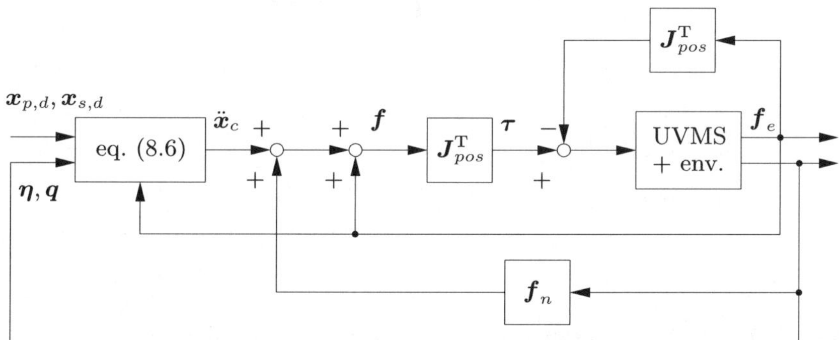


**Fig. 8.2.** Impedance Control Scheme

## 8.4 External Force Control

In this Section a force control scheme is presented to handle the strong limitations that are experienced in case of underwater systems. Based on the scheme proposed in [99], a force control scheme is analyzed that does not require exact dynamic compensation; however, the knowledge of part of the dynamic model can always be exploited when available. Extension of the original scheme to redundant systems is achieved via a task-priority inverse kinematics redundancy resolution algorithm [78] and suitable secondary tasks are defined to exploit all the degrees of freedom of the system.

**Force control scheme.** A sketch of the implemented scheme is provided in Figure 8.3. In our case the Inverse Kinematics (i.e., the block IK in the sketch) is solved at the differential level allowing to efficiently handle the system redundancy. The matrix $\boldsymbol{J}_k(\boldsymbol{R}_I^B)$ is the nonlinear, configuration dependent, matrix introduced in (2.58); its inverse, when resorting to the quaternion attitude representation, is always defined. The matrix $\boldsymbol{J}_{pos}(\boldsymbol{R}_I^B, \boldsymbol{q})$ has been defined in (2.67). Moreover, a stability analysis for the kinematically redundant case is provided. The Motion Control block can be any suitable motion control law for UVMS; thus, if a partial knowledge of the system is available, a model based control can be applied.

In Section 2.10, the mathematical model of an UVMS in contact with the environment is given.

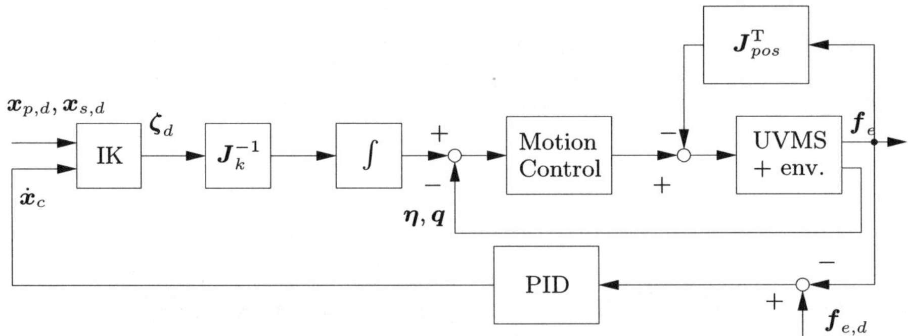


**Fig. 8.3.** External Force Control Scheme

### 8.4.1 Inverse Kinematics

Let $m$ be the number of degrees of freedom of the mission task. The system DOFs are $6 + n$, where 6 are the DOFs of the vehicle and $n$ is the number of manipulator's joints. When $6 + n > m$ the system is redundant with respect to the given task and Inverse Kinematics (IK) algorithms can be applied to exploit such redundancy. The IK algorithm implemented is based on a task-priority approach [78] (see also Section 6.5), that allows to manage the natural redundancy of the system while avoiding the occurrence of algorithmic singularities. In this phase we can define different secondary tasks to be fulfilled along with the primary task as long as they do not conflict. For example, we can ask the system not to change the vehicle orientation, or not to use the vehicle at all as long as the manipulator is working in a dexterous posture.

Let us define $\boldsymbol{x}_p$ as the primary task vector and $\boldsymbol{x}_s$ as the secondary task vector. The frame in which they are defined depends on the variables we are interested in: if the primary task is the position of the end-effector it should be normally defined in the inertial frame.

The task priority inverse kinematics algorithm is based on the following update law [78]

$$\begin{aligned}\boldsymbol{\zeta}_d &= \boldsymbol{J}_p^{\#}\left[\dot{\boldsymbol{x}}_{p,d} + \boldsymbol{\Lambda}_p(\boldsymbol{x}_{p,d} - \boldsymbol{x}_p)\right] + \\ &\quad + (\boldsymbol{I} - \boldsymbol{J}_p^{\#}\boldsymbol{J}_p)\boldsymbol{J}_s^{\#}\left[\dot{\boldsymbol{x}}_{s,d} + \boldsymbol{\Lambda}_s(\boldsymbol{x}_{s,d} - \boldsymbol{x}_s)\right] ,\end{aligned} \tag{8.7}$$

where $\boldsymbol{J}_p$ and $\boldsymbol{J}_s$ are the configuration-dependent primary and secondary task Jacobians respectively, the symbol $^{\#}$ denotes any kind of matrix inversion (e.g. Moore-Penrose), and $\boldsymbol{\Lambda}_p$ and $\boldsymbol{\Lambda}_s$ are positive definite matrices. It must be noted that $\boldsymbol{x}_p$, $\boldsymbol{x}_s$ and the corresponding Jacobians $\boldsymbol{J}_p$, $\boldsymbol{J}_s$ are functions of $\boldsymbol{\eta}_d$ and $\boldsymbol{q}_d$. The vector $\boldsymbol{\zeta}_d$ is defined in the body-fixed frame and a suitable integration must be applied to obtain the desired positions: $\boldsymbol{\eta}_d$, $\boldsymbol{q}_d$ (see (2.58)).

Let us assume that the primary task is the end-effector position that implies that $\boldsymbol{J}_p = \boldsymbol{J}_{pos}$. To introduce a force control action (8.7) is modified as follow: the vector $\dot{\boldsymbol{x}}_c$ is given as a reference value to the IK algorithm and the reference system velocities are computed as:

$$\begin{aligned}\boldsymbol{\zeta}_d &= \boldsymbol{J}_p^{\#}\left[\dot{\boldsymbol{x}}_{p,d} + \dot{\boldsymbol{x}}_c + \boldsymbol{\Lambda}_p(\boldsymbol{x}_{p,d} - \boldsymbol{x}_p)\right] + \\ &\quad + (\boldsymbol{I} - \boldsymbol{J}_p^{\#}\boldsymbol{J}_p)\boldsymbol{J}_s^{\#}\left[\dot{\boldsymbol{x}}_{s,d} + \boldsymbol{\Lambda}_s(\boldsymbol{x}_{s,d} - \boldsymbol{x}_s)\right] ,\end{aligned} \tag{8.8}$$

where

$$\dot{\boldsymbol{x}}_c = k_{f,p}\tilde{\boldsymbol{f}}_e - k_{f,v}\dot{\boldsymbol{f}}_e + k_{f,i}\int_0^t \tilde{\boldsymbol{f}}_e(\sigma)d\sigma \tag{8.9}$$

being $\tilde{\boldsymbol{f}}_e = \boldsymbol{f}_{e,d} - \boldsymbol{f}_e$ the force error. The direction in which force control is expected are included in the primary task vector, e.g., if the task requires to exert a force along $z$, the primary task includes the $z$ component of $\boldsymbol{x}$.

### 8.4.2 Stability Analysis

Pre-multiplying (8.8) by $\boldsymbol{J}_p \in \mathbb{R}^{3\times(6+n)}$ we obtain:

$$\boldsymbol{J}_p\boldsymbol{\zeta}_d = \dot{\boldsymbol{x}}_{p,d} + \dot{\boldsymbol{x}}_c + \boldsymbol{\Lambda}_p(\boldsymbol{x}_{p,d} - \boldsymbol{x}_p) \,. \tag{8.10}$$

Let us consider a regulation problem, i.e., the reference force $\boldsymbol{f}_{e,d}$ and the desired primary task $\boldsymbol{x}_{p,d}$ are constant. In addition, let us assume that $\boldsymbol{f}_{e,d} \in \mathcal{R}(\boldsymbol{K})$, where $\boldsymbol{K}$ is the stiffness matrix defined in (2.77), and that, as common in external force control approach, the motion controller guarantees perfect tracking, yielding $\boldsymbol{J}_p\boldsymbol{\zeta}_d = \dot{\boldsymbol{x}}$. We finally choose $\boldsymbol{\Lambda}_p = \lambda_p \boldsymbol{I}_m$. Thus, (8.10) can be rewritten as

$$\dot{\boldsymbol{x}} = \dot{\boldsymbol{x}}_c + \lambda_p(\boldsymbol{x}_{p,d} - \boldsymbol{x}) . \tag{8.11}$$

In view of the above assumptions the vector $\dot{\boldsymbol{x}}_c$ belongs to $\mathcal{R}(\boldsymbol{K})$; it is then simple to recognize that the motion component of the dynamics along the normal direction to the surface is decoupled from the motion components lying onto the contact plane. Therefore, it is convenient to multiply (8.11) by the two orthogonal projectors $\boldsymbol{n}\boldsymbol{n}^{\mathrm{T}}$ and $\boldsymbol{I} - \boldsymbol{n}\boldsymbol{n}^{\mathrm{T}}$ (see Figure 2.5), that gives the two decoupled dynamics

$$\boldsymbol{n}\boldsymbol{n}^{\mathrm{T}}\dot{\boldsymbol{x}} = \dot{\boldsymbol{x}}_c + \boldsymbol{n}\boldsymbol{n}^{\mathrm{T}}\lambda_p(\boldsymbol{x}_{p,d} - \boldsymbol{x}) \tag{8.12}$$

$$(\boldsymbol{I} - \boldsymbol{n}\boldsymbol{n}^{\mathrm{T}})\dot{\boldsymbol{x}} = (\boldsymbol{I} - \boldsymbol{n}\boldsymbol{n}^{\mathrm{T}})\lambda_p(\boldsymbol{x}_{p,d} - \boldsymbol{x}) \tag{8.13}$$

Equation (8.13) clearly shows convergence of the components of $\boldsymbol{x}$ tangent to the contact plane to the corresponding desired values when $\lambda_p > 0$.

To analyze convergence of (8.12), by differentiating (8.12) and by taking into account (2.76) and (8.9) we obtain the scalar equation

$$(1 + k_{f,v}k)\ddot{w} + (\lambda_p + k_{f,p}k)\dot{w} + k_{f,i}kw = k_{f,i}\boldsymbol{n}^{\mathrm{T}}(\boldsymbol{f}_d + k\boldsymbol{x}_\infty) , \tag{8.14}$$

where the variable $w = \boldsymbol{n}^{\mathrm{T}}\boldsymbol{x}$ is used for notation compactness and $k$, as defined in the modeling Chapter, is the environment stiffness.

Equation (8.14) shows that, with a proper choice of the control parameters, the component of $\boldsymbol{x}$ normal to the contact plane converges to the value

$$w_\infty = \boldsymbol{n}^{\mathrm{T}}(\frac{1}{k}\boldsymbol{f}_{e,d} + \boldsymbol{x}_\infty) .$$

In summary, the overall system converges to the equilibrium

$$\boldsymbol{x}_\infty = \boldsymbol{n}\boldsymbol{n}^{\mathrm{T}}(\frac{1}{k}\boldsymbol{f}_{e,d} + \boldsymbol{x}_\infty) + (\boldsymbol{I} - \boldsymbol{n}\boldsymbol{n}^{\mathrm{T}})\boldsymbol{x}_{p,d} ,$$

which can be easily recognized to ensure $\boldsymbol{f}_{e,\infty} = \boldsymbol{f}_{e,d}$.

### 8.4.3 Robustness

In the following, the robustness of the controller to react against unexpected impacts or errors in planning desired force/position directions is discussed.

The major drawback of hybrid control is that it is not robust to the occurrence of an impact in a direction where motion control has been planned. In such a case, in fact, the end effector is not compliant along that direction and strong interaction between manipulator and environment is experienced. This problem has been solved by resorting to the parallel approach [79] where the force control action overcomes the position control action at the contact.

The proposed scheme shows the same feature as the parallel control. In detail, if a contact occurs along a motion direction, we can see from (8.8) that the integral action of the force controller guarantees a null force error at steady state while a non-null position error $\boldsymbol{x}_d - \boldsymbol{x}$ is obtained. This implies that, in the direction where a null desired force is commanded, the manipulator reacts to unexpected impacts with a safe behavior. Moreover, in the directions in which a desired force is commanded, the desired position is overcome by the controller.

If $\boldsymbol{f}_{e,d} \notin R(\boldsymbol{K})$, i.e., the direction of $\boldsymbol{f}_{e,d}$ is not parallel to $\boldsymbol{n}$, the controller is trying to interact with the environment in directions in which the environment cannot generate reaction forces. A drift in that direction is then experienced.

### 8.4.4 Loss of Contact

Due to the floating base and the possible occurrence of external disturbances (such as, e.g., ocean current), it can happen that the end effector loses contact with the environment during the task fulfillment. In such a case the control action might become unsuitable. In fact, from (8.8) it can be noted that the desired force is interpreted as a motion reference velocity scaled by the force control gain. One way to handle this problem is the following: if the force sensor does not read any force value in the desired contact direction, the integrator in the force controller is reset and (8.8) is modified as follows:

$$\begin{aligned}\boldsymbol{\zeta}_d = \boldsymbol{J}_p^{\#} \left\{ \boldsymbol{H} \left[ \dot{\boldsymbol{x}}_{p,d} + \dot{\boldsymbol{x}}_c + \boldsymbol{\Lambda}_p(\boldsymbol{x}_{p,d} - \boldsymbol{x}_p) \right] + (\boldsymbol{I} - \boldsymbol{H})\, \dot{\boldsymbol{x}}_l \right\} + \\ (\boldsymbol{I} - \boldsymbol{J}_p \boldsymbol{J}_p^{\#}) \boldsymbol{J}_s^{\#} \left[ \dot{\boldsymbol{x}}_{s,d} + \boldsymbol{\Lambda}_s(\boldsymbol{x}_{s,d} - \boldsymbol{x}_s) \right] ,\end{aligned} \tag{8.15}$$

where $\boldsymbol{H} \in \mathrm{I\!R}^{m \times m}$ is a diagonal selection matrix with ones for the motion directions and zeros for the force directions, and $\dot{\boldsymbol{x}}_l$ is a desired velocity at which the end effector can safely impact the environment. Basically, the IK is handling the motion directions in the same way as when the contact occurs but, for the force direction, a reference velocity is given in a way to obtain again the contact. Notice that $\dot{\boldsymbol{x}}_c$ is not dropped out in case of loss of contact because it guarantees from unexpected contacts in the direction where motion control is expected (i.e., the directions in which the desired force is zero).

Matrix $\boldsymbol{H}$ can be interpreted as the selection matrix of a hybrid control scheme. Notice that this matrix is used only when there is *no contact* at the end effector and the properties of robustness discussed above still hold.

### 8.4.5 Implementation Issues

The implementation of the proposed force control scheme might benefit from some practical considerations:

- The vehicle and the manipulator are characterized by a different control bandwidth, due to the different inertia and actuator performance. Moreover limit cycles in underwater vehicles are usually experienced due to the thruster's characteristics. This implies that the use of the desired configuration in the kinematic errors computation in (8.8) would lead to coupling between the force and motion directions, since the Jacobian matrix is computed with respect to a position different from the actual position. In (8.8), thus, the real positions will be used to compute the errors.

- Force control tasks require accurate positioning of the end effector. On the other hand, the vehicle, i.e., the base of the manipulator, is characterized by large position errors. In (8.8), thus, it could be appropriate to decompose the desired end-effector velocity in a way to involve the manipulator alone in the fulfillment of the primary task. Let us define $\boldsymbol{J}_{p,man}(\boldsymbol{R}_B^I, \boldsymbol{q})$ as the Jacobian of the manipulator, it is possible to rewrite (8.8) as
$$\boldsymbol{\zeta}_d = \begin{bmatrix} \boldsymbol{0}_{6\times 1} \\ \boldsymbol{J}^{\#}_{p,man}\left[\dot{\boldsymbol{x}}_{p,d} + \dot{\boldsymbol{x}}_c + \boldsymbol{\Lambda}_p(\boldsymbol{x}_{p,d} - \boldsymbol{x}_p)\right] \end{bmatrix} + \\ +(\boldsymbol{I} - \boldsymbol{J}^{\#}_p \boldsymbol{J}_p)\boldsymbol{J}^{\#}_s\left[\dot{\boldsymbol{x}}_{s,d} + \boldsymbol{\Lambda}_s(\boldsymbol{x}_{s,d} - \boldsymbol{x}_s)\right] . \tag{8.16}$$
Notice that the same properties of (8.8) applies also for (8.16). A physical interpretation of (8.16) is the following: it is *asked* the manipulator to fulfill the primary task taking into account the movement of its base. At the same time the secondary task is fulfilled, with less strict requirements, with the whole system (e.g., the vehicle must move when the manipulator is working on the boundaries of its workspace).
- UVMSs are usually highly redundant. If a 6-DOF manipulator is used this means that 12 DOFs are available. It is then possible to define more tasks to be iteratively projected on the null space of the higher priority tasks. An example of 3 tasks could be: 1) motion/force control of the end effector, 2) increase manipulability measure of the manipulator, 3) limit roll and pitch orientation of the vehicle. See also Chap. 6.
- To decrease power consumption it is possible to implement IK algorithms with bounded reference values for the secondary tasks. Using some smooth functions, or fuzzy techniques, it is possible to activate the secondary tasks only when the relevant variables are out of a desired range. For the roll and pitch vehicle's angles, for example, it might be convenient to implement an algorithm that keeps them in a range, e.g., $\pm 10°$. See also Chap. 6.
- Force/moment sensor readings are usually corrupted by noise. The use of a derivative action in the control law, thus, can be difficult to implement. With the assumption of a frictionless and elastically compliant plane it can be observed a linear relation between $\dot{\boldsymbol{f}}_e$ and $\dot{\boldsymbol{x}}$. The force derivative action can then be substituted by a term proportional to $\dot{\boldsymbol{x}}$. The latter will be computed by differential kinematics from $\boldsymbol{\zeta}$ that is usually available from direct sensor readings or from the position readings via a numerical filter.

### 8.4.6 Simulations

To prove the effectiveness of the proposed force control scheme several simulations have been run under MATLAB© /SIMULINK© environment. The

UVMS simulated has 9 DOFs, 6 DOFs of the vehicle plus a 3-link manipulator mounted on it [231]. The controller has been implemented with a sampling frequency of 200 Hz. The vehicle is a box of dimensions $(2\times1\times0.5)$ m, and the vehicle fixed frame is located in the geometrical center of the body. The manipulator has a planar structure with 3 rotational joints. A sketch of the system seen from the vehicle $xz$ plane is shown in Figure 8.4. In the shown configuration it is $\boldsymbol{\eta} = [\,0 \quad 0 \quad 0 \quad 0 \quad 0 \quad 0\,]^{\mathrm{T}}$ m, deg, $\boldsymbol{q} = [\,45 \quad -90 \quad -45\,]^{\mathrm{T}}$ deg. The stiffness of the environment is $k = 10^4$ N/m. In the Appendix some details on the dynamic model are given.

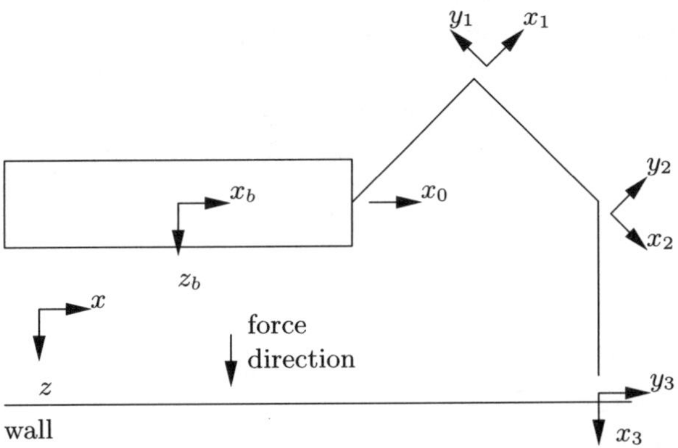


**Fig. 8.4.** Sketch of the simulated system for both controllers as seen from the $xz$ vehicle-fixed plane

The vehicle is supposed to start in a non dexterous configuration, so as to test the redundancy resolution capabilities of the proposed scheme. The initial configuration is:

$$\begin{aligned} \boldsymbol{\eta} &= [\,0 \quad 0 \quad -2 \quad 12 \quad 0 \quad 0\,]^{\mathrm{T}} \ \text{m, deg}\,, \\ \boldsymbol{q} &= [\,-90 \quad 15 \quad 0\,]^{\mathrm{T}} \ \text{deg} \end{aligned}$$

that corresponds to the end-effector position

$$\boldsymbol{x} = [\,x_E \quad y_E \quad z_E\,]^{\mathrm{T}} = [\,1.51 \quad -0.60 \quad 0.86\,]^{\mathrm{T}} \quad \text{m}\,.$$

Since the vehicle is far from the plane ($z = 1.005$ m), the manipulator is outstretched, i.e., close to a kinematic singularity.

The simulated task is to perform a force/position task at the end effector as primary task; specifically, the UVMS is required to move $-20$ cm along $x$ and apply a desired force of 200 N along $z$. The secondary tasks are to guarantee manipulator manipulability and to keep roll and pitch vehicle's angles in a safe range of $\pm10°$. In detail, the 3 task variables are:

$$\begin{aligned} \boldsymbol{x}_p &= \boldsymbol{x}\,, \\ \boldsymbol{x}_s &= [\,\phi \quad \theta\,]^{\mathrm{T}}\,, \\ x_t &= q_2\,, \end{aligned}$$

where $x_t$ expresses the third task to be projected in the null space of the higher priority tasks; in fact since the manipulator has a 3-link planar structure a measure of its manipulability is simply given by $q_2$, where $q_2 = 0$ corresponds to a kinematic singularity. In the initial position the 3 tasks are activated simultaneously and are performed by exploiting kinematic redundancy. Moreover an unexpected impact along $x$ is considered (for $x_E = 1.32\,\text{m}$).

A weighted pseudoinverse has been used to compute $\boldsymbol{J}_p^{\#}$ characterized by the weight matrix $\boldsymbol{W} = \text{blockdiag}\{10\boldsymbol{I}_6, \boldsymbol{I}_3\}$. To simulate an imperfect hovering of the vehicle a control law with *lower* gain for the vehicle was implemented; the performance of the simulated vehicle, thus, has an error that is of the same magnitude as that of a real vehicle in hovering. The motion controller implemented is the virtual decomposition adaptive based control presented in Section 7.9. Equation (8.16) has then been used to compute the desired velocities. The secondary task regarding the vehicle orientation has to be fulfilled only when the relevant variable is outside of a desired bound.

The control gains, in S.I. units, are:

$$
\begin{aligned}
k_{fp} &= 3 \cdot 10^{-3}\,, \\
k_{fi} &= 8 \cdot 10^{-3}\,, \\
k_{fv} &= 10^{-4}\,, \\
\boldsymbol{\Lambda}_p &= 0.6\boldsymbol{I}_3\,, \\
\boldsymbol{\Lambda}_s &= \boldsymbol{I}_3\,.
\end{aligned}
$$

The simulations have been run by adopting separate motion control schemes for the vehicle and the manipulator, since this is the case of many UVMSs. Better results would be obtained by resorting to a centralized motion control scheme in which dynamic coupling between vehicle and manipulator is compensated for. The initial value of the parameters has been chosen such that the gravity compensation at the beginning is different from the real one, adding an error bounded to about 10% for each parameter.

In Figure 8.5 the time history of the end-effector variables for the proposed force control scheme without exploiting the redundancy and without unexpected impact is shown. During the first 3 s the end effector is not in contact with the plane and the algorithm to handle loss of contact has been used. It can be noted that the primary task is successfully achieved.

In Figure 8.6 the same task has been achieved by exploiting the redundancy. The different behavior of the force can be explained by considering that the system impacts the plane in a different configuration with respect to the previous case because of the internal motion imposed by the redundancy resolution. This also causes a different end-effect velocity at the impact.

Figure 8.7 shows the time history of the secondary tasks in the two previous simulations, without exploiting redundancy (solid) and with the proposed control scheme (dashed). It can be recognized that without exploiting the redundancy the system performs the task in a non dexterous configuration, i.e., with a big roll angle and with the manipulator close to a kinematic

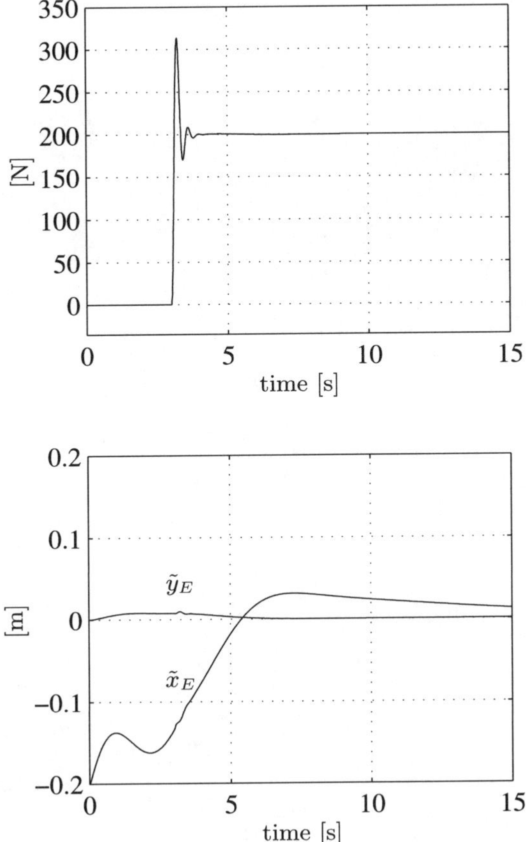


**Fig. 8.5.** External force control. Force along $z$ (top) and end-effector error components along the motion directions (bottom) without exploiting the redundancy and without unexpected impact

singularity. A suitable use of the system's redundancy allows to reconfigure the system and to achieve the secondary task.

Finally, in Figure 8.8 the time history of the primary task variables in case of an unplanned impact is shown. It can be noted that the undesired force along $x$, due to the unexpected impact, is recovered yielding a non-zero position error along that direction.

## 8.5 Explicit Force Control

In this Section, based on [119] two force control schemes for UVMS that overcome many of the above-mentioned difficulties associated with the un-

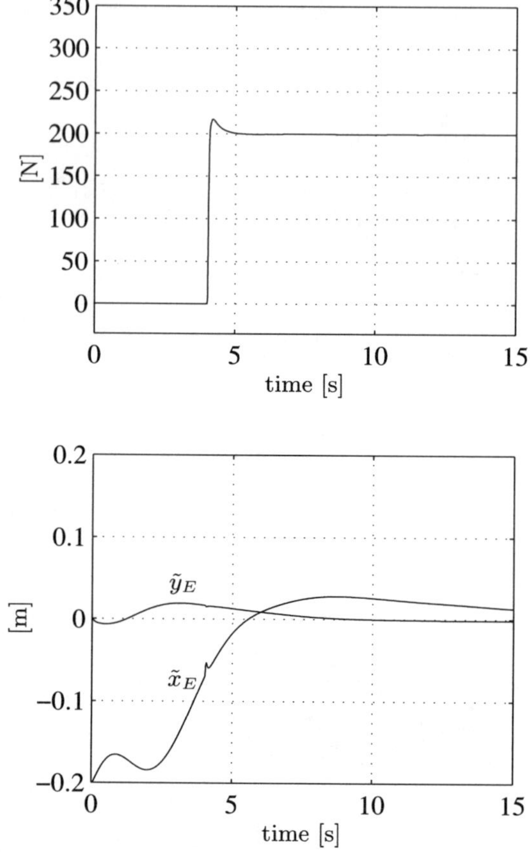


**Fig. 8.6.** External force control. Force along $z$ (top) and end-effector error components along the motion directions (bottom) exploiting the redundancy and without unexpected impact

derwater manipulation are presented. These force control schemes exploit the system redundancy by using a task-priority based inverse kinematics algorithm [78]. This approach allows us to satisfy various secondary criteria while controlling the contact force. Both these force control schemes require separate motion control schemes which can be chosen to suit the objective without affecting the performance of the force control. The possible occurrence of loss of contact due to vehicle movement is also analyzed. The proposed control schemes have extensively been tested in numerical simulation runs; the results obtained in a case study are reported to illustrate their performance.

Two different versions of the scheme are implemented based on different projections of the force error from the task space to the vehicle/joint space.

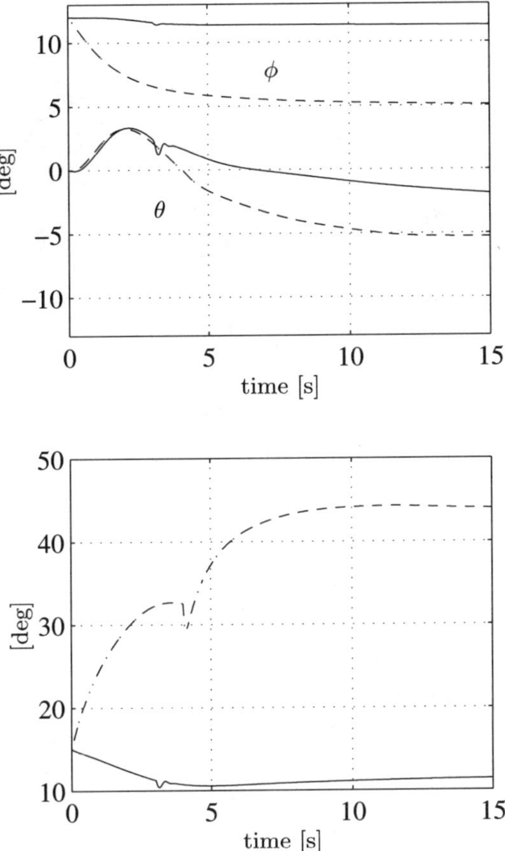


**Fig. 8.7.** External force control. Roll and pitch vehicle's angles (top) and $q_2$ as manipulability measure (bottom). Solid: without exploiting redundancy; Dashed: exploiting redundancy with the proposed scheme

The first scheme is obtained by using the transpose of the Jacobian to project the force error from the task space directly to the control input space, i.e., force/moments for the vehicle and torques for the manipulator, leading to an evident physical interpretation.

In the second scheme, instead, the force error is projected from the task space to the body-fixed velocities. This is done to avoid the need to directly access the control input; in many cases, in fact, a velocity controller is implemented on the manipulator and control torques are not accessible [119].

For both the control schemes, the kinematic control applied is the same as the algorithm exploited in the external force control scheme, already shown in Section 8.4.1.

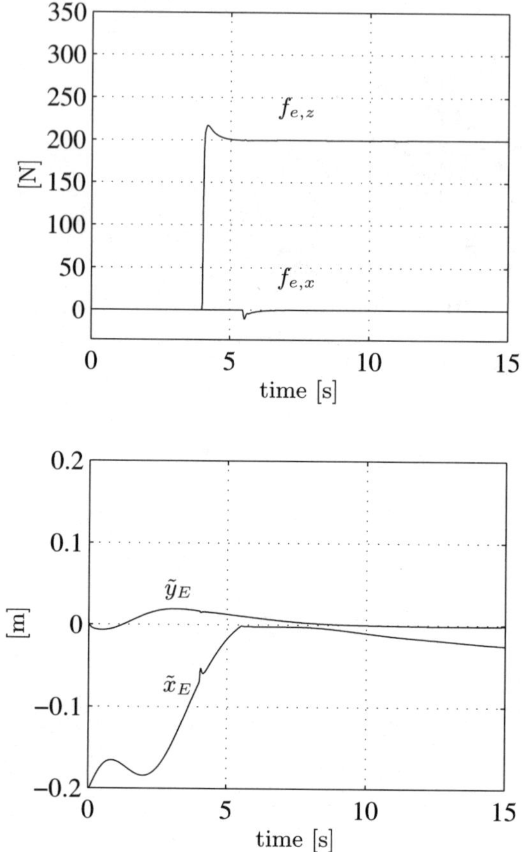


**Fig. 8.8.** External force control. Force along $z$ and $x$ (top) and end-effector error components along the motion directions (bottom) exploiting the redundancy and with unexpected impact along $x$

**Explicit force control, scheme 1.** A sketch of the implemented scheme is provided in Figure 8.9. In our case, the inverse kinematics is solved via a numerical algorithm based on velocity mapping to allow the handling of system redundancy efficiently (see previous Section and Section 6.5). It is possible to decompose the input torque as the sum of the torque output from the motion control and the torque output from the force control: $\boldsymbol{\tau} = \boldsymbol{\tau}^M + \boldsymbol{\tau}^F$. The force control action $\boldsymbol{\tau}^F$ is computed as:

$$\begin{aligned} \boldsymbol{\tau}^F &= \boldsymbol{J}_{pos}^{\mathrm{T}} \boldsymbol{u}_F \\ &= \boldsymbol{J}_{pos}^{\mathrm{T}} \left( -\boldsymbol{f}_e + k_{f,p} \tilde{\boldsymbol{f}}_e - k_{f,v} \dot{\boldsymbol{f}}_e + k_{f,i} \int_0^t \tilde{\boldsymbol{f}}_e(\sigma) d\sigma \right) , \end{aligned} \tag{8.17}$$

where $k_{f,p}$, $k_{f,v}$, $k_{f,i}$ are scalar positive gains, and $\tilde{\boldsymbol{f}}_e = \boldsymbol{f}_{e,d} - \boldsymbol{f}_e$ is the force error. Equation (8.17), thus, is a force control action in the task space that is further projected, via the transpose of the Jacobian, on the vehicle/joint space.

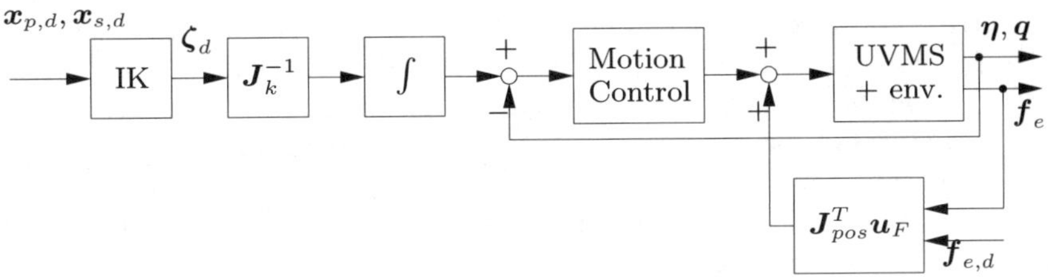


**Fig. 8.9.** Explicit force control scheme 1

**Explicit force control, scheme 2.** A sketch of the implemented scheme is provided in Figure 8.10. The force control action is composed of two loops; the action $-\boldsymbol{J}_{pos}^{\mathrm{T}} \boldsymbol{f}_e$ is aimed at compensating the end-effector contact force (included in the block labeled "UVMS + env."). The input of the motion control is a suitable integration of the velocity vector:

$$\boldsymbol{\zeta}_r = \boldsymbol{\zeta}_d + \boldsymbol{\zeta}_F = \boldsymbol{\zeta}_d + \boldsymbol{J}_{pos}^{\mathrm{T}} (k_{f,p}^* \tilde{\boldsymbol{f}}_e - k_{f,v}^* \dot{\boldsymbol{f}}_e + k_{f,i}^* \int_0^t \tilde{\boldsymbol{f}}_e(\sigma) d\sigma) ,$$

where $k_{f,p}^*$, $k_{f,v}^*$, $k_{f,i}^*$ are positive gains, and $\boldsymbol{\zeta}_d$ is the output of the inverse kinematics algorithm described in previous Section.

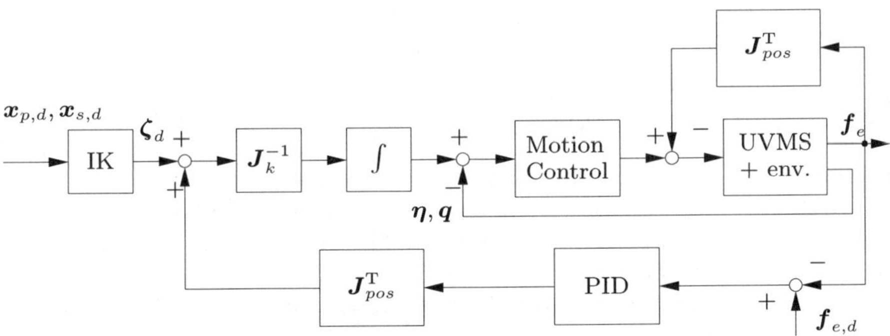


**Fig. 8.10.** Explicit force control scheme 2

### 8.5.1 Robustness

In the following, the robustness of the schemes to react against unexpected impacts or errors in planning desired force/position directions is discussed.

The use of the integral action in the force controller gives a higher priority to the force error with respect to the position error. In the motion control directions the desired force is null. An unexpected impact in a direction where the desired force is zero, thus, is handled by the controller yielding a safe behavior, which results in a non-null position error at steady state and a null force error, i.e., zero contact force.

If $\boldsymbol{f}_{e,d} \notin \mathcal{R}(\boldsymbol{K})$, i.e. the direction of $\boldsymbol{f}_{e,d}$ is not parallel to $\boldsymbol{n}$, the controller is commanded to interact with the environment in directions along which no reaction force exists. In this case, a drift motion in that direction is experienced.

These two force control schemes too, possess the safe behavior as the external force control scheme.

The possibility that the end effector loses contact with the environment is also taken into account. The same algorithm as presented in Section 8.4.4 were implemented.

### 8.5.2 Simulations

To test the effectiveness of the proposed force control schemes several numerical simulations were run under the MATLAB© /SIMULINK© environment. The controllers were implemented in discrete time with a sampling frequency of 200 Hz. The environmental stiffness is $k = 10^4$ N/m.

The simulated UVMS has 9 DOFs, 6 DOFs of the vehicle plus a 3-link manipulator mounted on it [231]. The vehicle is a box of dimensions $(2 \times 1 \times 0.5)$ m; the vehicle-fixed frame is located in the geometrical center of the body. The manipulator is a 3-link planar manipulator with rotational joints. Figure 8.4 shows a sketch of the system, seen from the vehicle's $xz$ plane, in the configuration

$$\begin{aligned} \boldsymbol{\eta} &= [0 \quad 0 \quad 0 \quad 0 \quad 0 \quad 0]^{\mathrm{T}} \ \text{m, deg}, \\ \boldsymbol{q} &= [45 \quad -90 \quad -45]^{\mathrm{T}} \quad \text{deg} \end{aligned}$$

corresponding to the end-effector position $\boldsymbol{x} = [2.41 \quad 0 \quad 1]^{\mathrm{T}}$ m.

A case study is considered aimed at the following objectives: as primary task, to perform force/motion control of the end effector (exert a force of 200 N along $z$, moving the end effector from 2.41 m to 2.21 m along $x$, while keeping $y$ at 0 m); as secondary task, to guarantee vehicle's roll and pitch angles being kept in the range $\pm 10°$; as tertiary task, to guarantee the manipulator's manipulability being kept in a safe range. In detail, the 3 task variables are:

$$\begin{aligned} \boldsymbol{x}_p &= \boldsymbol{x}, \\ \boldsymbol{x}_s &= [\phi \quad \theta]^{\mathrm{T}}, \\ x_t &= q_2, \end{aligned}$$

where $x_t$ expresses the third task; in fact, since the manipulator has a 3-link planar structure, a measure of its manipulability is simply given by $q_2$, where $q_2 = 0$ corresponds to a kinematic singularity.

Let us suppose that the system starts the mission in the non-dexterous configuration

$$\begin{aligned} \boldsymbol{\eta} &= [\,0 \quad 0 \quad -0.001 \quad 15 \quad 0 \quad -4\,]^{\mathrm{T}} \text{ m, deg}\,, \\ \boldsymbol{q} &= [\,46 \quad -89 \quad -44\,]^{\mathrm{T}} \qquad\qquad\quad \text{deg} \end{aligned}$$

corresponding to the end-effector position $\boldsymbol{x} = [\,2.45 \quad -0.42 \quad 0.91\,]^{\mathrm{T}}$ m.

A weighted pseudoinverse was used to compute $\boldsymbol{J}^{\#}$ characterized by the weight matrix $\boldsymbol{W} = \text{blockdiag}\{10\boldsymbol{I}_6, \boldsymbol{I}_3\}$. To simulate an imperfect hovering of the vehicle the control law was implemented with *lower* gains for the vehicle; the performance of the simulated vehicle, thus, has an error that is of the same magnitude of a real vehicle in hovering.

To accomplish the above task, firstly the force control scheme 1 is used together with the sliding mode motion control law described in Section 7.6; notice that non-perfect gravity and buoyancy compensation was assumed. Figure 8.11 reports the contact force and the end-effector error components obtained. Since the desired final position is given as a set-point, the initial end-effector position errors are large. In Figure 8.12, the position and orientation components of the vehicle are shown; it can be recognized that, despite the large starting value, the roll angle is kept in the desired range.

To take advantage of dynamic compensation actions, the force control scheme 2 is used to accomplish the same task as above together with the singularity-free adaptive control presented in Section 7.9; notice that only the restoring force terms have been considered to be compensated and a constant unknown error, bounded to $\pm 10\%$, of these parameters has been assumed. There are 4 parameters for each rigid body of the UVMS, giving 16 parameters in total. Moreover, during the task execution an unexpected impact occurs along $x$.

In Figure 8.13, the contact force and the end-effector error components are shown. It can be recognized that the unexpected impact, occurring at about 3 s, is safely handled: a transient force component along $x$ is experienced; nevertheless, at the steady state the desired force is achieved with null error.

It can be noted that the expected coupling between the force and the motion directions affects the $x$ direction also before the unplanned impact. This coupling, due to the structure and configuration of the manipulator, is not observed along $y$.

In Figure 8.14, the vehicle's position and orientation components are shown. It can be recognized that the vehicle moves by about 10 cm; nevertheless, the manipulator still performs the primary task accurately. Moreover, the large initial roll angle is recovered by exploiting the system redundancy since this task does not conflict with the higher priority task.

In the force control scheme labeled 1 the force error directly modifies the force/moments/torques acting on the UVMS, leading to an evident physical

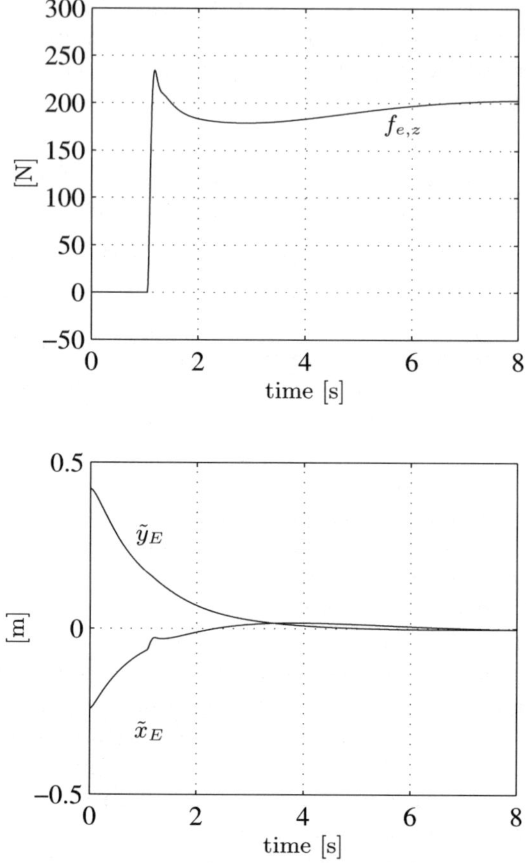


**Fig. 8.11.** Explicit force control scheme 1. Top: time history of the contact force. Bottom: time history of the end-effector errors along the motion directions

interpretation. In the force control scheme 2, instead, the force error builds a correction term acting on the body-fixed velocity references which fed the available motion control system of the UVMS.

The two proposed control schemes were tested in numerical simulation case studies and their performance is analyzed. Overall, the force control scheme 2 seems to be preferable with respect to the force control scheme 1 for two reasons. Firstly, it allows the adoption of adaptive motion control laws, thus making it possible dynamic compensation actions. Secondly, it naturally embeds the standard motion control of the UVMS, since it acts at the reference motion variables level.

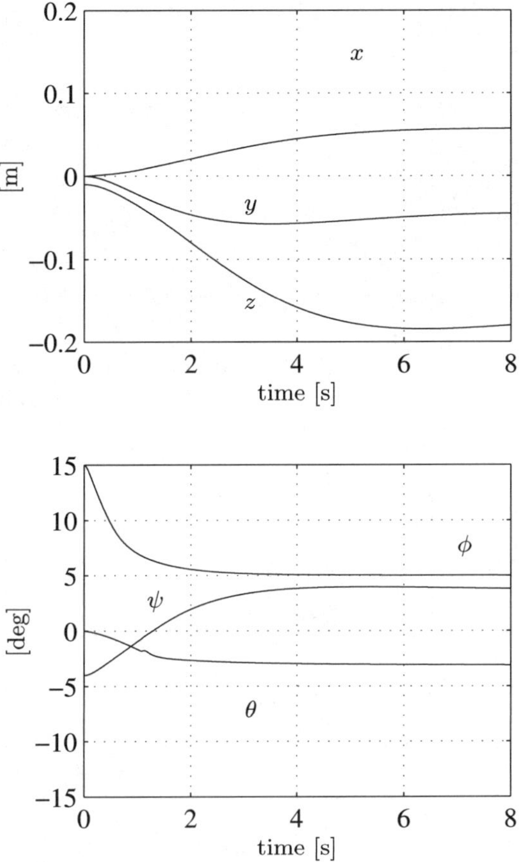


**Fig. 8.12.** Explicit force control scheme 1. Top: time history of the vehicle's position. Bottom: time history of the vehicle's orientation

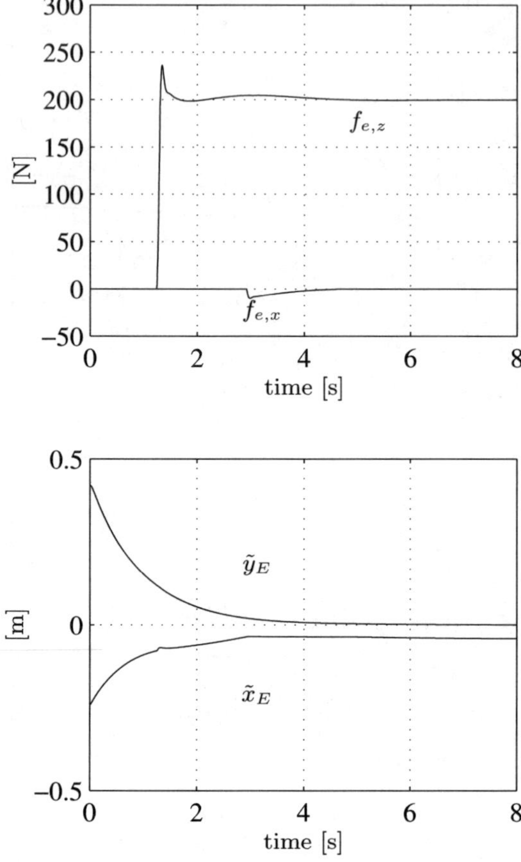


**Fig. 8.13.** Explicit force control scheme 2. Top: time history of the contact force in case of unexpected impact along $x$. Bottom: time history of the end-effector errors along the motion directions

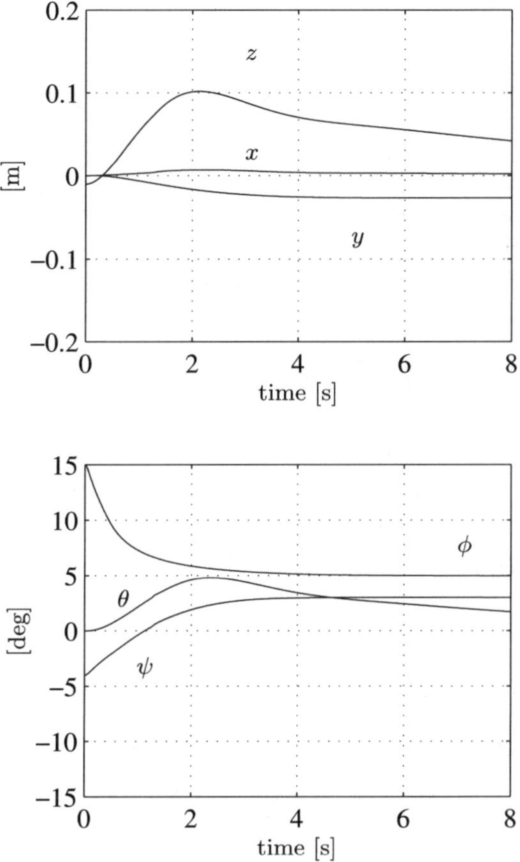


**Fig. 8.14.** Explicit force control scheme 2. Top: time history of the vehicle's position. Bottom: time history of the vehicle's orientation

## 8.6 Conclusions

Interaction control in the underwater environment is a very challenge task. Currently, only theoretical results exist that extended the industrial-based approaches to the UVMSs. Obviously, while these approaches worked in simulation their real effectiveness can be proven only with an exhaustive experimental analysis. Simulations, nevertheless, gave useful information about the need to equip the system with high bandwidth and low noise sensors in order to interact with stiff environment.

# 9. Coordinated Control of Platoons of AUVs

## 9.1 Introduction

The use of several AUVs to achieve fulfillment of a task might be of benefit in several situations: explorations of areas, de-mining [74], as in un the first Gulf's war, or interaction with the environment in which the team of AUVs may, e.g., push an object that one single AUV would not have the power to move. In Figure 1.6, a possible scenario of mine countermeasure using a platoon of AUVs under study at the SACLANT Undersea Research Center of the North Atlantic Treaty Organization (NATO) is given.

Considering surface marine vehicles, in [155] proposes a leader-follower formation control algorithm; experiments with one scale vehicle following a simulated leader are provided. Reference [141] presents a naval minesweeper platoon in which a supervisor vehicle is in charge of tasking in real-time the remaining vehicles. Concerning multi-robot systems in general, one of the first works is reported in [241], where a successful group behavior is achieved by considering only local sensing for each robot. The paper [291] is an interesting tutorial about control of multi-robot systems in a wide sense; it can be considered as a starting reading to understand the nomenclature and the research subjects on the topic. Since an exhaustive literature survey is huge it will be simply skipped in this context, focusing our attention on algorithm developed for the underwater environment.

In particular, the platoons of AUVs experience some specific characteristics;

- The AUVs mathematical models are generally nonholonomic, in fact, in case of exploration the use of torpedo-like vehicles is of common use;
- In case of AUVs equipped with control surfaces there is a substantial impossibility in hovering the vehicle due to the general absence of thrust at low velocity;
- The communication bandwidth is limited thus inhibiting a large exchange of data among the vehicles;
- Localization of the AUVs is not easy due to the absence of a single proprioceptive sensor that gives this information.

Reference [120, 47] reports interesting experimental results on the use of an underwater glider fleet for adaptive ocean sampling. The sampling, thus, is not achieved by resorting to deterministic, pre-programmed, paths but is made adaptive in order to extract the maximum possible information during one mission. The formation control is obtained using virtual bodies and artificial potential techniques [187, 218]; the Authors separate the problem into small area coverage (5 km) and synoptic area coverage for larger scale (5–100 km). For the former, experimental result performed with 3 gliders in 2003 are reported; the experiments were successful thus demonstrating the reliability of the approach in a real environment under the effect, e.g, of the ocean current.

In [54] a cooperative mission, whose experimental validation is planned for summer 2006, is described. The mission objective is to perform a side-scan survey searching for mines with mixed initiative interactions; the presence of a human operator in the loop, thus, is considered. In detail, several target mines will be deployed in a bay and the vehicles have to find and classify the targets in minimum-time.

In [275, 276], an effective decentralized control technique for platoons is proposed and simulated for underwater vehicles. Remarkably, the approach requires a limited amount of inter-vehicle communication that is independent from the platoon dimension; moreover, control of the platoon formation is achieved through definition of a suitable (global) task function without requiring the assignment of desired motion trajectories to the single vehicles. In Figure 1.7, one of the AUVs developed at the Virginia Tech is shown, these vehicles are very small in size and cheap with most of the components custom-engineered.

Reference [179] presents a behavior-based intelligent control architecture that computes discrete control actions. The control architecture separates the sensing aspect, called *perceptor* in the paper, from the control action, called *response controller* and it focuses on the latter implemented by means of a set of discrete event models. The design of a sampling mission for AUVs is also discussed in the paper.

In [158], the problem of formation control for AUVs is solved by using a leader-follower approach. For each vehicle a relative position with respect to the leader is given as a reference motion; simulation for two 3-dimensional nonholonomic vehicles are reported.

Reference [55] presents an algorithm to perform a search using multiple AUVs. Simulation results for the search of the minimum in a scalar fields are shown.

In [247] the communication issue for the underwater environment is taken into account. The vehicles operate in decentralized manner and communicate to achieve common control objectives. The network, thus, is time-varying, and the vehicles communicate with different subset of other vehicles along the mission. In particular, the case of disconnected network is addressed and possibly solved by resorting to a *fast enough* periodic fast switching. The communication constraint is taken into account also in [188] with numerical simulations of the effects of communication delays on the formation.

Reference [64] reports the development of an adaptive on-line planning algorithm for the exploration of an unknown oceanic environment by means of several AUVs. The aim is to have a map of the estimation of a desired variable such as, e.g., the water temperature, in a region. Each vehicle decides the next measurements on the basis of the current measurement and the past measurements of all the platoon such that the confidence level of the overall estimation is locally increased. The paper reports simulation results, an experimental set-up is currently under development based on self-designed low-cost vehicles [6, 7].

In [136] a Lypanuov-based technique is derived so that the underwater vehicles, supposed fully actuated, are steered along a set of spatial paths while keeping a given inter-vehicle formation. The path following for each vehicle is essentially decoupled while its advancing velocity is properly shaped so that the formation is properly achieved. Simulation results on a planar formation of 3 vehicles are provided.

In the following Section a specific control strategy, based on inverse kinematics algorithms for fixed-base manipulators, is discussed together with some simulation results; finally, the description of the experimental set-up of the Virginia Polytechnic Institute & State University is provided.

## 9.2 Kinematic Control of AUVs

In 2001 [48, 50, 277] B.E. Bishop and D.J. Stilwell for the first time recognize a formal similarity between the problem of robot redundancy resolution and the achievement of platoon's tasks at the differential level. The primary task is defined by the platoon's mean and variance and tasks such as obstacle avoidance and heading control are handled as secondary tasks by using the approach developed by Y. Nakamura [210]. G. Antonelli and S. Chiaverini, in [27, 28, 30], inherited this approach by proposing a different way to handle the redundancy with respect to the given task, specifically, the task priority redundancy resolution proposed by S. Chiaverini [78].

### Task Functions and Inverse Kinematics

While considering a platoon of $n$ vehicles, the aim is to control the value taken by a generic task function which suitably depends on the platoon state; an example of such function is the one expressing the mean value of all the vehicles' positions as a synthetic data about the platoon location.

To keep the notation compact it is useful to define the vector $\boldsymbol{p} \in \mathbb{R}^{3n}$

$$\boldsymbol{p} = \begin{bmatrix} \boldsymbol{\eta}_{1,1} \\ \vdots \\ \boldsymbol{\eta}_{1,n} \end{bmatrix}$$

where $\eta_{1,i}$ is the 3 dimensional position of the $i$ th vehicle, and

$$\boldsymbol{v} = \begin{bmatrix} \dot{\boldsymbol{\eta}}_{1,1} \\ \vdots \\ \dot{\boldsymbol{\eta}}_{1,n} \end{bmatrix} .$$

By defining as $\boldsymbol{\sigma} \in \mathbb{R}^m$ the task variable to be controlled, it is:

$$\boldsymbol{\sigma} = \boldsymbol{f}(\boldsymbol{\eta}_{1,1}, \ldots, \boldsymbol{\eta}_{1,n}) = \boldsymbol{f}(\boldsymbol{p}) \tag{9.1}$$

with

$$\dot{\boldsymbol{\sigma}} = \sum_{i=1}^{n} \frac{\partial \boldsymbol{f}(\boldsymbol{p})}{\partial \boldsymbol{\eta}_{1,i}} \dot{\boldsymbol{\eta}}_{1,i} = \boldsymbol{J}(\boldsymbol{p})\boldsymbol{v} , \tag{9.2}$$

where $\boldsymbol{J} \in \mathbb{R}^{m \times 3n}$ is the configuration-dependent task Jacobian matrix[1].

An effective way to generate motion references for the AUVs starting from desired values $\boldsymbol{\sigma}_d(t)$ of the task function is to act at the differential level by inverting the (locally linear) mapping (9.2); in fact, this problem has been widely studied in robotics (see, e.g., [262] for a tutorial or Chapter 6 for an UVMS's application). For low-rectangular matrices —which is usually the case in platoon formation control, being $3n \gg m$— the problem admits infinite solutions and such *redundancy* is often exploited to optimize some criterion. A typical requirement is to pursue minimum-norm velocity, leading to the least-squares solution:

$$\boldsymbol{v}_d = \boldsymbol{J}^{\dagger} \dot{\boldsymbol{\sigma}}_d = \boldsymbol{J}^{\mathrm{T}} \left( \boldsymbol{J}^{\mathrm{T}} \boldsymbol{J} \right)^{-1} \dot{\boldsymbol{\sigma}}_d . \tag{9.3}$$

At this point, the vehicle motion controllers need reference position trajectories besides the velocity references; these can be obtained by time integration of $\boldsymbol{v}_d$. However, discrete-time integration of the vehicles' reference velocities would result in a numerical drift of the reconstructed vehicles' positions; the drift can be counteracted by a so-called Closed Loop Inverse Kinematics (CLIK) version of the algorithm, namely,

$$\boldsymbol{v}_d(t_k) = \boldsymbol{J}^{\dagger} \left( \dot{\boldsymbol{\sigma}}_d + \boldsymbol{\Lambda}(\boldsymbol{\sigma}_d - \boldsymbol{\sigma}) \right) \Big|_{t=t_k} \tag{9.4}$$

$$\boldsymbol{p}_d(t_k) = \boldsymbol{p}_d(t_{k-1}) + \boldsymbol{v}_d(t_k) \Delta t , \tag{9.5}$$

where $t_k$ is the $k$-th time sample, $\Delta t$ is the sampling period, and $\boldsymbol{\Lambda}$ is a suitable constant positive definite matrix of gains. It must be remarked that the loop is closed on algorithmic quantities at the *input* of the motion controllers and does not involve measurement of quantities at the *output* of the motion controllers; in other words, the dynamics of the vehicles' motion controllers is out of the loop in (9.4).

[1] The symbol $\boldsymbol{J}$ has been used in Equation (2.68) for a specific Jacobian matrix, in the remainder of the Chapter, however, it will denote a generic task Jacobian matrix.

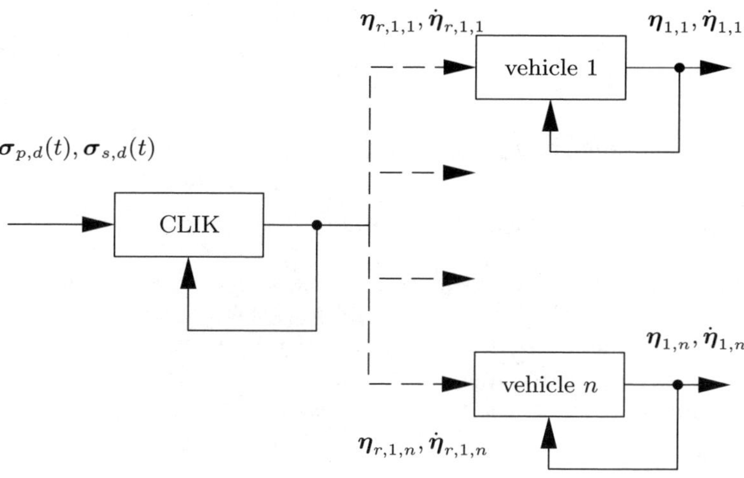


**Fig. 9.1.** Sketch of the centralized kinematic control strategy for platoons

The structure of such a centralized platoon control algorithm is sketched in Figure 9.1. It must be remarked that each vehicle is equipped with a dynamic low level controller in charge of following the reference trajectory. This structure can be implemented on a platoon of AUVs without modifying their dynamic controllers that can be tuned, thus, at their best performance. The dynamic low level controller is not affected by the algorithm implemented in the CLIK. It is sufficient that it implements a kind of input interpolation to match the different sampling frequency of the two loops.

In view of the underwater communication limitations, it must be remarked that only one-way communication channel must be strictly set-up; communication among robots, thus, is not required. Moreover, the communication channel is in charge of updating the vehicles' references; it is not, thus, inside the control loop with obvious advantages in terms of stability and performance.

A high-level supervisor might benefit from sensor measurements. Those can be vehicle-fixed sensors (e.g., sonar measurements or video camera) or other distributed sensors (e.g., baseline acoustic measurements). These variables can be acquired from different sources at different sampling frequencies and merged using a Kalman filtering approach. In this case, however, communication from each vehicle to the centralized algorithm may be required.

### Task Function for Platoon Average Position

An example of task function for platoon formation control is one expressing the mean value of all the vehicles' positions as a synthetic data about the platoon location. In this case, the 3-dimensional task function $\sigma_a$ is simply given by:

$$\boldsymbol{\sigma}_a = \boldsymbol{f}_a(\boldsymbol{\eta}_{1,1}, \dots, \boldsymbol{\eta}_{1,n}) = \frac{1}{n}\sum_{i=1}^{n} \boldsymbol{\eta}_{1,i} = \overline{\boldsymbol{p}}, \tag{9.6}$$

whose Jacobian matrix $\boldsymbol{J}_a \in \mathbb{R}^{3\times 3n}$ can be easily derived [28].

### Task Function for Platoon Variance

Together with the platoon average position, it is of interest considering the variance of all the vehicles' positions as a synthetic data on their spreading around the average position. It is clear that by controlling these two task variables it is possible to influence the shape of the platoon. The task function for platoon variance $\boldsymbol{\sigma}_v \in \mathbb{R}^3$ is defined as

$$\boldsymbol{\sigma}_v = \frac{1}{n}\sum_{i=1}^{n} (\boldsymbol{\eta}_{1,i} - \overline{\boldsymbol{p}})^2 . \tag{9.7}$$

In [28] the Jacobian $\boldsymbol{J}_v$ is provided; moreover, different tasks, such as keeping the platoon in a bounded geometrical area or in a rigid formation, are discussed.

### Merging Several Tasks

Different task functions may be used at the same time by stacking the corresponding single-task variables in an overall task vector. As a result, the corresponding single-task Jacobians are stacked too in an overall task Jacobian, and the inverse solution (9.3) acts by simply adding the partial vehicles' velocities that would be obtained if (locally at the current configuration) each task were executed alone. It is simple to recognize that this approach is poorly effective since conflicting tasks would generate counteracting partial vehicles' velocities.

A possible technique to handle this problem has been proposed in [50], which consists in assigning a relative priority to the single task functions, thus resorting to the task-priority inverse kinematics introduced in [196, 210] for ground-fixed redundant manipulators. Nevertheless, as discussed in [78], just in case of conflicting tasks it is necessary to devise singularity-robust algorithms that ensure proper behavior of the inverse velocity mapping. For this reason, according to [78], G. Antonelli and S. Chiaverini propose to modify the CLIK solution (9.4) into

$$\boldsymbol{v}_d(t_k) = \boldsymbol{J}_p^{\dagger}\Big(\dot{\boldsymbol{\sigma}}_{p,d} + \boldsymbol{\Lambda}_p(\boldsymbol{\sigma}_{p,d}-\boldsymbol{\sigma}_p)\Big) + \left(\boldsymbol{I} - \boldsymbol{J}_p^{\dagger}\boldsymbol{J}_p\right)\boldsymbol{J}_s^{\dagger}\Big(\dot{\boldsymbol{\sigma}}_{s,d} + \boldsymbol{\Lambda}_s(\boldsymbol{\sigma}_{s,d}-\boldsymbol{\sigma}_s)\Big)\,, \tag{9.8}$$

where the subscript $p$ denotes primary-task quantities, the subscript $s$ denotes secondary-task quantities, and $\boldsymbol{\Lambda}_p$, $\boldsymbol{\Lambda}_s$ are suitable positive definite matrices. The extension to a generic number of tasks can be easily derived.

Equation (9.8) has a nice geometrical interpretation. Each task is projected onto the vehicles' velocity space by the use of the corresponding pseudoinverse, i.e., as if it were acting alone; then, before adding the two contributions, the lower-priority (secondary) task is further projected onto the null space of the higher-priority (primary) task so as to remove those velocity components that would conflict with it.

A different technique to manage conflicting tasks in the framework of the task priority approach has been proposed in [29] (see Section 6.6), where a fuzzy inference system is used to merge multiple secondary tasks given to underwater vehicle-manipulator systems.

## Obstacle Avoidance Strategy

In [50] obstacle avoidance has been developed in the frame of task priority inverse kinematics following the approach of Y. Nakamura [210].

In the following, based on the work developed in 1991 at the Università degli Studi Napoli Federico II for industrial manipulators [76], the obstacle avoidance is chosen as the primary task by selecting:

$$\boldsymbol{\sigma}_o = \begin{bmatrix} \|\boldsymbol{\eta}_1 - \boldsymbol{\eta}_o\| \\ \vdots \\ \|\boldsymbol{\eta}_n - \boldsymbol{\eta}_o\| \end{bmatrix} \tag{9.9}$$

where $\boldsymbol{\eta}_o$ is the coordinate vector of an obstacle, and the other tasks as secondary tasks. Defining as $\boldsymbol{r}_i$ the versor of $\boldsymbol{\eta}_i - \boldsymbol{\eta}_o$ it can be recognized that the pseudoinverse of $\boldsymbol{J}_o \in \mathbb{R}^{n \times 3n}$ is simply given by:

$$\boldsymbol{J}_o^{\dagger} = \begin{bmatrix} \boldsymbol{r}_1 & & \mathbf{0}_{3\times 1} \\ & \ddots & \\ \mathbf{0}_{3\times 1} & & \boldsymbol{r}_n \end{bmatrix} \tag{9.10}$$

and the null-space projection matrix, considering the planar case for simplicity, is given by:

$$\boldsymbol{N}_o = \begin{bmatrix} \begin{matrix} 1 - r_{1,x}^2 & -r_{1,x}r_{1,y} \\ -r_{1,x}r_{1,y} & 1 - r_{1,y}^2 \end{matrix} & & \boldsymbol{O}_{2\times 2} \\ & \ddots & \\ \boldsymbol{O}_{2\times 2} & & \begin{matrix} 1 - r_{n,x}^2 & -r_{n,x}r_{n,y} \\ -r_{n,x}r_{n,y} & 1 - r_{n,y}^2 \end{matrix} \end{bmatrix}.$$

In this approach, the obstacle avoidance task is also used for each vehicle with respect to the other vehicles of the platoon. This prevents from the possible occurrence of accidents during, e.g., formation changing or obstacle avoidance. In case of simultaneous proximity of an obstacle and another vehicle, the safer behavior is obtained by projecting the secondary-task velocities in the null space of *both* the obstacle avoidance tasks. This will simply zero the required velocity for the vehicle that is commanded to go closer to another vehicle and the obstacle at the same time.

The developed theory consider holonomic vehicle, in [49] nonholonomic vehicles are explicitly considered as composing the platoon. Recently, in [278], a partially decentralized version of the kinematic control is proposed to properly consider the limited bandwidth of the platoon.

### 9.2.1 Simulations

The kinematic-control-based algorithm has been tested in numerical simulation case studies. The objective of the simulations is to show the performance of the sole CLIK algorithm, while assuming a bounded tracking error of the vehicle controllers. The platoon is composed of 10 vehicles; the sampling time of the CLIK is 0.5 s and the total simulation duration is 100 s. The primary task is the platoon average position, while the secondary task is the variance. In detail, the average is commanded to be constant along $y$ and increases linearly at a velocity of 1 m/s along $x$:

$$\boldsymbol{\sigma}_a = \begin{bmatrix} t \\ 0 \end{bmatrix} \quad \mathrm{m}$$

while the desired variance is

$$\boldsymbol{\sigma}_v = \begin{bmatrix} 30 \\ 30 \end{bmatrix} \quad \mathrm{m}^2 .$$

The CLIK's gains have been set to:

$$\begin{aligned} \boldsymbol{\Lambda}_a &= 0.1\boldsymbol{I}_2 \\ \boldsymbol{\Lambda}_v &= 0.05\boldsymbol{I}_2 . \end{aligned}$$

In case of presence of obstacles or excessive proximity among the vehicles, the primary task is the obstacle avoidance, while mean and variance are, respectively, second and third task. A safe distance of 4 m among the vehicles is required. In Figure 9.2 it can be observed that, in absence of obstacles, the vehicles reconfigure themselves in order to fulfill mean and variance task.

The same initial condition with the same mean and variance tasks have been used to run a second simulation with the presence of an obstacle. One of the vehicles needs to slow down its velocity, avoid the vehicles in its proximity, avoid the obstacle and finally join the other vehicles to fulfill the mean-variance task (Figure 9.3).

Figure 9.4 shows the same mission (same required mean and variance starting from the same initial conditions) with the presence of 3 obstacles. It can be recognized that the mission is safely executed.

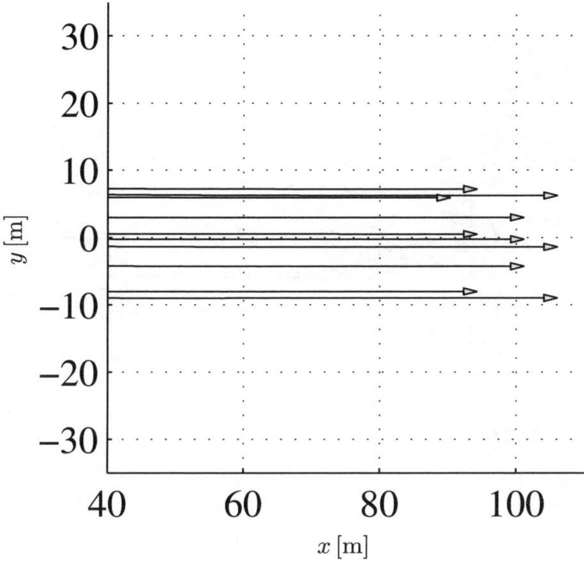


**Fig. 9.2.** Kinematic control of multi-AUVs. In absence of obstacles the platoon satisfies the mean and variance task

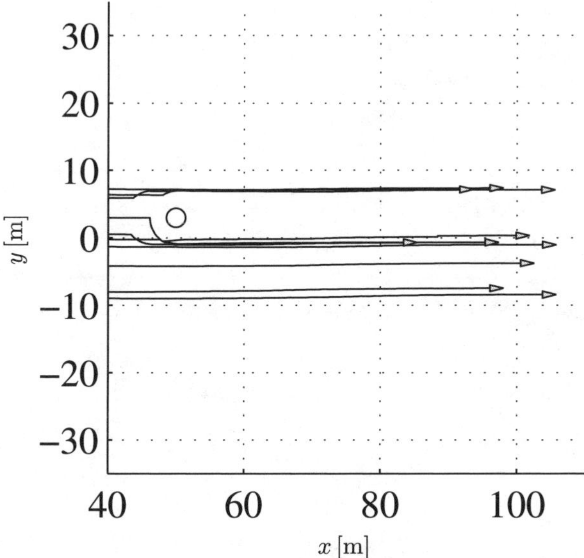


**Fig. 9.3.** Kinematic control of multi-AUVs. An obstacle is encountered and the primary task becomes its avoidance

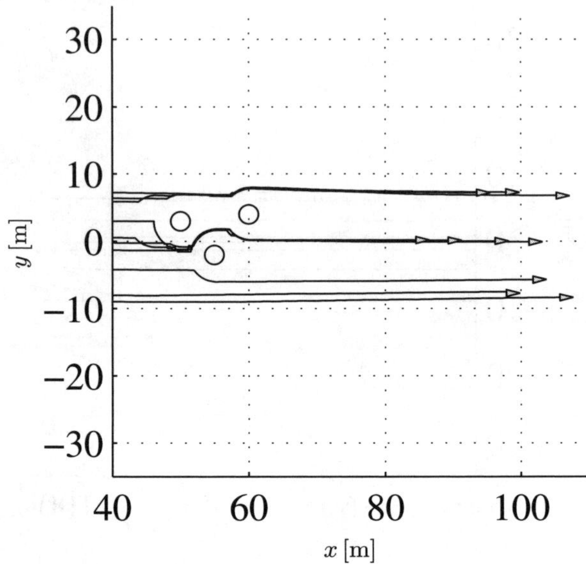


**Fig. 9.4.** In the framework of kinematic control of multi-AUVs it is possible to safely handles several situations

## 9.3 Experimental Set-Up at the Virginia Tech

Figure 9.5 reports the small AUVs developed at the Virginia Polytechnic Institute & State University to perform adaptive sampling with multiple underwater vehicles. The AUVs have been designed in order to be very small in size and relatively cheap [132].

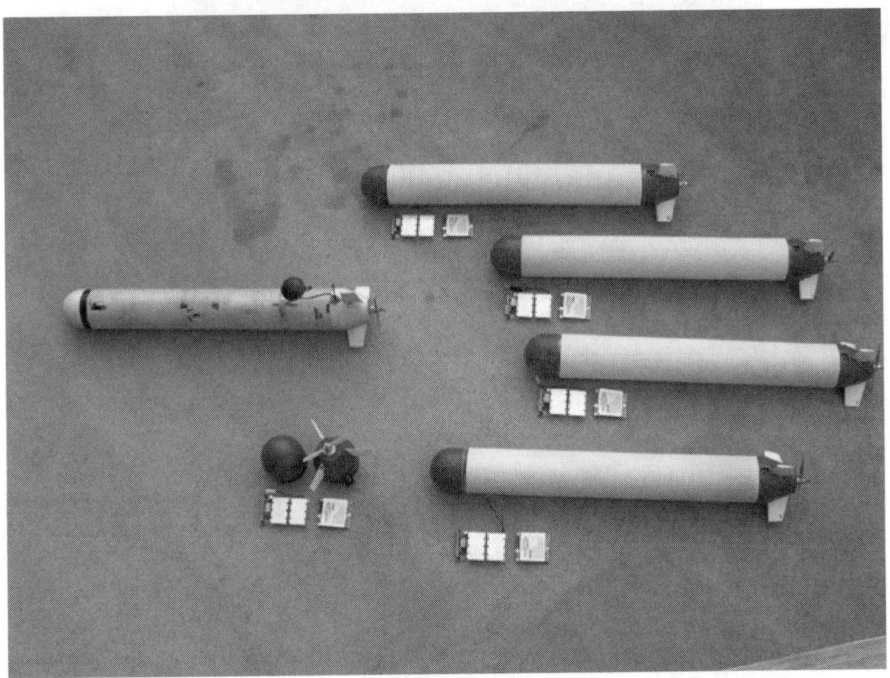

**Fig. 9.5.** Platoon of AUVs developed at the Virginia Polytechnic Institute & State University (courtesy of D. Stilwell)

One of the use of the platoon will be the adaptive sampling of environmental processes, i.e., the mapping and tracking of marine processes. In [66], C.J. Cannell and D.J. Stilwell compare two estimation methods in different conditions, moreover, the benefit achieved using more than one vehicles is also discussed. Reporting from [66], Figure 9.6 shows simulation results of two AUVs estimating the center of mass of a neutral dye tracer.

In [229] some preliminary experimental results have been reported by J. Petrich, C.A. Woolsey and D.J. Stilwell using one of the vehicles. However, since absolute positions are required to prove the proposed algorithm, the vehicle is commanded to move at the surface and its position is measured with a sampling time of 30 s.

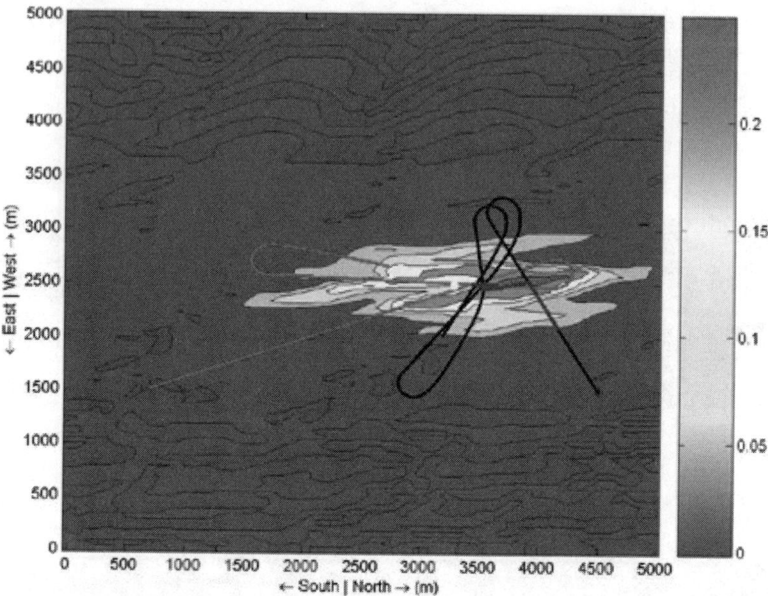


**Fig. 9.6.** Simulation of two AUVs estimating the center of mass of a neutral dye tracer (courtesy of D. Stilwell, Virginia Polytechnic Institute & State University)

## 9.4 Conclusions

In this Chapter a very quick overview on control strategy for multiple AUVs is given. The so called kinematic control has been detailed reporting simulation results. Some preliminary experimental results, performed at the Virginia Tech University have been shown.

Some interesting issues concern fault tolerant capabilities of the platoon, scalability of the algorithm, and local sensing aspects.

# 10. Concluding Remarks

*"Cheshire-cat —Alice began— would you tell me, please, which way I ought to go from here?". "That depends a good deal on where you want to get to", said the cheshire-cat.*

*Lewis Carroll, "Alice's adventures in wonderland".*

The underwater environment is hostile, the communication is difficult and external disturbances can strongly affect the dynamics of the system involved in the mission. Moreover, the sensing devices are not fully reliable in the underwater environment. This is particularly true for the positioning sensors. The vehicle actuating systems, usually composed of thrusters, is highly nonlinear and subject to limit cycles. Some very simple operations for a ground robotic system, such as *to hold its position* can become difficult for UVMSs. This difficulty dramatically increases for complex missions.

Moreover, also if one wants to neglect the technical difficulties and focus on the mathematical model he deals with characteristics difficult to handle. The equations of motion are coupled and nonlinear; the hydrodynamic effects are not completely known leading to the practical impossibility to identify the dynamic parameters. Together with the complexity of the mathematical model this is a serious obstacle versus the implementation of model-based control laws.

In this framework, the control of the UVMS must be *robust*, in a wide sense, reliable and of simple implementation. With this in mind the techniques for motion and force control of UVMSs needs to be developed.

The kinematic control approach is important for accomplishing an autonomous robotic mission. Differently from industrial robotics, where an off-line trajectory planning is a possible approach, in case of unknown, unstructured environment, the trajectory has to be be generated in real-time. The possible occurrence of kinematic singularities, thus, is not avoided with a *task space control law*. Moreover, UVMSs are usually redundant with respect to the given task; a kinematic control approach, thus, can exploit such a redundancy.

The motion control of UVMSs must deal with the strong constraints discussed along the book. In this work, different approaches were discussed. Control of an underwater robotic system can not be reduced at the sole *translation* of similar approaches designed for ground-fixed manipulator; as an example, it is interesting to stress the results achieved in Chapter 3: certain control laws devoted at the study of 6-DOFs dynamic control of the vehicle can deteriorate the transient due to a wrong adaptive or integral action. In the Author's opinion, this result is particularly significant.

The following step needs to be the experimental validation of the several control laws developed for UVMSs and tested only in simulation.

*Nec plus ultra.*

# A. Mathematical models

## A.1 Introduction

This Appendix reports the parameters of the different mathematical models used along the book. For all the models the gravity vector, expressed in the inertial frame, and the water density are given by:

$$
\begin{aligned}
\boldsymbol{g}^I &= [\,0 \quad 0 \quad 9.81\,]^{\mathrm{T}}\,\mathrm{m/s}^2 \\
\rho &= 1000\,\mathrm{kg/m}^3 .
\end{aligned}
$$

## A.2 Phoenix

The data of the vehicle Phoenix, developed at the Naval Postgraduate School (Monterey, CA, USA), are given in [145]. These have been used as data of the vehicle carrying a manipulator with 2, 3 and 6 DOFs in the simulations of the dynamic control laws.

$$
\begin{array}{rcllll}
L & = & 5.3 & \mathrm{m} & \text{vehicle length} \\
m & = & 5454.54 & \mathrm{kg} & \text{vehicle weight} \\
\boldsymbol{r}_G^B & = & [\,0 \quad 0 \quad 0.061\,]^{\mathrm{T}} & \mathrm{m} & \\
\boldsymbol{r}_B^B & = & [\,0 \quad 0 \quad 0\,]^{\mathrm{T}} & \mathrm{m} & \\
W & = & 53400 & \mathrm{N} & \\
B & = & 53400 & \mathrm{N} & \\
I_x & = & 2038 & \mathrm{Nms}^2 & \\
I_y & = & 13587 & \mathrm{Nms}^2 & \\
I_z & = & 13587 & \mathrm{Nms}^2 & \\
I_{xy} & = & -13.58 & \mathrm{Nms}^2 & \\
I_{yz} & = & -13.58 & \mathrm{Nms}^2 & \\
I_{xz} & = & -13.58 & \mathrm{Nms}^2 &
\end{array}
$$

where:

$$
\boldsymbol{I}_{O_b} = \begin{bmatrix} I_x & I_{xy} & I_{xz} \\ I_{xy} & I_y & I_{yz} \\ I_{xz} & I_{yz} & I_z \end{bmatrix} .
$$

The inertia matrix, including the added mass terms, is given by:

$$
\boldsymbol{M}_v = \begin{bmatrix}
6.019 \cdot 10^3 & 5.122 \cdot 10^{-8} & -1.180 \cdot 10^{-2} \\
5.122 \cdot 10^{-8} & 9.551 \cdot 10^3 & 3.717 \cdot 10^{-6} \\
-1.180 \cdot 10^{-2} & 3.717 \cdot 10^{-6} & 2.332 \cdot 10^4 \\
-3.200 \cdot 10^{-5} & -3.802 \cdot 10^2 & -1.514 \cdot 10^{-4} \\
3.325 \cdot 10^2 & -3.067 \cdot 10^{-5} & 2.683 \cdot 10^3 \\
-6.731 \cdot 10^{-5} & -4.736 \cdot 10^2 & -4.750 \cdot 10^{-4}
\end{matrix} \cdots
$$

$$
\cdots \begin{matrix}
-3.200 \cdot 10^{-5} & 3.325 \cdot 10^2 & -6.731 \cdot 10^{-5} \\
-3.802 \cdot 10^2 & -3.067 \cdot 10^{-5} & -4.736 \cdot 10^2 \\
-1.514 \cdot 10^{-4} & 2.683 \cdot 10^3 & -4.750 \cdot 10^{-4} \\
4.129 \cdot 10^3 & 1.358 \cdot 10^1 & 8.467 \cdot 10^1 \\
1.358 \cdot 10^1 & 4.913 \cdot 10^4 & 1.357 \cdot 10^1 \\
8.467 \cdot 10^1 & 1.357 \cdot 10^1 & 2.069 \cdot 10^4
\end{bmatrix}.
$$

The hydrodynamic derivatives are given by:

$$
\begin{array}{llllll}
X_{p|p|} = & 2.8 \cdot 10^3 & X_{q|q|} = & -5.9 \cdot 10^3 & X_{r|r|} = & 1.6 \cdot 10^3 \\
X_{pr} = & 2.9 \cdot 10^2 & X_{\dot{u}} = & -5.6 \cdot 10^2 & X_{\omega q} = & -1.5 \cdot 10^4 \\
X_{vp} = & -2.2 \cdot 10^2 & X_{vr} = & 1.5 \cdot 10^3 & X_{v|v|} = & 7.4 \cdot 10^2 \\
X_{\omega|\omega|} = & 2.4 \cdot 10^3 & & & & \\
Y_{\dot{p}} = & 47 & Y_{\dot{r}} = & 4.7 \cdot 10^2 & Y_{pq} = & 1.6 \cdot 10^3 \\
Y_{qr} = & -2.6 \cdot 10^3 & Y_{\dot{v}} = & -4.1 \cdot 10^3 & Y_{up} = & 2.2 \cdot 10^2 \\
Y_{r} = & 2.2 \cdot 10^3 & Y_{vq} = & 1.8 \cdot 10^3 & Y_{\omega p} = & 1.7 \cdot 10^4 \\
Y_{\omega r} = & -1.4 \cdot 10^3 & Y_{v} = & -1.4 \cdot 10^3 & Y_{v\omega} = & 9.5 \cdot 10^2 \\
Z_{\dot{q}} = & -2.7 \cdot 10^3 & Z_{p|p|} = & 51 & Z_{pr} = & 2.6 \cdot 10^3 \\
Z_{r|r|} = & -2.9 \cdot 10^3 & Z_{\dot{\omega}} = & -1.8 \cdot 10^4 & Z_{uq} = & -1 \cdot 10^4 \\
Z_{vp} = & -3.6 \cdot 10^3 & Z_{vr} = & 3.3 \cdot 10^3 & Z_{u\omega} = & -4.2 \cdot 10^3 \\
Z_{v|v|} = & -9.5 \cdot 10^2 & & & & \\
K_{\dot{p}} = & -2 \cdot 10^3 & K_{\dot{r}} = & -71 & K_{pq} = & -1.4 \cdot 10^2 \\
K_{qr} = & 3.5 \cdot 10^4 & K_{\dot{v}} = & 47 & K_{up} = & -4.3 \cdot 10^3 \\
K_{ur} = & -3.3 \cdot 10^2 & K_{vq} = & -2 \cdot 10^3 & K_{\omega p} = & -51 \\
K_{\omega r} = & 5.5 \cdot 10^3 & K_{uv} = & 2.3 \cdot 10^2 & K_{v\omega} = & -1.4 \cdot 10^4 \\
M_{\dot{q}} = & -3.5 \cdot 10^4 & M_{p|p|} = & 1.1 \cdot 10^2 & M_{pr} = & 1 \cdot 10^4 \\
M_{r|r|} = & 6 \cdot 10^3 & M_{\dot{\omega}} = & -2.7 \cdot 10^3 & M_{uq} = & -2.7 \cdot 10^4 \\
M_{vp} = & 4.7 \cdot 10^2 & M_{vr} = & 6.7 \cdot 10^3 & M_{u\omega} = & 7.4 \cdot 10^3 \\
M_{v|v|} = & -1.9 \cdot 10^3 & & & & \\
N_{\dot{p}} = & -71 & N_{\dot{r}} = & -7.1 \cdot 10^3 & N_{pq} = & -4.4 \cdot 10^4 \\
N_{qr} = & 5.6 \cdot 10^3 & N_{\dot{v}} = & 4.7 \cdot 10^2 & N_{up} = & -3.3 \cdot 10^2 \\
N_{ur} = & -6.3 \cdot 10^3 & N_{vq} = & -3.9 \cdot 10^3 & N_{\omega p} = & -6.7 \cdot 10^3 \\
N_{\omega r} = & 2.9 \cdot 10^3 & N_{uv} = & -5.5 \cdot 10^2 & N_{v\omega} = & -2 \cdot 10^3
\end{array}
$$

On the sway, heave, pitch and yaw degree of motion the damping due to the vortex shedding is also considered. By defining as $b(x)$ the vehicle breadth and $h(x)$ the vehicle height this term is modeled as a discretized version of the following:

$$
\begin{aligned}
Y &= -\frac{1}{2}\rho \int_{tail}^{nose} \left[C_{dy}h(x)(v+xr)^2 + C_{dz}b(x)(\omega-xq)^2\right] \frac{(v+xr)}{U_{cf}(x)}dx \\
Z &= \frac{1}{2}\rho \int_{tail}^{nose} \left[C_{dy}h(x)(v+xr)^2 + C_{dz}b(x)(\omega-xq)^2\right] \frac{(\omega-xq)}{U_{cf}(x)}dx \\
M &= -\frac{1}{2}\rho \int_{tail}^{nose} \left[C_{dy}h(x)(v+xr)^2 + C_{dz}b(x)(\omega-xq)^2\right] \frac{(\omega+xq)}{U_{cf}(x)}xdx \\
N &= -\frac{1}{2}\rho \int_{tail}^{nose} \left[C_{dy}h(x)(v+xr)^2 + C_{dz}b(x)(\omega-xq)^2\right] \frac{(v+xr)}{U_{cf}(x)}xdx
\end{aligned}
$$

where the cross-flow velocity is computed as:

$$U_{cf}(x) = \sqrt{(v+xr)^2 + (\omega-xq)^2}.$$

In the simulations $h(x)$ and $b(x)$ have been considered constant, in detail $h(x) = 0.5\,\mathrm{m}$ and $b(x) = 1\,\mathrm{m}$. Finally, $C_{dy} = C_{dz} = 0.6$.

## A.3 Phoenix+6DOF SMART 3S

The vehicle, whose model is widely known and used in literature [127, 145], is supposed to carry a SMART-3S manipulator manufactured by COMAU whose main parameters are reported in Table A.1.

**Table A.1.** mass [kg], Denavit-Hartenberg parameters [m,rad], radius [m], length [m] and viscous friction [Nms] of the manipulator mounted on the underwater vehicle

| | mass | $a$ | $d$ | $\theta$ | $\alpha$ | radius | length | viscous frict. |
|---|---|---|---|---|---|---|---|---|
| link 1 | 80 | 0.15 | 0 | $q_1$ | $-\pi/2$ | 0.2 | 0.85 | 30 |
| link 2 | 80 | 0.61 | 0 | $q_2$ | 0 | 0.1 | 1 | 20 |
| link 3 | 30 | 0.11 | 0 | $q_3$ | $-\pi/2$ | 0.15 | 0.2 | 5 |
| link 4 | 50 | 0 | 0.610 | $q_4$ | $\pi/2$ | 0.1 | 0.8 | 10 |
| link 5 | 20 | 0 | −0.113 | $q_5$ | $-\pi/2$ | 0.15 | 0.2 | 5 |
| link 6 | 25 | 0 | 0.103 | $q_6$ | 0 | 0.1 | 0.3 | 6 |

The manipulator is supposed to be mounted under the vehicle in the middle of its length, the vector positions of the origin of the frames $i-1$ to frame/center-of-mass/center-of-buoyancy of link $i$ are not reported for brevity. The dry friction has not been considered to avoid chattering behavior and increase output readability. All the links are modeled as cylinders. Their volumes, thus, are computed as $\delta_i = \pi * L_i r_i^2$, where $L_i$ and $r_i$ are the link lengths and radius, respectively. The modeling of the hydrodynamic effects, using the strip theory, benefit from the simplified geometric assumption on the link shapes [127, 256].

For each of the cylinder the mass is computed as:

$$
\boldsymbol{M}_i = \begin{bmatrix} m_i + \rho\delta_i & 0 & 0 & * & * & * \\ 0 & m_i + \rho\delta_i & 0 & * & * & * \\ 0 & 0 & m_i + 0.1m_i & * & * & * \\ * & * & * & I_{x,i} + \pi\rho L_i^3 r_i^2/12 & 0 & 0 \\ * & * & * & 0 & I_{y,i} + \pi\rho L_i^3 r_i^2/12 & 0 \\ * & * & * & 0 & 0 & I_{z,i} \end{bmatrix},
$$

where the link inertia are given in Table A.2. In this case, the cylinder length is considered along the $z_i$ axis.

**Table A.2.** Link inertia [Nms$^2$] of the manipulator mounted on the underwater vehicle

| | $I_{x,i}$ | $I_{y,i}$ | $I_{z,i}$ | $I_{xy,i}$ | $I_{xz,i}$ | $I_{yz,i}$ |
|---|---|---|---|---|---|---|
| link 1 | 100 | 30 | 100 | 0 | 0 | 0 |
| link 2 | 20 | 80 | 80 | 0 | 0 | 0 |
| link 3 | 2 | 0.5 | 2 | 0 | 0 | 0 |
| link 4 | 50 | 9 | 50 | 0 | 0 | 0 |
| link 5 | 5 | 4 | 5 | 0 | 0 | 0 |
| link 6 | 5 | 5 | 3 | 0 | 0 | 0 |

The linear skin and quadratic drag coefficients are given by $D_s = 0.4$ and $C_d = 0.6$, respectively. The lift coefficient $C_l$ is considered null.

## A.4 ODIN

The mathematical model of ODIN, an AUV developed at the ASL, University of Hawaii, has been used to test several, experimentally validated, control laws ([34, 35, 83, 216, 232, 233, 251, 320]):

$$
\begin{array}{rcllr}
r & = & 0.3 & \text{m} & \text{vehicle radius} \\
m & = & 125 & \text{kg} & \text{vehicle weight} \\
\boldsymbol{r}_G^B & = & [0 \quad 0 \quad 0.05]^{\mathrm{T}} & \text{m} & \\
\boldsymbol{r}_B^B & = & [0 \quad 0 \quad 0]^{\mathrm{T}} & \text{m} & \\
W & = & 1226 & \text{N} & \\
I_x & = & 8 & \text{Nms}^2 & \\
I_y & = & 8 & \text{Nms}^2 & \\
I_z & = & 8 & \text{Nms}^2 & \\
I_{xy} & = & 0 & \text{Nms}^2 & \\
I_{yz} & = & 0 & \text{Nms}^2 & \\
I_{xz} & = & 0 & \text{Nms}^2 &
\end{array}
$$

The buoyancy $B$ is computed considering the vehicle as spherical. The hydrodynamic derivatives are given by:

$$
\begin{array}{rclrcl}
X_{\dot{u}} & = & -62.5 & X_{u|u|} & = & -48 \\
Y_{\dot{v}} & = & -62.5 & Y_{v|v|} & = & -48 \\
Z_{\dot{\omega}} & = & -62.5 & Z_{\omega|\omega|} & = & -48 \\
K_p & = & -30 & K_{p|p|} & = & -80 \\
M_q & = & -30 & M_{q|q|} & = & -80 \\
N_r & = & -30 & N_{r|r|} & = & -80
\end{array}
$$

It can be observed that a diagonal structures of the matrices $\boldsymbol{M}_A$ and $\boldsymbol{D}$ is obtained.

According to simple geometrical considerations, the following TCM is observed:

$$
\boldsymbol{B}_v = \begin{bmatrix}
s & -s & -s & s & 0 & 0 & 0 & 0 \\
s & -s & -s & -s & 0 & 0 & 0 & 0 \\
0 & 0 & 0 & 0 & -1 & -1 & -1 & -1 \\
0 & 0 & 0 & 0 & l_1 s & l_1 s & -l_1 s & -l_1 s \\
0 & 0 & 0 & 0 & l_1 s & -l_1 s & -l_1 s & l_1 s \\
l_2 & -l_2 & l_2 & -l_2 & 0 & 0 & 0 & 0
\end{bmatrix}
$$

with $s = \sin(\pi/4)$, $l_1 = 0.381\,\text{m}$ and $l_2 = 0.508\,\text{m}$.

## A.5 9-DOF UVMS

Details of the mathematical model of the vehicle carrying a 3-link manipulator used in the simulations of the interaction control chapter can be found in [231], its main characteristics are reported in the following. The vehicle considered is a box of $2 \times 1 \times 0.5\,\text{m}$ characterized by:

$$
\begin{array}{rcll}
L & = & 2 & \text{m} \quad \text{vehicle length} \\
m & = & 1050 & \text{kg} \quad \text{vehicle weight} \\
\boldsymbol{r}_G^B & = & [0 \;\; 0 \;\; 0]^{\mathrm{T}} & \text{m} \\
\boldsymbol{r}_B^B & = & [0 \;\; 0 \;\; 0]^{\mathrm{T}} & \text{m} \\
W & = & 10300 & \text{N} \\
B & = & 7900 & \text{N} \\
I_x & = & 66 & \text{Nms}^2 \\
I_y & = & 223 & \text{Nms}^2 \\
I_z & = & 223 & \text{Nms}^2 \\
I_{xy} & = & 0 & \text{Nms}^2 \\
I_{yz} & = & 0 & \text{Nms}^2 \\
I_{xz} & = & 0 & \text{Nms}^2
\end{array}
$$

The hydrodynamic derivatives are given by:

$$
\begin{array}{rcl}
X_{\dot{u}} & = & -307 \\
Y_{\dot{v}} & = & -1025 \\
Z_{\dot{\omega}} & = & -1537 \\
K_{\dot{p}} & = & -12 \\
M_{\dot{q}} & = & -36 \\
N_{\dot{r}} & = & -36
\end{array}
$$

The drag coefficient is given by $C_d = 1.8$. All the hydrodynamic effects have been computed by assuming a regular shape for the vehicle and the link (a box and a cylinder, respectively).

In Tables A.3 and A.4 the mass, length, radius and inertia of the 3-link manipulator carried by the vehicle are reported.

**Table A.3.** mass [kg], length [m] and radius [m] of the 9-DOF UVMS

| | mass | length | radius |
|---|---|---|---|
| link 1 | 33 | 1 | 0.1 |
| link 2 | 19 | 1 | 0.08 |
| link 3 | 12 | 1 | 0.07 |

**Table A.4.** Link inertia [Nms$^2$] of the 3-link manipulator mounted on the underwater vehicle

| | $I_{x,i}$ | $I_{y,i}$ | $I_{z,i}$ | $I_{xy,i}$ | $I_{xz,i}$ | $I_{yz,i}$ |
|---|---|---|---|---|---|---|
| link 1 | 11 | 11 | 1.65 | 0 | 0 | 0 |
| link 2 | 6.3 | 6.3 | 0.75 | 0 | 0 | 0 |
| link 3 | 4 | 4 | .4 | 0 | 0 | 0 |

# References


1. Alekseev Y.K., Kostenko V.V. and Shumsky A.Y. (1994) Use of Identification and Fault Diagnostic Methods for Underwater Robotics. In: MTS/IEEE Techno-Ocean '94, Brest, France, 489–494
2. Alessandri A., Caccia M., Indiveri G. and Veruggio G. (1998) Application of LS and EKF Techniques to the Identification of Underwater Vehicles. In: 1998 IEEE International Conference on Control Applications, Trieste, Italy, 1084–1088
3. Alessandri A., Caccia M. and Veruggio G. (1998) A Model-Based Approach to Fault Diagnosis in Unmanned Underwater Vehicles. In: MTS/IEEE Techno-Ocean '98, Nice, France, 825–829
4. Alessandri A., Caccia M. and Veruggio G. (1999) Fault Detection of Actuator Faults in Unmanned Underwater Vehicles. Control Engineering Practice, 7:357–368
5. Alessandri A., Gibbons A., Healey A.J. and Veruggio G. (1999) Robust Model-Based Fault Diagnosis for Unmanned Underwater Vehicles Using Sliding Mode-Observers. In: Proceedings International Symposium Unmanned Untethered Submersible Technology, Durham, New Hampshire, 352-359
6. Alvarez A., Caffaz A., Caiti A., Casalino G., Clerici E., Giorgi F., Gualdesi L. and Turetta A. (2004) Design and realization of a very low cost prototypal autonomous vehicle for coastal oceanographic missions. In: IFAC Conference on Control Applications in Marine Systems - CAMS 2004, Ancona, Italy, 471–476
7. Alvarez A., Caffaz A., Caiti A., Casalino G., Clerici E., Giorgi F., Gualdesi L., Turetta A. and Viviani R. (2005) Folaga: a very low cost autonomous underwater vehicle for coastal oceanography. In: Proceddings 16th IFAC World Congress, Praha, Czech Republic
8. Antonelli G. (2003) Underwater Robots. Motion and Force Control of Vehicle-Manipulator Systems. Springer Tracts in Advanced Robotics, Springer-Verlag, Heidelberg, Germany
9. Antonelli G. (2003) A New Adaptive Control Law for the Phantom ROV. In: 7th IFAC Symposium on Robot Control, Wroclaw, Poland, 569–574
10. Antonelli G. (2002) A survey of fault detection/tolerance strategies for AUVs and ROVs. In: Fault diagnosis and tolerance for mechatronic systems. Recent advances, Caccavale F., Villani L., (Eds.), Springer Tracts in Advanced Robotics, Springer-Verlag, 109–127
11. Antonelli G.[1], Caccavale F. and Chiaverini S. (1999) A Modular Scheme for Adaptive Control of Underwater Vehicle-Manipulator Systems. In: 1999 American Control Conference, San Diego, California, 3008–3012


[1] Note that, as customary in Italy, for the papers concerning the Author of this monograph, G. Antonelli and when *all* co-Authors are Italians, the list is in alphabetical order regardless of the importance of single contributions

12. Antonelli G., Caccavale F. and Chiaverini S. (2001) A Virtual Decomposition Based Approach to Adaptive Control of Underwater Vehicle-Manipulator Systems. In: 9th Mediterranean Conference on Control and Automation, Dubrovnik, Croatia
13. Antonelli G., Caccavale F., Chiaverini S. and Fusco G. (2001) A Novel Adaptive Control Law for Autonomous Underwater Vehicles. In: IEEE International Conference on Robotics and Automation, Seoul, Korea, 447–451
14. Antonelli G., Caccavale F., Chiaverini S. and Fusco G. (2001) On the Use of Integral Control Actions for Autonomous Underwater Vehicles. In: European Control Conference, Porto, Portugal, 1186–1191
15. Antonelli G., Caccavale F., Chiaverini S. and Fusco G. (2002) A Modular Control Law for Underwater Vehicle-Manipulator Systems Adapting on a Minimum Set of Parameters. In: IFAC World Congress, Barcelona, Spain
16. Antonelli G., Caccavale F., Chiaverini S. and Fusco G. (2003) A Novel Adaptive Control Law for Underwater Vehicles. IEEE Transactions on Control Systems Technology, 11(2):221–232
17. Antonelli G., Caccavale F., Chiaverini S. and Villani L. (1998) An Output Feedback Algorithm for Position and Attitude Tracking Control of Underwater Vehicles. In: 1998 IEEE Conference on Decision and Control, Tampa, Florida, 4567–4572
18. Antonelli G., Caccavale F., Chiaverini S. and Villani L. (2000) Tracking Control for Underwater Vehicle-Manipulator Systems with Velocity Estimation. IEEE Journal of Oceanic Engineering, 25:399–413
19. Antonelli G., Caccavale F., Chiaverini S. and Villani L. (2000) Control of Underwater Vehicle-Manipulator Systems Using Only Position and Orientation Measurements. 6th IFAC Symposium on Robot Control, Wien, Austria, 463–468
20. Antonelli G. and Chiaverini S. (1998) Task-Priority Redundancy Resolution for Underwater Vehicle-Manipulator Systems. In: IEEE International Conference on Robotics and Automation, Leuven, Belgium, 768–773
21. Antonelli G. and Chiaverini S. (1998) Singularity-Free Regulation of Underwater Vehicle-Manipulator Systems. In: 1998 American Control Conference, Philadelphia, PA, 399–403
22. Antonelli G. and Chiaverini S. (1998) Adaptive Tracking Control of Underwater Vehicle-Manipulator Systems. In: 1998 IEEE International Conference on Control Applications, Trieste, Italy, 1089–1093
23. Antonelli G. and Chiaverini S. (2000) SIMURV. A simulation package for Underwater Vehicle-Manipulator Systems. In: 3rd IMACS Symposium on Mathematical Modelling, Wien, Austria, 533–536
24. Antonelli G. and Chiaverini S. (2000) Fuzzy Inverse Kinematics for Underwater Vehicle-Manipulator Systems. 7th International Symposium on Advances in Robot Kinematics, Piran-Portorož, Slovenia. In: Advances in Robot Kinematics, J. Lenarčič and M.M. Stanišić (Eds.), Kluwer Academic Publishers, Dordrecht, the Netherlands, 249–256
25. Antonelli G. and Chiaverini S. (2000) A Fuzzy Approach to Redundancy Resolution for Underwater Vehicle-Manipulator Systems. In: $5^{th}$ IFAC Conference on Manoeuvring and Control of Marine Craft, Aalborg, Denmark
26. Antonelli G. and Chiaverini S. (2002) A Fuzzy Approach to Redundancy Resolution for Underwater Vehicle-Manipulator Systems. Control Engineering Practice, in press
27. Antonelli G. and Chiaverini S. (2003) Obstacle Avoidance for a Platoon of Autonomous Underwater Vehicles. In: 6 th IFAC Conference on Manoeuvring and Control of Marine Craft, Girona, Spain, 143–148

28. Antonelli G. and Chiaverini S. (2003). Kinematic Control of a Platoon of Autonomous Vehicles. In: Proceedings 2003 IEEE International Conference on Robotics and Automation. Taipei, Taiwan, 1464–1469
29. Antonelli G. and Chiaverini S. (2003) Fuzzy Redundancy Resolution and Motion Coordination for Underwater Vehicle-Manipulator Systems. IEEE Transactions on Fuzzy Systems, 11(1)109–120
30. Antonelli G. and Chiaverini S. (2004). Fault Tolerant Kinematic Control of Platoons of Autonomous Vehicles. In: Proceedings 2003 IEEE International Conference on Robotics and Automation, New Orleans, Louisiana, 3313–3318
31. Antonelli G., Chiaverini S. and Sarkar N. (1999) Explicit Force Control for Underwater Vehicle-Manipulator Systems with Adaptive Motion Control Law. In: IEEE Hong Kong Symposium on Robotics and Control, Hong Kong, 361–366
32. Antonelli G., Chiaverini S. and Sarkar N. (1999) An Explicit Force Control Scheme for Underwater Vehicle-Manipulator Systems. In: IEEE/RSJ International Conference on Intelligent Robots and Systems, Kyongju, Korea, 136–141
33. Antonelli G., Chiaverini S. and Sarkar N. (2001) External Force Control for Underwater Vehicle-Manipulator Systems. IEEE Transactions on Robotics and Automation, 17:931–938
34. Antonelli G., Chiaverini S., Sarkar N. and West M. (1999) Singularity-free Adaptive Control for Underwater Vehicles in 6 dof. Experimental Results on ODIN. In: IEEE International Symposium on Computational Intelligence in Robotics and Automation, Monterey, California, 64–69
35. Antonelli G., Chiaverini S., Sarkar N. and West M. (2001) Adaptive Control of an Autonomous Underwater Vehicle: Experimental Results on ODIN. IEEE Transactions on Control System Technology, 9:756–765
36. Antonelli G., Sarkar N. and Chiaverini S. (1999) External Force Control for Underwater Vehicle-Manipulator Systems. In: 1999 IEEE Conference on Decision and Control, Phoenix, Arizona, 2975–2980
37. Antonelli G., Sarkar N. and Chiaverini S. (2002) Explicit Force Control for Underwater Vehicle-Manipulator Systems. Robotica, 20:251–260
38. Arimoto S. (1996) Control Theory of Nonlinear Mechanical Systems. A Passivity-based and Circuit-theoretic Approach. Clarendon Press
39. Babcock IV P.S. and Zinchuk J.J. (1990) Fault Tolerant Design Optimization: Application to an Autonomous Underwater Vechile Navigation System. In: Proceedings Symposium on Autonomous Underwater Vehicle Technology, Washington DC, 34–43
40. Bachmayer L., Whitcomb L.L. and Grosenbaugh M.A. (2000) An Accurate Four-Quadrant Nonlinear Dynamical Model for Marine Trhusters: Theory and Experimental Validation. IEEE Journal of Oceanic Engineering, 25:146–159
41. Balasuriya A. and T. Ura (2000) A Study of the Control of an Underactuated Underwater Robotic Vehicle. In: IEEE/RSJ International Conference on Intelligent Robots and Systems, Takamatsu, Japan, 849–854
42. Barnett D. and McClaran S. (1996) Architecture of the Texas A& M Autonomous Underwater Vehicle Controller. In: Proceedings Symposium on Autonomous Underwater Vehicle Technology, Monterey, California, 231–237
43. Beale G.O. and Kim J.H. (2000) A Robust Approach to Reconfigurable Control. In: $5^{th}$ IFAC Conference on Manoeuvring and Control of Marine Craft, Aalborg, Denmark, 197–202
44. Ben-Israel and Greville T.N.E (1974) Generalized Inverse: Theory and Application. Wiley, New York
45. Berghuis H. and Nijmeijer H. (1993) A Passivity Approach to Controller-Observer Design for Robots. IEEE Transactions on Robotics and Automation, 9:740–754

46. Bellingham J.G., Goudey C.A., Consi T.R., Bales J.W., Atwood D.K., Leonard J.J. and Chryssostomidis C. (1994) A second generation survey AUV. In: Proceedings Symposium on Autonomous Underwater Vehicle Technology, Cambridge, Massachusetts, 148–155
47. Bhatta P, Fiorelli E., Lekien F. Leonard N.E., Paley D.A., Zhang F., Bachmayer R., Davis R.E, Fratantoni D.M. and Sepulchre R. (2005) Coordination of an Underwater Glider Fleet for Adaptive Ocean Sampling. In: Proceedings IARP International Workshop on Underwater Robotics, Genova, Italy, 61–69
48. Bishop B.E. (2003) On the Use of Redundant Manipulator Techniques for Control of Platoons of Cooperating Robotic Vehicles. IEEE Transcations of Systems, Man and Cybernetics, part. A, 33(5):608–615
49. Bishop B.E. (2005) Control of Platoons of Nonholonomic Vehicles Using Redundant Manipulators Analogs. In: IEEE International Conference on Robotics and Automation, Barcelona, Spain, 4617–4622
50. Bishop B.E. and Stilwell D.J. (2001) On the Application of Redundant Manipulator Techniques to the Control of Platoons of Autonomous Vehicles. In: Proceddings 2001 IEEE International Conference on Control Applications, México City, Mexico, 823–828
51. Boeke M., Aust E. and Schultheiss G.F. (1991) Underwater Operation of a 6-Axes Robot Based on Offline Programming and Graphical Simulation. In: IEEE International Conference on Robotics and Automation, Sacramento, California, 1342–1347
52. Bono R., Bruzzone G., Bruzzone G. and Caccia M. (1999) ROV Actuator Fault Diagnosis Through Servo-Amplifiers' Monitoring: an Operational Experience. In: MTS/IEEE Techno-Ocean '99, Seattle, Washington, 1318–1324
53. Bono R., Caccia M. and Verruggio G. (1994) Software Architecture and Pool Tests of a Small Virtual AUV. In: IFAC Symposium on Robot Control, Capri, Italy, 1031–1036
54. Borges de Sousa J. and Healey A.J. (2005) A Case Study in Mixed Initiative Coordination and Control of AUVs. In: Proceedings IARP International Workshop on Underwater Robotics, Genova, Italy, 103–111
55. Borges de Sousa J., Johansson K.H., Speranzon A. and Da Silva J.E. (2005) A Control Architecture for Multiple Submarines in Coordinated Search Missions. In: IFAC World Congress, Praha, Czech Republic
56. Bozzo G.M., Maddalena D. and Terribile A. (1991) Laboratory and Sea Tests of the Supervisory Controlled Underwater Telemanipulation System. In: IEEE International Conference on Robotics and Automation, Sacramento, California, 1336–1341
57. Brady M., Hellerbach J.M., Johnson T.L., Lozano-Perez T., Mason M.T. (1982) Robot Motion: Planning and Control, MIT Press
58. Brooks R.A. (1986) A Robust Layered Control System for a Mobile Robot. IEEE Journal of Robotics and Automation, 2(1):14–23
59. Broome D. and Wang Q. (1991) Adaptive Control of Underwater Robotic Manipulators. In: International Conference on Advanced Robotics, Pisa, Italy, 1321–1326
60. Caccavale F., Natale C., Siciliano B. and Villani L. (1998) Resolved-acceleration control of robot manipulators: A critical review with experiments. Robotica, 16:565–573
61. Caccia M., Bono R., Bruzzone Ga., Bruzzone Gi., Spirandelli E. and Verruggio G. (2001) Experiences on Actuator Fault Detection, Diagnosis and Accomodation for ROVs. In: Proceedings International Symposium Unmanned Untethered Submersible Technology, Durham, New Hampshire

62. Caccia M., Indiveri G. and Veruggio G. (2000) Modeling and Identification of Open-Frame Variable Configuration Underwater Vehicles. IEEE Journal of Oceanic Engineering, 25:227-240
63. Caccia M. and Verruggio G. (2000) Guidance and Control of a Reconfigurable Unmanned Underwater Vehicle. Control Engineering Practice, 8:21–37
64. Caiti A., Munafò A. and Viviani R. (2005) Cooperating autonomous underwater vehicles to estimate ocean environmental parameters. In: Proceedings IARP International Workshop on Underwater Robotics, Genova, Italy, 87–94
65. Cancilliere F.M. (1994) Advanced UUV Technology. In: MTS/IEEE Techno-Ocean '94, Brest, France, 147–151
66. Cannell C.J. and Stilwell D.J. (2005) A Comparison of Two Approaches for Adaptive Sampling of Environmental Processes Using Autonomous Underwater Vehicles. In: MTS/IEEE Techno-Ocean '05, Washington, DC
67. Canudas de Wit C., Olguin Diaz E., and Perrier M. (1998) Robust Nonlinear Control of an Underwater Vehicle/Manipulator System with Composite Dynamics. In: IEEE International Conference on Robotics and Automation, Leuven, Belgium, 452–457
68. Canudas de Wit C., Olguin Diaz E., and Perrier M. (1998) Control of Underwater Vehicle/Manipulator with Composite Dynamics. In: 1998 American Control Conference, Philadelphia, PA, 389–393
69. Canudas de Wit C., Olguin Diaz E., and Perrier M. (2000) Nonlinear Control of an Underwater Vehicle/Manipulator System with Composite Dynamics. IEEE Transactions on Control System Technology, 8:948–960
70. Canudas de Wit C., Siciliano B., and Bastin G. (Eds.) (1996) Theory of Robot Control, Springer-Verlag, Berlin, Germany
71. Casalino G., Angeletti D., Bozzo T. and Marani G. (2001) Dexterous Underwater Object Manipulation via Multirobot Cooperating Systems. In: IEEE International Conference on Robotics and Automation, Seoul, Korea, 3220–3225
72. Casalino G. and Turetta A. (2003) Coordination and Control of Multiarm, Non-Holonomic Mobile Manipulators. In: IEEE/RSJ International Conference on Intelligent Robots and Systems, Las Vegas, Nevada, 2203–2210
73. Casalino G., Turetta A. and Sorbara A. (2005) Computationally Distributed Control and Coordination Architectures for Underwater Reconfigurable Free-Flying Multi-Manipulators. In: Proceedings IARP International Workshop on Underwater Robotics, Genova, Italy, 145–152
74. Cecchi D., Caiti A, Fioravanti S., Baralli F and Bovio E. (2005) Target Detection Using Multiple Autonomous Underwater Vehicles. In: Proceedings IARP International Workshop on Underwater Robotics, Genova, Italy, 161–168
75. Chen X., Marco D., Smith S., An E., Ganesan K. and Healey T. (1997) 6 DOF Nonlinear AUV Simulation Toolbox. In: Proceedings MTS/IEEE Ocean 97, Halifax, Canada
76. Chiacchio P., Chiaverini S., Sciavicco L. and Siciliano B. (1991) Closed-Loop Inverse Kinematics Schemes for Constrained Redundant Manipulators with Task Space Augmentation and Task Priority Strategy. The International Journal of Robotics Research, 10(4)410–425
77. Chiaverini S. (1993) Estimate of the Two Smallest Singular Values of the Jacobian Matrix: Application to Damped Least-Squares Inverse Kinematics. Journal of Robotic Systems, 10:991–1008
78. Chiaverini S. (1997) Singularity-Robust Task-Priority Redundancy Resolution for Real-Time Kinematic Control of Robot Manipulators. IEEE Transactions on Robotics and Automation, 13:398–410

79. Chiaverini S. and Sciavicco L. (1993) The Parallel Approach to Force/Position Control of Robotic Manipulators. IEEE Transactions on Robotics and Automation, 9:361–373
80. Chiaverini S. and Siciliano B. (1999) The Unit Quaternion: A Useful Tool for Inverse Kinematics of Robot Manipulators. Systems Analysis, Modelling and Simulation, 35:45–60
81. Choi H.T., Hanai A., Choi S.K. and Yuh J. (2003) Development of an Underwater Robot: ODIN III. In: IEEE/RSJ International Conference on Intelligent Robots and Systems, Las Vegas, Nevada, 836–841
82. Choi S.K. and Yuh J. (1993) Design of Advanced Underwater Robotic Vehicle and graphic Workstation. In: IEEE International Conference on Robotics and Automation, Atlanta, Georgia, 99–105
83. Choi S.K. and Yuh J. (1996) Experimental Study on a Learning Control System with Bound Estimation for Underwater Robots. In: Autonomous Robots, Yuh J., Ura T., Bekey G.A. (Eds.), Kluwer Academic Publishers, 187–194
84. Choi S.K. and Yuh J. (1996) Experimental Study on a Learning Control System with Bound Estimation for Underwater Robots. In: 1996 IEEE International Conference on Robotics and Automation, Minneapolis, Minnesota, 2160–2165
85. Chou J.C.K. (1992) Quaternion Kinematic and Dynamic Differential Equations. IEEE Transactions on Robotics and Automation, 1:53–64
86. Chung G.B., Eom K.S., Yi B.-J., Suh I.H., Oh S.RS., and Cho Y.J. (2000) Disturbance Observer-Based Robust Control for Underwater Robotic Systems with Passive Joints. In: IEEE International Conference on Robotics and Automation, San Francisco, California, 1775–1780
87. Chung G.B., Eom K.S., Yi B.-J., Suh I.H., Oh S.RS., Chung W.K., and Kim J.O. (2001) Disturbance Observer-Based Robust Control for Underwater Robotic Systems with Passive Joints. In: Journal of Advanced Robotics, 15:575–588
88. Conte G. and Serrani A. (1994) Robust Lyapunov-Based Design for Autonomous Underwater Vehicles. In: IFAC Symposium on Robot Control, Capri, Italy, 321–326
89. Conte G. and Serrani A. (1997) Robust Lyapunov-Based Design for Autonomous Underwater Vehicles. 5th IFAC Symposium on Robot Control, Nantes, France, 321–326
90. Conte G. and Serrani A. (1998) Global Robust Tracking with Disturbance Attenuation for Unmanned Underwater Vehicles. In: 1998 IEEE International Conference on Control Applications, Trieste, Italy, 1094–1098
91. Corradini M.L. and Orlando G. (1997) A Discrete Adaptive Variable-Structure Controller for MIMO Systems, and Its Application to an Underwater ROV. IEEE Transactions on Control Systems Technology, 5:349–359
92. Coste-Manière E., Peuch A., Perrier M., Rigaud V., Wang H.H., Rock S.M. and Lee M.J. (1996) Joint Evaluation of Mission Programming for Underwater Robots. In: IEEE International Conference on Robotics and Automation, Minneapolis, Minnesota, 2492-2497
93. Cristi R., Pappulias F.A. and Healey A. (1990) Adaptive Sliding Mode Control of Autonomous Underwater Vehicles in the Dive Plane. IEEE Journal of Oceanic Engineering, 15:152–160
94. Cui Y., Podder T. and Sarkar N. (1999) Impedance Control of Underwater Vehicle-Manipulator Systems (UVMS). In: IEEE/RSJ International Conference on Intelligent Robots and Systems, Kyongju, Korea, 148–153
95. Cui Y. and Sarkar N. (2000) A Unified Force Control Approach to Autonomus Underwater Manipulation. In: IEEE International Conference on Robotics and Automation, San Francisco, California, 1263–1268

96. Cui Y. and Sarkar N. (2001) A unified force control approach to autonomous underwater manipulation. Robotica, 19(3):255–266
97. da Cunha J.P.V.S., Costa R.R. and Hsu L. (1995) Design of a High Performance Variable Structure Position Control of ROV's. IEEE Journal of Oceanic Engineering, 20:42–55
98. De Angelis C.M. and Whitney J.E. (2000) Adaptive Calibration of an Autonomous Underwater Vehicle Navigation System. In: MTS/IEEE Techno-Ocean '00, Providence, Rhode Island, 1273–1275
99. Dégoulange E. and Dauchez P. (1994) External Force Control of an Industrial PUMA 560 Robot. Journal of Robotic Systems, 11:523–540
100. Demin X. and Lei G. (2000) Wavelet Transform and its Application to Autonomous Underwater Vehicle Control System Fault Detection. In: Proceedings 2000 International Symposium Underwater Technology, Tokyo, Japan, 99–104
101. De Schutter J. and Van Brussel H. (1988) Compliant Robot Motion II. A Control Approach Based on External Control Loops. International Journal of Robotics Research, 7:18–33
102. Deuker B., Perrier M. and Amy B. (1998) Fault-Diagnosis of Subsea Robots Using Neuro-Symbolic Hybrid Systems. In: MTS/IEEE Techno-Ocean '98, Nice, France, 830–834
103. Driankov D., Hellendoorn H. and Reinfrank M. (1995) An Introduction to Fuzzy Control. Springer-Verlag, Berlin, Germany
104. Dubowsky S. and Papadopoulos E. (1993) The Kinematics, Dynamics, and Control of Free-Flying and Free-Floating Space Robotic Systems. IEEE Transactions on Robotics and Automation, 9:531–543
105. Dunnigan M.W. and Russell G.T. (1994) Reduction of the Dynamic Coupling Between a Manipulator and ROV Using Variable Structure Control. In: International Conference on Control, 1578–1583
106. Dunnigan M.W. and Russell G.T. (1998) Evaluation and Reduction of the Dynamic Coupling Between a Manipulator and an Underwater Vehicle. IEEE Journal of Oceanic Engineering, 23:260–273
107. Egeland O. (1987) Task-Space Tracking with Redundant Manipulators. IEEE Journal Robotics and Automation, 3:471–475
108. Egeland O. and Godhavn J.-M. (1994) Passivity-Based Adaptive Attitude Control of a Rigid Spacecraft. In: IEEE Transactions on Automatic Control, 39:842–846
109. Egeland O. and Sagli J.R. (1990) Kinematics and control of a Space manipulator using the Macro-Micro Manipulator Concept. In: 1990 IEEE Conference on Decision and Control, Honolulu, Hawaii, 3096–3101
110. Egeland O. and Sagli J.R. (1993) Coordination of Motion in a Spacecraft/Manipulator System. The International Journal of Robotics Research, 12:366–379
111. Eiani-Cherif S., Lebret G. and Perrier M. (1997) Identification and Control of a Submarine Vehicle. 5th IFAC Symposium on Robot Control, Nantes, France, 307–312
112. Faltinsen O.M. (1990) Sea Loads on Ships and Offshore Structures, Cambridge University Press
113. Farrell J., Berger T. and Appleby B.D. (1993) Using Learning Techniques to Accomodate Unanticipated Faults. Control Systems Magazine, 13(3):40–49
114. Featherstone R. (1983) The Calculation of Robot Dynamics Using Articulated-Body Inertias. The International Journal of Robotics Research, 2:13–30
115. Featherstone R. (1987) Robot Dynamics Algorithms, Kluwer, Boston

116. Feezor M.D., Sorrel F.Y., Blankinship P.R. and Bellingham J.G. (2001) Autonomous Underwater Vehicle Homing/Docking via Electromagnetic Guidance. IEEE Journal of Oceanic Engineering, 26(4):515–521
117. Ferguson J.S. (1998) The Theseus Autonomous Underwater Vehicle. Two Successful Missions. In: Proceedings 1998 International Symposium Underwater Technology, Tokyo, Japan, 109–114
118. Ferguson J.S., Pope A., Butler B., Verrall R. (1999) Theseus AUV - Two Record Breaking Missions. Sea Technology Magazine, 65–70
119. Ferretti G., Magnani G., and Rocco P. (1997) Toward the Implementation of Hybrid Position/Force Control in Industrial Robots. IEEE Transactions on Robotics and Automation, 16:838–845
120. Fiorelli E., Leonard N.E., Bhatta P, Paley D.A., Bachmayer R. and Fratantoni D.M. (2004) Multi-AUV Control and Adaptive Sampling in Monterey Bay. In: IEEE Autonomous Underwater Vehicles 2004: Workshop on Multiple AUV Operations (AUV04)
121. Fjellstad O. and Fossen T.I. (1994) Singularity-Free Tracking of Unmanned Underwater Vehicles in 6 DOF. In: 1994 IEEE Conference on Decision and Control, Lake Buena Vista, Florida, 1128–1133
122. Fjellstad O. and Fossen T.I. (1994) Position and Attitude Tracking of AUVs: A Quaternion Feedback Approach. IEEE Journal of Oceanic Engineering, 19:512–518
123. Fjellstad O. and Fossen T.I. (1994) Quaternion Feedback Regulation of Underwater Vehicles. In: 1994 IEEE International Conference on Control Applications, Glasgow, United Kingdom, 857–862
124. Fossen T.I. (1991) Adaptive Macro-Micro Control of Nonlinear Underwater Robotic Systems. In: International Conference on Advanced Robotics, Pisa, Italy, 1569–1572
125. Fossen T.I. and Balchen J. (1991) The NEROV Autonomous Underwater Vehicle. In: MTS/IEEE Techno-Ocean '91 Conference, Honolulu, Hawaii
126. Fossen T.I. and Sagatun S.I. (1991) Adaptive Control of Nonlinear Systems: A case Study of Underwater Robotic Systems. Journal of Robotic Systems, 8:393–412
127. Fossen T.I. (1994) Guidance and Control of Ocean Vehicles, John Wiley & Sons, Chichester, United Kingdom
128. Fossen T.I. (2002) Marine Control Systems: Guidance, Navigation and Control of Ships, Rigs and Underwater Vehicles, Marine Cybernetics AS, Trondheim, Norway
129. Fossen T.I. and Fjellstad O. (1995) Robust Adaptive Control of Underwater Vehicles: A Comparative Study. In: IFAC Workshop on Control Applications in Marine Systems, Trondheim, Norway, 66–74
130. Fossen T.I. and Sagatun S.I. (1991) Adaptive Control of Nonlinear Underwater Robotic Systems. In: IEEE International Conference on Robotics and Automation, Sacramento, California, 1687–1694
131. Fraisse P., Lapierre L., Dauchez P. (2000) Position/Force Control of an Underwater Vehicle Equipped with a Robotic Manipulator. In: 6th IFAC Symposium on Robot Control, Wien, Austria, 475–479
132. Gadre A., Mach J.E., Stilwell D.J., and Wick C.E. (2003) Design of a Prototype Miniature Autonomous Underwater Vehicle. In: IEEE/RSJ International Conference on Intelligent Robots and Systems, Las Vegas, Nevada, 842–846
133. Gadre A. and Stilwell D.J. (2005) A Complete Solution to Underwater Navigation in the Presence of Unknown Currents Based on Range Measurements from a Single Location. In: IEEE/RSJ International Conference on Intelligent Robots and Systems, Edmonton, Canada, 1743–1748

134. Garcia R., Puig J., Ridao P. and Cufi X. (2002) Augmented State Kalman Filtering for AUV Navigation. In: IEEE International Conference on Robotics and Automation, Washington, DC, 4010–4015
135. Gautier M. (1990) Numerical Calculation of the Base Inertial Parameters of Robots. In: 1990 IEEE International Confernece on Robotics and Automation, Cincinnati, Ohio, 1020–1025
136. Ghabcheloo R., Pascoal A., Silvestre C., Carvalho D. (2005) Coordinated Motion Control of Multiple Autonomous Underwater Vehicle. In: Proceedings IARP International Workshop on Underwater Robotics, Genova, Italy, 41–50
137. Goheen K.R. (1991) Modeling Methods for Underwater Robotic Vehicle Dynamics. Journal of Robotic Systems, 8:295–317
138. Goheen K.R. and Jeffreys R. (1993) Multivariable Self-Tuning Autopilots for Autonomous and Remotely Operated Underwater Vehicles. IEEE Journal of Oceanic Engineering, 15:144–151
139. Grenon G., An P.E., Smith S.M. and Healey A.J. (2001) Enhancement of the Inertial Navigation System for the Morpheus Autonomous Underwater Vehicles. IEEE Journal of Oceanic Engineering, 26:548–560
140. Hamilton K., Lane D., Taylor N. and Brown K. (2001) Fault Diagnosis on Autonomous Robotic Vehicles with RECOVERY: an Integrated Heterogeneous-Knowledge Approach. In: IEEE International Conference on Robotics and Automation, San Francisco, California, 1251–1256
141. Healey A.J. (2001) Application of Formation Control for Multi-Vehicle Robotic Minesweeping. In: Proceddings 40 th IEEE Conference on Decision and Control, Orlando, Florida, 1497–1502
142. Healey A.J. (1992) A Neural Network Approach to Failure Diagnostics for Underwater Vehicles. In: Proceedings IEEE Oceanic Engineering Society Symposium on Autonomous Underwater Vehicles, Washington D.C., 131–134
143. Healey A.J. (1998) Analytical Redundancy and Fuzzy Inference in AUV Fault Detection and Compensation. In: Proceedings Oceanology 1998, Brighton, 45–50
144. Healey A.J., Bahrke F., Navarrete J. (1992) Failure Diagnostics for Underwater Vehicles: A Neural Network Approach. In: IFAC Conference on Maneuvering and Control of Marine Craft, Aalborg, Denmark, 293–306
145. Healey A.J. and Lienard D. (1993) Multivariable Sliding Mode Control for Autonomous Diving and Steering of Unmanned Underwater Vehicles. IEEE Journal of Oceanic Engineering, 18:327–339
146. Healey A.J. and Marco D.B. (1992) Experimental Verification of Mission Planning by Autonomous Mission Execution and Data Visualization Using the NPS AUV II. In: Proceedings 1992 Symposium Autonomous Underwater Vehicle Technology, Washington, DC, 65–72
147. Healey A.J., Rock S.M., Cody S., Miles D. and Brown J.P. (1995) Toward an Improved Understanding of Thruster Dynamics for Underwater Vehicles. IEEE Journal of Oceanic Engineering, 20:354–361
148. Healey A.J. and Kim Y. (1997) Control of Small Robotic Vehicles in Unexploded Ordnance Clearance. In: IEEE International Conference Robotics and Automation, Albuquerque, New Mexico
149. Hogan N. (1985) Impedance Control: An Approach to Manipulation. (Parts I-III). Transactions ASME Journal of Dynamic Systems, Measurement, and Control, 107:1–24
150. Hooker W. and Margulies G. (1965) The Dynamical Attitude Equations for an n-Body Satellite. Journal Astronaut Science, XII(4), 123

151. Hornfeld W. and Frenzel E. (1998) Intelligent AUV On-Board Health Monitoring Software (INDOS). In: MTS/IEEE Techno-Ocean '98, Nice, France, 815–819
152. Hsu L., Costa R.R., Lizarrade F. and Soares daCunha J.P.V. (2000) Avaliação Experimental da Modelagem e Simulação da Dinâmica de um Veículo Submarino de Operação Remota (in Portuguese). Revista Controle & Automacão, 11:82–93
153. Hsu L., Costa R.R., Lizarrade F. and Soares daCunha J.P.V. (2000) Dynamic Positioning of Remotely Operated Underwater Vehicles. IEEE Robotics and Automation Magazine, 21–31
154. Hutchison B. (1991) Velocity Aided Inertial Navigation. In: Proceedings Sensor Navigation Issues for UUVs, Cambridge, Massachusetts, 8–9
155. Ihle I.-A. F. , Skjetne R. and Fossen T.I. (2004) Nonlinear formation control of marine craft with experimental results. In: IEEE Conference on Decision and Control, Paradise Island, Bahamas, 1:680–685
156. Ioi K. and Itoh K. (1990) Modelling and Simulation of an Underwater Manipulator. Advanced Robotics, 4:303–317
157. Indiveri G. (1998) Modelling and Identification of Underwater Robotic Systems, Ph.D. Thesis in Electronic Engineering and Computer Science, University of Genova, Genova, Italy
158. Ikeda T., Mita T. and Yamakita M. (2005) Formation Control of Autonomous Underwater Vehicles. In: IFAC World Congress, Praha, Czech Republic
159. Ishimi K., Ohtsuki Y., Manabe T. and Nakashima K. (1991) Manipulation System for Subsea Operation. In: IEEE International Conference on Robotics and Automation, Sacramento, California, 1348–1353
160. Ishibashi S., Shimizu E. and Masanori I. (2001) Motion Planning for a Manipulator Equipped on an Underwater Robot. In: 27th Annnual Conference of the IEEE Industrial Electronics Society, Denver, Colorado, 558–563
161. Ishibashi S., Shimizu E. and Ito M. (2002) The Motion Planning for Underwater Manipulators Depend on Genetic Algorithm. In: IFAC World Congress, Barcelona, Spain
162. Ishitsuka M., Sagara S. and Ishii K. (2004) Dynamics Analysis and resolved Acceleration Control of an Autonomous Underwater Vehicle Equipped with a manipulator. In: 2004 International Symposium on Underwater Technology, Taipei, Taiwan, 277–281
163. Jun B.H., Lee P.M. and Lee J. (2004) Manipulability Analysis of Underwater Robotics Arms on ROV and Application to Task-Oriented Joint Configuration. In: MTS/IEEE Techno-Ocean '04, Kobe, Japan, 1548–1553
164. Jun B.H., Lee J. and Lee P.M. (2005) A Repetitive Periodic Motion Planning of Articulated Underwater Robots Subjet to Drag Optimization. In: IEEE/RSJ International Conference on Intelligent Robots and Systems, Edmonton, Canada, 3868–3873
165. Kajita H. and Kosuge K. (1997) Force Control of Robot Floating on the Water Utilizing Vehicle Restoring Force. In: IEEE/RSJ International Conference on Intelligent Robots and Systems, Grenoble, France, 162–167
166. Kapellos D.S. and Espiau B. (1997) Design of Control Procedures for a Free-floating Underwater Manipulation System. In: IEEE International Conference Robotics and Automation, Albuquerque, New Mexico, 2158–2165
167. Kato N., Kojima J., Kato Y., Matumoto S. and Asakawa K. (1998) Optimization of Configuration of Autonomous Underwater Vehicles for Inspection of Underwater Cables. In: IEEE International Conference on Robotics and Automation, Leuven, Belgium, 1045-1050

168. Kato N. and Lane D.M. (1996) Co-ordinated Control of Multiple Manipulators in Underwater Robots. In: IEEE International Conference on Robotics and Automation, Minneapolis, Minnesota, 2505–2510
169. Kawamura S. and Sakagami N. (2002) Analysis on Dynamics of Underwater Robot Manipulators Basing on Iteratived Learning Control and Time-Scale Transformation. In: IEEE International Conference on Robotics and Automation, Washington, DC, 1088–1094
170. Khalil H.K. (1996) Nonlinear Systems, 2nd Edition, Prentice-Hall, Upper Saddle River, New Jersey
171. Khatib O. (1987) A Unified Approach for Motion and Force Control of Robot Manipulators: The Operational Space Formulation. IEEE Journal of Robotics and Automation, 3:43–53
172. Khatib O., Yokoi K., Chang K., Ruspini D., Holmberg R. and Casal A. (1996) Vehicle/Arm Coordination and Multiple Mobile Manipulator Decentralized Cooperation. IEEE/RSJ International Conference on Intelligent Robots and Systems, Osaka, Japon, 546–553.
173. Kim J.H. and Beale G.O. (2001) Fault Detection and Classification in Underwater Vehicle Using the $T^2$ Statistic. In: 9 th Mediterranean Conference on Control and Automation, Dubrovnik, Croatia
174. Kim J., Chung W.K. and Yuh J. (2003) Dynamic Analysis and Two-Time Scale Control for Underwater Vehicle-Manipulator Systems. In: IEEE/RSJ International Conference on Intelligent Robots and Systems, Las Vegas, Nevada, 577–582
175. Kim T. and Yuh J. (2001) A Novel Neuro-Fuzzy Controller for Autonomous Underwater Vehicle. In: IEEE International Conference on Robotics and Automation, Seoul, Korea, 2350–2355
176. Kim J., Han J., Chung W.K. and Yuh J. (2005) Accurate Thruster Modeling with Non-Parallel Ambient Flow for Underwater Vehicles. In: IEEE/RSJ International Conference on Intelligent Robots and Systems, Edmonton, Canada, 1737–1742
177. Kirkwood W.J., Shane F., Gashler D., Au D., Thomas H., Sibenac M., O'Reilly T.C., Konvalina T., McEwen R., Bahlavouni A., Tervalon N. and Bellingham J.G. (2001) Development of a Long Endurance Autonomous Underwater Vehicle for Ocean Science Exploration. In: MTS/IEEE Conference and Exhibition Oceans '01, Honolulu, Hawaii, 1504–1512
178. Koh T.H., Lau M.W.S., Low E., Seet G., Swei S. Cheng P.L. (2002) A Study of the Control of an Underactuated Underwater Robotic Vehicle. In: Proceedings 2002 IEEE/RSJ International Conference on Intelligent Robots and Systems, Lausanne, Switzerland, 2049–2054
179. Kumar R. and Stover J.A. (2000) A Behavior-Based Intelligent Control Architecture with Application to Coordination of Multiple Underwater Vehicles. IEEE Transcations of Systems, Man and Cybernetics, 30(6):767–784
180. Lane D.M., Bartolini G., Cannata G., Casalino G., Davies J.B.C., Veruggio G., Canals M. and Smith C. (1998) Advanced Manipulation for Deep Underwater Sampling: the AMADEUS Research Project. In: 1998 IEEE International Conference on Control Applications, Trieste, Italy, 1068–1073
181. Larkum T. and Broome D. (1994) Advanced Controller for an Underwater Manipulator. In: 1994 IEEE International Conference on Control Applications, Glasgow, United Kingdom, 1081–1086
182. Leabourne K.N., Rock S.M. and Lee M.J. (1998) Model Development of an Underwater Manipulator for Coordinated Arm-Vehicle Control. In: MTS/IEEE Techno-Ocean '98, Nice, France, 941–946

183. Lee W. and Kang G. (1998) A Fuzzy Model-Based Controller of an Underwater Robotic vehicle Under the Influence of Thruster Dynamics. In: IEEE International Conference on Robotics and Automation, Leuven, Belgium, 756-761
184. Lee C.S.G., Wang J. and Yuh J. (2001) Self-Adaptive Neuro-Fuzzy Sitemaps with Fast Parameters Learning for Autonomous Underwater Vehicle Control. In: Journal of Advanced Robotics, 15(5):589–608
185. Lee P.M. and Yuh J. (1999) Application of Non-regressor Based Adaptive Control to an Underwater Mobile Platform-mounted Manipulator. In: 1999 IEEE International Conference on Control Applications, Kohala Coast, Hawaii, 1135–1140
186. Lee M. and Choi H.S. (2000) A Robust Neural Controller for Underwater Robot Manipulators. IEEE Transactions on Neural Networks, 11:1465–1470
187. Leonard N.E. and Fiorelli E. (2001) Virtual leaders, artificial potentials and coordinated control of groups. In: 2001 IEEE Conference on Decision and Control, Orlando, Forida, 2968–2973
188. Leonessa A and Morel Y. (2005) Indirect Collaborative Control of Autonomous Vehicles with Limited Communication Bandwidth. In: Proceedings IARP International Workshop on Underwater Robotics, Genova, Italy, 233–240
189. Lévesque B. and Richard M.J. (1994) Dynamic Analysis of a Manipulator in a Fluid Environment. The International Journal of Robotics Research, 13:221–231
190. Liceaga E.C., Qiao H., Liceaga-Castro J. (1991) Modelling and Control of a Marine Robot Arm. In: 1991 IEEE Conference on Decision and Control, Brighton, England, 704–705
191. Liégeois A. (1977) Automatic Supervisory Control of the Configuration and Behavior of Muldibody Mechanisms. IEEE Transcations of Systems, Man and Cybernetics, 7:868–871
192. Lin C.C., Chang C.C., Jeng M.D. and Wang J.H.(2001) Development of an Intelligent Underwater Robotic System. In: MTS/IEEE Techno-Ocean '01, 1036–1040
193. Lizarralde F., Wen J.T., and Hsu L. (1995) Quaternion-Based Coordinated Control of a Subsea Mobile Manipulator with Only Position Measurements. In: 1995 IEEE Conference on Decision and Control, New Orleans, Louisiana, 3996–4001
194. Loncarić J. (1987) Normal Forms of Stiffness and Compliance Matrices. IEEE Journal of Robotics and Automation, 3:567–572
195. Maciejewski A.A. and Klein C.A. (1985) Obstacle Avoidance for Kinematically Redundant Manipulators in Dynamically Varying Environments. International Journal of Robotics Research, 4:109–117
196. Maciejewski A.A. (1988) Numerical Filtering for the Operation of Robotic Manipulators through Kinematicaly Singular Configurations. Journal of Robotic Systems, 5:527–552
197. Mahesh H. and Yuh J. (1991) Control of Underwater Robots in Working Mode. In: IEEE International Conference on Robotics and Automation, Sacramento, California, 2630–2635
198. Mahesh H., Yuh J. and Lakshmi R. (1991) A Coordinated Control of an Underwater Vehicle and Robotic Manipulator. Journal of Robotic Systems, 8:339–370
199. Mangoubi R.S., Appleby B.D., Verghese G.C. and VanderVelde W.E. (1995) A Robust Failure Detection and Isolation Algorithm. In: Proceedings 34th Conference Decision & Control, New Orleans, Louisiana, 2377–2382
200. Marchand É., Chaumette F., Spindler F. and Perrier M. (2001) Controlling the Manipulator of an Underwater ROV Using a Coarse Calibrated Pan/Tilt Ca-

mera. In: IEEE International Conference on Robotics and Automation, Seoul, Korea, 2773–2778
201. Marco D.B. and Healey A.J. (2001) Command, Control, and Navigation Experimental Results with the NPS ARIES AUV. IEEE Journal Oceanic Engineering, 26:466–476
202. Mason M.T. (1981) Compliance and Force Control for Computer Controlled Manipulators. IEEE Transactions on Systems, Man, and Cybernetics, 11:418–432
203. McLain T.W. and Rock S.M. (1988) Development and Experimental Validation of an Underwater Manipulator Hydrodynamic Model. The International Journal of Robotics Research, 17:748–759
204. McLain T.W., Rock S.M. and Lee M.J. (1995) Experiments in the Coordination of Underwater Manipulator and Vehicle Control. In: MTS/IEEE Techno-Ocean '95, 1208–1215
205. McLain T.W., Rock S.M. and Lee M.J. (1996) Coordinated Control of an Underwater Robotic System. In: Video Proceedings IEEE International Conference on Robotics and Automation, Minneapolis, Minnesota
206. McLain T.W., Rock S.M. and Lee M.J. (1996) Experiments in the Coordinated Control of an Underwater Arm/Vehicle System. In: Autonomous Robots, Yuh J., Ura T., Bekey G.A. (Eds.), Kluwer Academic Publishers, 213–232
207. McMillan S. and Orin D.E. (1995) Efficient Computation of Articulated-Body Inertias Using Successive Axial Screws. IEEE Transactions on Robotics and Automation, 11:606-611
208. McMillan S., Orin D.E. and McGhee R.B. (1994) Efficient Dynamic Simulation of an Unmanned Underwater Vehicle with a Manipulator. In: IEEE International Conference on Robotics and Automation, San Diego, California, 1133-1140
209. Nakamura Y. and Hanafusa H. (1986) Inverse Kinematic Solutions with Singularity Robustness for Robot Manipulator Control. Transactions ASME Journal of Dynamic Systems, Measurements and Control, 108:163–171
210. Nakamura Y., Hanafusa H. and Yoshikawa T. (1987) Task-Priority Based Redundancy Control of Robot Manipulators. International Journal Robotics Research, 6:3–15
211. Natale C., Siciliano B. and Villani L. (1999) Spatial Impedance Control of Redundant Manipulators. In: IEEE International Conference on Robotics and Automation, Detroit, Michigan, 1788–1793
212. Nelson E., McClaran S. and Barnett D. (1996) Development and Validation of the Texas A&M University Autonomous Underwater Vehicle Controller. In: Proceedings Symposium on Autonomous Underwater Vehicle Technology, Monterey California, 203–208
213. Nemec B. (1997) Force Control of Redundant Robots. In: IFAC Symposium on Robot Control, Nantes, France, 215–220
214. Newman J.N. (1977) Marine Hydrodynamics, MIT Press, Cambridge, MA
215. Nie J., Yuh J., Kardash E., and Fossen T.I. (1998) On-Board Sensor-Based Adaptive Control of Small UUVS in Very Shallow Water. In: IFAC Conference on Control Applications in Marine Systems, Fukuoka, Japan, 201–206
216. Nie J., Yuh J., Kardash E., and Fossen T.I. (2000) On-Board Sensor-Based Adaptive Control of Small UUV's in Very Shallow Water. International Journal of Adaptive Control Signal Process, 13:441–451
217. Oda M. and Ohkami Y. (1997) Coordinated Control of Spacecraft Attitude and Space Manipulators. Control Engineering Practice, 5:11–21

218. Ögren P. and Leonard N.E. (2003). Obstacle avoidance in Formation. In: Proceedings 2003 IEEE International Conference on Robotics and Automation. Taipei, Taiwan, 2492–2497
219. Oh Y., Chung W.K., Youm Y., Suh I.H. (1998) Motion/Force Decomposition of Redundant Manipulator and Its Application to Hybrid Impedance Control. In: IEEE International Conference on Robotics and Automation, Leuven, Belgium, 1441–1446
220. Olguin Diaz E. (1999) Modélisation et Commande d'un Système Véhicule/Manipulateur Sous-Marin (in French), Thèse pour le grade de Docteur de l'Institut National Polytechnique de Grenoble, Grenoble, France
221. Oliveira P. and Pascoal A. (1998) Navigation Systems Design: an Application of Multi-Rate Filtering Theory. In: MTS/IEEE Techno-Ocean '98, Nice, France, 1348–1353
222. Orrick A., McDermott M., Barnett D., Nelson E. and Williams G. (1994) Failure Detection in an Autonomous Underwater Vehicle. In: Proceedings Symposium on Autonomous Underwater Vehicle Technology, Cambridge, Massachusetts, 377–382
223. Ortega R. and Spong M.W. (1989) Adaptive Motion Control of Rigid Robots: a Tutorial. Automatica, 25:877–888
224. Padir T. and Koivo A.J. (2003) Modeling of Two Underwater Vehicles with Manipulators On-Board. IEEE Transcations of Systems, Man and Cybernetics, 33:1259–1364
225. Pascoal A., Kaminer I. and Oliveira P. (1998) Navigation System Design Using Time-Varying Complementary Filters. In: 1998 IEEE International Conference on Control Applications, Trieste, Italy, 1099–1104
226. Peng Z. and Adachi N. (1993) Compliant Motion Control of Kinematically Redundant Manipulators. IEEE Transactions on Robotics and Automation, 9:831–837
227. Pereira J. and Duncan A. (2000) System identification of Underwater Vehicles. In: Proceedings 2000 International Symposium Underwater Technology, Tokyo, Japan, 419–424
228. Perrault D. and Nahon M. (1998) Fault-Tolerant Control of an Autonomous Underwater Vehicle. In: MTS/IEEE Techno-Ocean '98, Nice, France, 820–824
229. Petrich J., Woolsey C.A. and Stilwell D.J. (2005) Identification of a Low-Complexity Flow Field Model for AUV Applications. In: MTS/IEEE Techno-Ocean '05, Washington, DC
230. Pierrot F., Benoit M. and Dauchez P. (1998) SamoS: A Pythagorean Solution for Omnidirectional Underwater Vehicles. J.Lenarčič and M.L. Husty (eds.), In: Advances in Robot Kinematics: Analysis and Control, Kluwer Academic Publishers, the Netherlands
231. Podder T.K. (1998) Dynamic and Control of Kinematically Redundant Underwater Vehicle-Manipulator Systems. Autonomous Systems Laboratory Technical Report, ASL 98-01, University of Hawaii, Honolulu, Hawaii
232. Podder T.K., Antonelli G. and Sarkar N. (2000) Fault Tolerant Control of an Autonomous Underwater Vehicle Under Thruster Redundancy: Simulations and Experiments. In: IEEE International Conference on Robotics and Automation, San Francisco, California, 1251–1256
233. Podder T.K., Antonelli G. and Sarkar N. (2001) An Experimental Investigation into the Fault-Tolerant Control of an Autonomous Underwater Vehicle. Journal of Advanced Robotics, 15:501–520
234. Podder T.K. and Sarkar N. (1999) Fault Tolerant Decomposition of Thruster Forces of an Autonomous Underwater Vehicle. In: IEEE International Conference on Robotics and Automation, Leuven, Belgium, 84–89

235. Podder T.K. and Sarkar N. (2000) Dynamic Trajectory Planning for Autonomous Underwater Vehicle-Manipulator Systems. In: IEEE International Conference on Robotics and Automation, San Francisco, California, 3461–3466
236. Podder T.K. and Sarkar N. (2001) Fault Tolerant Control of an Autonomous Underwater Vehicle Under Thruster Redundancy. Robotics and Autonomous Systems, 34:39–52
237. Quinn A.W. and Lane D. (1994) Computational Issues in Motion Planning for Autonomous Underwater Vehicle with Manipulators. In: IEEE Symposium on Autonomus Underwater Vehicle Technology, Boston, 255–262
238. Rae G.J.S. and Dunn S.E. (1994) On-Line Damage Detection for Autonomous Underwater Vehicle. In: Proceedings Symposium on Autonomous Underwater Vehicle Technology, Cambridge, Massachusetts, 383–392
239. Raibert M.H. and Craig J.J. (1981) Hybrid Position/Force Control of Manipulators. Transactions ASME Journal of Dynamic Systems, Measurement, and Control, 102:126–133
240. Rauch H.E. (1994) Intelligent Fault Diagnosis and Control Reconfiguration. In: IEEE International Symposium on Intelligent Control, Chicago, Illinois, 6–12
241. Reynolds C. (1987) Flocks, herd and schools: A distributed behavioral model. In: Proceedings Computer Graphics, 21(4)25–34
242. Richard M.J. and Lévesque B. (1996) Stochastic Dynamical Modelling of an Open-Chain Manipulator in a Fluid Environment. Mechanism and Machine Theory, 31:561–572
243. Riedel J.S. (2000) Shallow Water Stationkeeping of an Autonomous Underwater Vehicle: The Experimental Results of a Disturbance Compensation Controller. In: MTS/IEEE Techno-Ocean '00, Providence, Rhode Island, 1017–1024
244. Riedel J.S. and Healey A.J. (1998) Shallow Water Station Keeping of AUV Using Multi-Sensor Fusion for Wave Disturbance Prediction and Compensation. In: MTS/IEEE Techno-Ocean '98, Kobe, Japan, 1064–1068
245. Ryu J.H., Kwon D.S. and Lee P.M. (2001) Control of Underwater Manipulators Mounted on a ROV Using Base Force Information. In: IEEE International Conference on Robotics and Automation, Seoul, Korea, 3238–3243
246. Roberson R.E. and Schwertassek R. (1988) Dynamics of Multibody Systems, Springer-Verlag, Berlin
247. Roberson D.G. and Stilwell D.J. (2005) Control fo an Autonomous Underwater Vehicle Platoon with a Switched Communication Netwrok. In: 2005 American Control Conference, Portland, Oregon, 4333–4338
248. Salisbury J.K. (1980) Active Stiffness Control of a Manipulator in Cartesian Coordinates. In: 1980 IEEE Conference on Decision and Control, Albuquerque, New Mexico, 95–100
249. Sarkar N. and Podder T.K. (1999) Motion Coordination of Underwater Vehicle Manipulator Systems Subject to Drag Optimization. In: IEEE International Conference on Robotics and Automation, Detroit, Michigan, 387–392
250. Sarkar N. and Podder T.K. (2001) Coordinated Motion Planning and Control Autonomous Underwater Vehicle-Manipulator Systems Subject to Drag Optimization. IEEE Journal Oceanic Engineering, 26:228–239
251. Sarkar N., Podder T.K., and Antonelli G. (2002) Fault-accommodating thruster force allocation of an AUV considering thruster redundancy and saturation. IEEE Transactions on Robotics and Automation, 18:223–233
252. Sarkar N., Yuh J., and Podder T.K. (1999) Adaptive Control of Underwater Vehicle-Manipulator Systems Subject to Joint Limits. In: IEEE/RSJ International Conference on Intelligent Robots and Systems, Kyongju, Korea, 142–147

253. Sarpkaya T. and Isaacson M. (1981) Mechanics of Wave Forces on Offshore Structures, Van Nostrand Reinhold Co., New York, New York
254. Sciavicco L. and Siciliano B. (2000) Modeling and Control of Robot Manipulators. Springer-Verlag, London, United Kingdom
255. Schjølberg I. and Fossen T.I. (1994) Modelling and Control of Underwater Vehicle-Manipulator Systems. In: $3^{rd}$ Conference on Marine Craft Manoeuvring and Control, Southampton, United Kingdom, 45–57
256. Schjølberg I. (1996) Modeling and Control of Underwater Robotic Systems, PhD thesis for the Doktor ingeniør degree, Norwegian University of Science and Technology, Trondheim, Norway
257. Schubak G.E. and Scott D.S. (1995) A Techno-Economic Comparison of Power Systems for Autonomous Underwater Vehicles. IEEE Journal Oceanic Engineering, 20:94–100
258. Schultz A.C. (1992) Adaptive Testing of Controllers for Autonomous Vehicles. In: Proceedings Symposium on Autonomous Underwater Vehicle Technology, Washington, DC, 158–164
259. Serrani A. (1997) Modellistica, Simulazione e Controllo Nonlineare di Veicoli Sottomarini (in Italian), Dottorato di Ricerca in Sistemi Artificiali, Università degli Studi di Ancona, Ancona, Italy, 1997
260. Serrani A. and Zanoli S.M. (1998) Designing Guidance and Control Schemes for AUVs Using an Integrated Simulation Environment. In: IEEE International Conference on Robotics and Automation, Leuven, Belgium, 1079–1083
261. Shepperd S. W. (1978) Quaternion from Rotation Matrix. Journal of Guidance and Control, 1:223–224
262. Siciliano B. (1990) Kinematic Control of Redundant Robot Manipulators: A Tutorial. Journal of Intelligent Robotic Systems, 3:201–212
263. Siciliano B. and Slotine J.J. (1991), A General Framework for Managing Multiple Tasks in Highly Redundant Robotic Systems, In: International Conference on Advanced Robotics, Pisa, Italy, 1211–1216
264. Slotine J.J. (1984) Sliding Controller Design for Non-Linear Systems. International Journal of Control, 40:421–434
265. Slotine J.J. and Benedetto M.D.D. (1990) Hamiltonian Adaptive Control of Spacecraft. IEEE Transactions on Automatic Control, 35:848–852
266. Slotine J.J. and Li W. (1987) Adaptive Strategies in Constrained Manipulation. In: IEEE International Conference on Robotics and Automation, Raleigh, NC, 595–601
267. Slotine J.J. and Li W. (1987) On the Adaptive Control of Robot Manipulators. International Journal of Robotics Research, 6:49–59
268. Slotine J.J. and Li W. (1991) Applied Nonlinear Control, Prentice-Hall International Editions, USA
269. Smallwood D.A. and Whitcomb L.L. (2001) Toward Model Based Dynamic Positioning of Underwater Robotic Vehicles. In: MTS/IEEE Techno-Ocean '01, 1106–1114
270. Smallwood D.A. and Whitcomb L.L. (2002) The Effect of Model Accuracy and Thruster Saturation on Tracking Performance of Model Based Controllers for Underwater Robotic Vehicles: Experimental Results. In: IEEE International Conference on Robotics and Automation, Washington, DC, 1081–1087
271. Smallwood D.A. and Whitcomb L.L. (2003) Adaptive Identification of Dynamically Positioned Underwater Robotic Vehicles. IEEE Transactions on Control System Technology, 11(4):505–515
272. SNAME, The Society of Naval Architects and Marine Engineers (1950) Nomenclature for Treating the Motion of a Submerged Body Through a Fluid. In: Technical and Research Bulletin, 1–5

273. Solvang B., Deng Z., Lien T.K. (2000) A Methodological Framework for Developing ROV-Manipulator Systems for Underwater Unmanned Intervention. In: MTS/IEEE Techno-Ocean '01, 1085–1091
274. Stevens B. and Lewis F. (1992) Aircraft Control and Simulations, John Wiley & Sons Ltd., Chichester, United Kingdom
275. Stilwell D.J. (2002) Decentralized Control Synthesis for a Platoon of Autonomous Vehicles. In: IEEE International Conference on Robotics and Automation, Washington, DC, 744–749
276. Stilwell D.J. and Bishop B.E. (2000) Platoons of Underwater Vehicles. IEEE Control Systems Magazine, 45–52
277. Stilwell D.J. and Bishop B.E. (2002) Redundant Manipulator Techniques for Path Planning and Control of a Platoon of Autonomous Vehicles. In: Conference on Decision and Control, Las Vegas, Nevada, 2093–2098
278. Stilwell D.J. and Bishop B.E. (2005) Redundant Manipulator Techniques for Partially Decentralized Path Planning and Control of a Platoon of Autonomous Vehicles. IEEE Transcations of Systems, Man and Cybernetics, part. B, 35(4):842–848
279. Stokey R.P. (1994) Software Design Technique for the Man Machine Interface to a Complex Underwater Vehicle. In: MTS/IEEE Techno-Ocean '94, Brest, France, 119–124
280. Sun Y.C. and Cheah C.C. (2003) Adaptive Setpoint Control for Autonomous Underwater Vehicles. In: 2003 IEEE Conference on Decision and Control, Maui, Hawaii, 1262–1267
281. Sun Y.C. and Cheah C.C. (2004) Adaptive Setpoint Control of Underwater Vehicle-Manipulator Systems. In: 2004 IEEE Conference on Robotics, Automation and Mechatronics, Singapore, 434–439
282. Sun Y.C. and Cheah C.C. (2004) Coordinated Control of Multiple Cooperative Underwater Vehicle-Manipulator Systems Holding a Common Load. In: MTS/IEEE Techno-Ocean '04, Kobe, Japan, 1542–1547
283. Tacconi G. and Tiano A. (1989) Reconfigurable Control of an Autonomous Underwater Vehicle. In: Proceedings International Symposium Unmanned Untethered Submersible Technology, 486–493
284. Takai M., Fujii T. and Ura T. (1995) A Model Based Diagnosis System for Autonomous Underwater Vehicles using Artifical Neural Networks. In: Proceedings International Symposium Unmanned Untethered Submersible Technology, Durham, New Hampshire, 243–252
285. Tarn T.J. and Yang S.P. (1997) Modeling and Control for Underwater Robotic Manipulators - An example. In: IEEE International Conference Robotics and Automation, Albuquerque, New Mexico, 2166–2171
286. Tarn T.J., Shoults G.A. and Yang S.P. (1996) A Dynamic Model for an Underwater Vehicle with a Robotic Manipulator using Kane's Method. Autonomous Robots, 3:269–283
287. Terribile A., Prendin W. and Lanza R. (1994) An innovative Electromechanical Underwater Telemanipulator - Present Status and Future Development. In: MTS/IEEE Techno-Ocean '94, II:188–191
288. Tong G. and Jimao Z. (1998) A Rapid Reconfiguration Strategy for UUV Control. In: Proceedings 1998 International Symposium Underwater Technology, Tokyo, Japan, 478–483
289. Tsukamoto C.L., Lee W., Yuh J., Choi S.K. and Lorentz J. (1997) Comparison Study on Advanced Thrusters Control of Underwater Robots. In: IEEE International Conference Robotics and Automation, Albuquerque, New Mexico, 1845–1850

290. Uliana M., Andreucci F. and B. Papalia (1997) The Navigation System of an Autonomous Underwater Vehicle for Antarctic Exploration. In: MTS/IEEE Techno-Ocean '97, Halifax, Canada, 403–408
291. Uny Cao Y., Fukunaga A.S. and Kanhg A.B. (1997) Cooperative Mobile Robotics: Antecedents and Directions. Autonomous Robots, 4:7–27
292. Ura T. (2003) Steps to Intelligent AUVs. In: 6 th IFAC Conference on Manoeuvring and Control of Marine Craft, Girona, Spain
293. Utkin V.I. (1977) Variable Structure Systems with Sliding Modes. IEEE Transactions in Automatic and Control, 22:212–222
294. Valavanis K.P., Gracanin D., Matijasevic M., Kolluru R. and Demetriou G.A. (1997) Control Architecture for Autonomous Underwater Vehicles. IEEE Control Systems, 48–64
295. Vasudevan C. and Ganesan K. (1996) Case-Based path Planning for Autonomous Underwater Vehicles. In: Underwater Robots, Yuh, Ura and Bekey (Eds.), Kluwer Academic Publishers, Boston, 1–15
296. Walker N.W. and Orin D.E. (1982) Efficient Dynamic Computer Simulation of Robotic Mechanisms. Journal of Dynamic Systems, Measurements, and Control, 104:205–211
297. Wang H.H., Rock S.M. and Lee M.J. (1995) Experiments in Automatic Retrieval of Underwater Objects with an AUV. In: MTS/IEEE Techno-Ocean '95, 366–373
298. Wen J.T.Y. and Delgado K.K. (1994) The Attitude Control Problem. IEEE Transactions on Automatic Control, 36:1148–1162
299. Whitcomb L.L. (2000) Underwater Robotics: Out of the Research Laboratory and Into the Field. In: 2000 IEEE International Conference on Robotics and Automation, San Franscisco, California, 709–716
300. Whitcomb L.L. and Yoerger D.R. (1999) Development, Comparison, and Preliminary Experimental Validation of Nonlinear Dynamic Thruster Models. IEEE Journal of Oceanic Engineering, 24:481–494
301. Whitney D.E. (1969) Resolved motion rate control of manipulators and human prostheses. IEEE Transactions of Man, Machine Systems, 10:47–53
302. Whitney D.E. (1977) Force Feedback Control of Manipulators Fine Motions. Transactions ASME Journal of Dynamic Systems, Measurement, and Control, 99:91–97
303. Whitney D.E. (1987) Historical Perspective and State of the Art in Robot Force Control. International Journal of Robotics Research, 6:3–14
304. Xiaoping Y., Bachmann E.R. and Arslan S. (2000) An Inertial Navigation System for Small Autonomous Underwater Vehicles. In: IEEE International Conference on Robotics and Automation, San Francisco, California, 1781–1786
305. Xu Y. and Shum H.Y. (1994) Dynamic Control and Coupling of a Free-Flying Space Robot System. Journal of Robotic Systems, 11:573–589
306. Yang K.C., Yuh J. and Choi S.K. (1998) Experimental Study of Fault-Tolerant System Design for Underwater Robots. In: IEEE International Conference on Robotics and Automation, Leuven, Belgium, 1051–1056
307. Yang K.C., Yuh J. and Choi S.K. (1999) Fault-Tolerant System Design of an Autonomous Underwater Vehicle – ODIN: an experimental study. International Journal of System Science, 30(9):1011–1019
308. Yavnai A. (1996) Architecture for an Autonomous Reconfigureable Intelligent Control System (ARICS). In: Proceedings Symposium on Autonomous Underwater Vehicle Technology, Monterey, California, 238–245
309. Yoerger D.R., Cooke J.G., and Slotine J.J. (1990) The Influence of Thruster Dynamics on Underwater Vehicle Behavior and their Incorporation into Control System Design. IEEE Journal of Oceanic Engineering, 15:167–178

310. Yoerger D.R., Slotine J.J. (1985) Robust Trajectory Control of Underwater Vehicles. IEEE Journal of Oceanic Engineering, 10:462–470
311. Yoerger D.R., Slotine J.J. (1991) Adaptive Sliding Control of an Experimental underwater Vehicle. IEEE International Conference on Robotics and Automation, Sacramento, California, 2746–2751
312. Yoshikawa T. (1985) Manipulability of Robotic Mechanisms. International Journal Robotics Research, 4:3–95
313. Young K-K.D. (1978) Controller Design for a Manipulator Using Theory of Variable Structure Systems. IEEE Transactions on Systems, Man and Cybernetics, 8:210–218
314. Yuh J. (1990) Modeling and Control of Underwater Robotic Vehicles. IEEE Transactions on Systems, Man, and Cybernetics, 20:1475–1483
315. Yuh J. (1995) Development in Underwater Robotics. In: IEEE International Conference on Robotics and Automation, Nagoya, Japan, 1862–1867
316. Yuh J. (1996) An Adaptive and Learning Control System for Underwater Robots. In: Proceedings 13th IFAC World Congress, San Francisco, California, A:145–150
317. Yuh J. (2003) Exploring the Mysterious Underwater World with Robots. In: 6 th IFAC Conference on Manoeuvring and Control of Marine Craft, Girona, Spain
318. Yuh J., Choi S.K., Ikehara C., Kim G.H., McMurty G., Ghasemi-Nejhad M., Sarkar N. and Sugihara K. (1998) Design of a semi-autonomous underwater vehicle for intervention missions (SAUVIM). In: Proceedings 1998 International Symposium on Underwater Technology, 63–68
319. Yuh J. and Gonugunta K.V (1993) Learning Control of Underwater Robotic Vehicles. In: IEEE International Conference on Robotics and Automation, Atlanta, Georgia, 106–111
320. Yuh J., Nie J. and Lee C.S.G. (1999) Experimental Study on Adaptive Control of Underwater Robots. In: IEEE International Conference on Robotics and Automation, Detroit, Michigan, 393–398
321. Yuh J. and West M. (2001) Underwater Robotics. Journal of Advanced Robotics, 15:609–639
322. Yuh J., Zhao S. and Lee P.M. (2001) Application of Adaptive Disturbance Observer Control to an Underwater Manipulator. In: IEEE International Conference on Robotics and Automation, Seoul, Korea, 3244–3249
323. Zhao S. and Yuh J. (2005) Experimental Study on Advanced Underwater Robot Control. IEEE Transactions on Robotics, 21(4):695–703
324. Zheng X. (1992) Layered Control of a Practical AUV. In: Proceedings Symposium on Autonomous Underwater Vehicle Technology, Washington, DC, 142–147
325. Zhu W.H., Xi Y.G., Zhong Z.J., Bien Z. and De Schutter J. (1997) Virtual Decomposition Based Control for Generalized High Dimensional Robotic Systems with Complicated Structure. IEEE Transactions on Robotics and Automation, 13:411-436

Printing: Krips bv, Meppel
Binding: Stürtz, Würzburg

RECEIVED
OCT 16 2006